ON THE DUAL USES OF SCIENCE AND ETHICS
PRINCIPLES, PRACTICES, AND PROSPECTS

ON THE DUAL USES OF SCIENCE AND ETHICS
PRINCIPLES, PRACTICES, AND PROSPECTS

edited by

Brian Rappert and Michael J. Selgelid

Australian
National
University

E PRESS

Centre for
Applied Philosophy
and **Public Ethics**
An Australian Research Council Funded Special Research Centre

Practical Ethics and Public Policy Monograph 4
Series Editor: Michael J. Selgelid

ANU
E PRESS

Published by ANU E Press
The Australian National University
Canberra ACT 0200, Australia
Email: anuepress@anu.edu.au
This title is also available online at http://epress.anu.edu.au

---

National Library of Australia Cataloguing-in-Publication entry

Title: On the dual uses of science and ethics : principles, practices, and prospects /
          editors, Brian Rappert, Michael J. Selgelid.

ISBN: 9781925021332 (paperback) 9781925021349 (ebook)

Subjects: Science and civilization--Moral and ethical aspects.
          Science--Moral and ethical aspects.
          Science and the humanities.

Other Authors/Contributors:

          Rappert, Brian, editor.

          Selgelid, Michael J., editor.

Dewey Number: 501

Cover design and layout by ANU E Press

Contents

Introduction

Part I: Dual Use in Context

Part II: Ethical Frameworks and Principles

Part III: Ethical Practices

Part IV: Ethical Futures

Conclusion

Contributors

Valentina Bartolucci

Valentina Bartolucci is a lecturer at the University of Pisa, Italy, and at the Marist college campus of Florence, Italy. Valentina is also Visiting Fellow at the University of Derby (UK). She maintains a close connection with the School for Social and International Studies at the University of Bradford, UK. Valentina has a degree in peace studies from the University of Pisa, an MA in conflict management and human rights from the Sant' Anna School of Advanced Studies, and a PhD from the University of Bradford. Dr Bartolucci has previously worked as a junior technical expert for the Italian Ministry of Foreign Affairs in Morocco, and as a consultant for various international organisations (in Italy, Burkina-Faso and India). She is currently offering consultancies on issues linked to terrorism, organised crime, migration, and American and Middle Eastern foreign policies for various national and international organisations as well as government departments around Europe. In recognition of her impact internationally, she was recently made a Member of the Aspen Institute (group 'I protagonisti italiani nel mondo'). In 2011, she was nominated for the Who's Who Academic Excellence. Apart from her doctoral thesis on the governmental discourse on terrorism, Dr Bartolucci has written numerous articles in the fields of politics, critical discourse analysis, security and surveillance. Her research mainly addresses the foreign and security policies of the United States and other states, terrorism and counterterrorism, discourses on threat and emergency, and strategic communication as anti-terrorism. She has recently been awarded with a Fulbright Research Scholar Bursary for 2013–14 to be spent at Arizona State University working on 'public diplomacy as anti-terrorism'.

Louise Bezuidenhout

Louise Bezuidenhout completed her PhD in cardiothoracic surgery at the University of Cape Town, South Africa, and has worked as a postdoctoral scientist in the department of cardiovascular sciences at the University of Edinburgh, UK. In 2008 she was the recipient of an EU scholarship for the Erasmus Mundus Masters of Bioethics program run by the Catholic University of Leuven, Radboud University Nijmegen and the University of Padova. These dual interests in science and bioethics led her to enrol for a PhD in sociology at the University of Exeter under Professor Brian Rappert, which she completed in 2013. Her research focused on the need for contextual sensitivity in life-science ethics debates, particularly examining how scientists in developing countries interact with topical ethical issues such as dual use. She is currently a research fellow at the University of Witwatersrand in South Africa.

Koos van der Bruggen

Koos van der Bruggen (born 1951) studied political science (international relations) at Radboud University Nijmegen (the Netherlands). He wrote his PhD thesis on criteria for the ethical judgment of nuclear deterrence at the Vrije Universiteit (Free University), Amsterdam. Dr van der Bruggen has been working at the Rathenau Institute, the Dutch organisation for parliamentary technology assessment, for more than 12 years. He was involved in research and debates on social and ethical aspects of technological developments (biomedical technology, genetics and military technology). More recently, van der Bruggen has been working at the Royal Netherlands Academy of Arts and Sciences (developing a code of conduct for scientists regarding issues of biosecurity) and Delft University of Technology (research on biosecurity and dual use). He was also the secretary of the Committee of Inquiry into the Dutch involvement in the Iraq War of 2003 (Davids Committee, 2009). At the moment he is working as an independent researcher on peace and security issues. Dr van der Bruggen has published (in Dutch and English) on just-war issues, on issues of ethics and technology and on biological weapons.

Steve Clarke

Steve Clarke is a Senior Research Fellow in the Centre for Applied Philosophy and Public Ethics at Charles Sturt University in Australia. He is also a James Martin Research Fellow in the Institute for Science and Ethics in the Oxford Martin School and Faculty of Philosophy at the University of Oxford, and a Research Fellow in the Uehiro Centre for Practical Ethics, also at Oxford. He holds a PhD in philosophy from Monash University and has previously held appointments at the University of Melbourne, the University of Cape Town and La Trobe University. Steve is a broad-ranging philosopher who has published more than 60 papers in refereed journals and scholarly collections. His papers have appeared in such journals as the *American Philosophical Quarterly*, the *British Journal for the Philosophy of Science*, the *Journal of Medicine and Philosophy*, *Religious Studies* and *Synthese*. He is the author of two books (*Metaphysics and the Disunity of Scientific Knowledge*, Ashgate, 1998; and *The Justification of Religious Violence*, Wiley-Blackwell, 2014) and the co-editor of three books. The latest of these is *Religion, Intolerance and Conflict: A Scientific and Conceptual Investigation*, edited by Steve Clarke, Russell Powell and Julian Savulescu (Oxford University Press, 2013).

Nancy D. Connell

Nancy D. Connell is Professor in the Division of Infectious Disease in the Department of Medicine at Rutgers New Jersey Medical School (RNJMS) and the Rutgers Biomedical Health Sciences. Dr Connell, a Harvard University PhD in microbiology, has made her major research focus the antibacterial drug discovery in respiratory pathogens such as *M. tuberculosis* and *B. anthracis*. She is Director of the Biosafety Level Three (BSL3) Facility of RNJMS's Center for the Study of Emerging and Re-emerging Pathogens and chairs the university's Institutional Biosafety Committee. Dr Connell has been continuously funded by the National Institutes of Health (NIH) and other agencies since 1993 and serves on numerous NIH study sections and review panels. She has served on a number of committees of the National Academy of Sciences—for example, the Committee on Advances in Technology and the Prevention of their Application to Next Generation Biowarfare Agents (2004), Trends in Science and Technology Relevant to the Biological Weapons Convention; an international workshop (2010), and the Committee to Review the Scientific Approaches used in the FBI's Investigation of the 2001 *Bacillus anthracis* mailings (2011). Current work is with the National Academies of Sciences Education Institute on Responsible Science, held in Jordan (2012) and Malaysia (2013).

Michael Crowley

Michael Crowley is the Project Coordinator of the Bradford Non-Lethal Weapons Research Programme (BNLWRP) and is also a senior research associate for the Omega Research Foundation. He has worked for 20 years on arms control, security and human rights issues, including as executive director of the Verification Research, Training and Information Centre (VERTIC). He also acted as chairman of the Bio-Weapons Prevention Project. Prior to this he worked at the Omega Research Foundation exploring options for effective restriction of the development, trade and use of security equipment and technology employed in torture and ill treatment. He has also managed the Arms Trade Treaty project at the Arias Foundation in Costa Rica and worked as senior arms trade analyst at BASIC. He has also held several research and policy positions with Amnesty International, both in the UK section and at the International Secretariat. He holds a BSc in genetics from Liverpool University, and an MRes and PhD from Bradford University, where his doctoral thesis explored the regulation of riot-control agents, incapacitating agents and related means of delivery.

Malcolm Dando

Malcolm Dando is Professor of International Security at the University of Bradford. A biologist by training, his main research interest is in the preservation of the prohibitions embodied in the Chemical Weapons Convention and the Biological Weapons Convention at a time of rapid scientific and technological change in the life sciences. His recent publications include *Deadly Cultures: Biological Weapons Since 1945* (Harvard University Press, 2006), which he edited with Mark Wheelis and Lajos Rozsa.

Thomas Douglas

Thomas Douglas is a Uehiro Research Fellow in the Oxford Uehiro Centre for Practical Ethics, Faculty of Philosophy, University of Oxford, and a Junior Golding Fellow at Brasenose College, Oxford. He trained in medicine (MBChB, Otago) and philosophy (DPhil, Oxford) and works in applied and normative ethics. His current research addresses the ethics of producing dangerous knowledge and of using medical technologies for non-medical purposes, particularly crime prevention. He has previously worked on organ donation policy, reproductive decision-making, slippery-slope arguments and compensatory justice.

Dr Nicholas G. Evans

Dr Nicholas G. Evans is an Adjunct Research Assistant at the Centre for Applied Philosophy and Public Ethics, Charles Sturt University, Canberra. His dissertation focused on the ethics of censoring dual-use research in the life sciences through an examination of the history of nuclear science. Nicholas is the author of a number of articles on military ethics, the philosophy of science, and ethics, and has taught philosophy, military ethics and physics at universities around Australia. His research interests include emerging military technologies, public health ethics, concepts of responsibility and autonomy, and friendship.

John Forge

John Forge is an Honorary Associate in the Unit for History and Philosophy of Science, Sydney University, having previously worked at the Universities of New South Wales, Wollongong, Griffith and Macquarie. His research ranged from the philosophy of physical science, especially explanation, in the early years, to science and ethics later on. Forge has authored numerous articles and six books, the latest being *The Responsible Scientist* (Pittsburgh, 2008), winner of the David Harold Tribe Prize for Philosophy and the Eureka Prize for Research Ethics; *Designed to Kill: The Case against Weapons Research* (Springer, 2012); and *On the Morality of Weapons Research* (Ashgate, 2014).

Alexander Kelle

Alexander Kelle is a political scientist by training and a Senior Policy Officer in the Office of Strategy and Policy of the Organisation for the Prohibition of Chemical Weapons in the Hague, Netherlands. He contributed to this volume while senior lecturer in politics and international relations at the Department of Politics, Languages and International Studies at the University of Bath, UK. Before moving to Bath, he held positions at Queen's University Belfast, the University of Bradford, Stanford University, Goethe University Frankfurt and the Peace Research Institute Frankfurt. His past research has addressed international security cooperation—with a view to chemical and biological weapons—dual-use governance of synthetic biology, and the foreign and security policies of Western liberal democracies.

Seumas Miller

Seumas Miller is a Professorial Research Fellow at the Centre for Applied Philosophy and Public Ethics (CAPPE) (an Australian Research Council Special Research Centre) at Charles Sturt University (Canberra) and the 3TU Centre for Ethics and Technology at Delft University of Technology (the Hague) (joint position). He is the foundation director of CAPPE (2000–07). He has authored or co-authored more than 150 academic articles and 15 books, including (with M. Selgelid) *Ethical and Philosophical Consideration of the Dual Use Dilemma in the Biological Sciences* (Springer, 2008), *Terrorism and Counter-Terrorism: Ethics and Liberal Democracy* (Blackwell, 2009), *The Moral Foundations of Social Institutions: A Philosophical Study* (Cambridge University Press, 2010) and (with I. Gordon) *Investigative Ethics: Ethics for Police Detectives and Criminal Investigators* (Blackwell, 2013).

Brian Rappert

Brian Rappert is a Professor of Science, Technology and Public Affairs and Head of the Department of Sociology and Philosophy at the University of Exeter. His long-term interest has been the examination of how choices can be and are made about the adoption and regulation of security-related technologies— particularly in conditions of uncertainty and disagreement. Recent books by Rappert include *Controlling the Weapons of War* (Routledge, 2006), *Biotechnology, Security and the Search for Limits* (Palgrave, 2007), *Technology & Security* (ed., Palgrave, 2007), Biosecurity (co-ed., Palgrave, 2009) and *Experimental Secrets* (UPA, 2009). More recently he has been interested in the social, ethical and epistemological issues associated with researching and writing about secrets, as in his book *Experimental Secrets* (2009) and *How to Look Good in a War* (2012).

Andreas Alois Reis

Andreas Alois Reis (MD, MSc) is Technical Officer in the Department of Ethics and Social Determinants of Health at the World Health Organisation (WHO) headquarters in Geneva. After medical education in internal medicine in Germany, France and Chile, he pursued studies in health economics. His main area of work is public health ethics, with a focus on distributive justice and equitable access to health services, and ethical aspects of infectious diseases such as HIV, pandemic influenza and tuberculosis. He has lectured and organised training for WHO in more than 50 countries, and is currently serving on the editorial board of *Public Health Ethics*.

David B. Resnik

David B. Resnik (JD, PhD) is a bioethicist and Institutional Review Board (IRB) Chair at the National Institute of Environmental Health Science, National Institutes of Health. He has published eight books and 200 articles on ethical, legal, social and philosophical issues in science, medicine and technology, and is Associate Editor of the journal *Accountability in Research*.

Michael J. Selgelid

Professor Michael J. Selgelid is Director of the Centre for Human Bioethics at Monash University in Melbourne, Australia. He was previously Director of the World Health Organisation Collaborating Centre for Bioethics at The Australian National University. His research focus is public health ethics—with emphasis on ethical issues associated with biotechnology and infectious disease. He is editor of a book series in public health ethics analysis for Springer, Co-Editor of *Monash Bioethics Review*, and Associate Editor of the *Journal of Medical Ethics*. He co-authored (with Seumas Miller) *Ethical and Philosophical Consideration of the Dual-Use Dilemma in the Biological Sciences* (Springer, 2008).

Michael Smithson

Michael Smithson is a Professor in the Research School of Psychology at The Australian National University in Canberra. He is the author of six books, co-editor of two, and his other publications include more than 140 refereed journal articles and book chapters. His primary research interests are in judgment and decision-making under uncertainty, statistical methods for the social sciences, and applications of fuzzy-set theory to the social sciences.

Judi Sture

Judi Sture is the Head of the Graduate School at the University of Bradford, UK, where she leads two doctoral research training programs. She lectures in research ethics and research methodology and is closely involved in devising and developing postgraduate and ethics policy and practice at the university and beyond. As a member of the Wellcome Trust Dual-Use Bioethics Group and associate member of the Bradford Disarmament Research Centre, she is engaged with colleagues from a number of UK and overseas universities in developing a bioethics approach to counter-biosecurity threats in the life sciences. Judi holds a BSc(Hons) in archaeology (University of Bradford), in which she specialised in the study of human skeletal remains, and a PhD (University of Durham) in biological anthropology, focusing on environmental associations with human birth defects. Her research in biological anthropology continues, including further work on developmental defects. She is currently engaged in analysis of skeletal remains held at the Museum of London and is working with the British Association for Biological Anthropology and Osteoarchaeology on developing ethical practice in the profession.

Emmanuelle Tuerlings

Dr Emmanuelle Tuerlings has extensive experience on dual-use research, biosecurity and global health security. She carries out training and consultancy with a particular interest in public health. She was a scientist to the Biorisk Reduction for Dangerous Pathogens (BDP) program at WHO headquarters, Geneva, from 2004 to 2011, where she was leading the project 'Responsible Life Sciences Research for Global Health Security'. Before joining WHO, she was based at the Harvard Sussex Program, University of Sussex, UK. She also worked with several international and non-governmental organisations on issues related to dual-use biological technologies and their governance. She holds an MSc and a doctorate in science and technology policy from the University of Sussex (UK).

Suzanne Uniacke

Suzanne Uniacke is Director of the Centre for Applied Philosophy and Public Ethics, Charles Sturt University, Canberra. She has held positions in philosophy departments in England and Australia and visiting research fellowships at St Andrews, Harvard and Stirling universities. Professor Uniacke has published extensively in ethics, applied philosophy and philosophy of law and was editor of the *Journal of Applied Philosophy* from 2001 to 2013.

Simon Whitby

Simon Whitby works at the interface between life and associated science and national security communities to address the threat of deliberate disease in the context of rapidly advancing dual-use science and technology. Whitby's work contributes to the discourse on dual-use biosecurity and bioethics and thus on raising awareness at governmental, civil society, life science and industry levels about the ethical, legal and social implications of life science and associated science research. He has been actively engaged in building a worldwide capability in dual-use bioethics to engage life and associated science communities in awareness-raising programs about the importance of responsible conduct of life-science research. Significantly, he has developed a novel and innovative online distance-learning masters-level train-the-trainer program as well as short courses in applied dual-use biosecurity/bioethics.

Jim Whitman

Jim Whitman is Professor of Global Governance in the Department of Peace Studies, University of Bradford, and Director of Postgraduate Studies for the School of Social and International Studies. His latest book, *Governance Challenges of Nanotechnology*, will be published by Palgrave in 2014.

Acknowledgments

A Wellcome Trust award, 'Building A Sustainable Capacity in Dual-Use Bioethics', supported a January 2010 workshop titled 'Promoting Dual Use Ethics' at The Australian National University that underpinned the development of this volume. A grant by the Research Councils UK (ES/K011308/1) enabled Brian Rappert and Malcolm Dando to undertake work for their chapters. An award from the Economic and Social Research Council covered the production costs associated with making this volume Open Access through ANU E Press.

Abbreviations

AAAS	American Association for the Advancement of Science
AMA	American Medical Association
ANU	The Australian National University
APS	American Phytopathological Society
ASM	American Society for Microbiology
BBSRC	Biotechnology and Bioscience Research Council
BDP	Biorisk Reduction for Dangerous Pathogens program
BIO	Biotechnology Industry Organization
BMA	British Medical Association
BNLWRP	Bradford Non-Lethal Weapons Research Programme
BSL	biosafety level
BSLIII	Biosafety Level Three facility
Bt	*Bacillus thuringiensis*
BTWC	Biological and Toxin Weapons Convention
BW	biological weapons
BWC	Biological Weapons Convention
CAPPE	Centre for Applied Philosophy and Public Ethics
CBA	cost–benefit analysis
CBRN	chemical, biological, radiological and nuclear
CBW	chemical and biological weapons
CISSM	Center for International and Security Studies at Maryland
COB	'cat out of the bag'
CSAIL	Computer Science and Artificial Intelligence Laboratory
CSIS	Centre for Strategic and International Studies
CWC	Chemical Weapons Convention
DARPA	Defense Advanced Research Projects Agency
DDE	doctrine of double effect
DHHS	Department of Health and Human Services (US)
DHS	Department of Homeland Security (US)
DIY	do-it-yourself
DOC	Department of Commerce (US)
EGE	European Group on Ethics
ELSI	ethical, legal and social implications
ENSN	European Neuroscience and Society Network

EU	European Union
EVC	expected-value criterion
EVP	expected-value procedure
FBI	Federal Bureau of Investigation (US)
FDA	Food and Drug Administration (US)
fMRI	functional magnetic resonance imaging
GE	genetic engineering
HAC	holistic arms control
HSP	Harvard Sussex Project
IACUCs	institutional animal care and use committees
IAEA	International Atomic Energy Agency
IAP	InterAcademy Panel
IASB	Industry Association Synthetic Biology
IBC	institutional biosafety committee
ICPS	International Consortium for Polynucleotide Synthesis
ICRC	International Committee of the Red Cross
INN	International Neuroethics Network
IPR	intellectual property rights
IRBs	institutional review boards
ISAAA	International Service for the Acquisition of Agri-biotech Applications
ISU	Implementation Support Unit (UN)
IUBMB	International Union of Biochemistry and Molecular Biology
MAS	marker-assisted selection
MIT	Massachusetts Institute of Technology
MOOTW	military operations other than war
MRC	Medical Research Council (UK)
MTA	material transfer agreement
NATO	North Atlantic Treaty Organisation
NCD	Neuroscience Center at Dartmouth
NEST	New and Emerging Science and Technology
NET	Neuroethics New Emerging Team
NGO	non-governmental organisation
NIEHS	National Institute of Environmental Health Sciences (US)
NIH	National Institutes of Health (US)
NMHP	no means to harm principle
NNI	National Nanotechnology Initiative (US)

NPG	Nature Publishing Group
NRA	*National Research Act (US)*
NRC	Nuclear Regulatory Commission
NRC	National Research Council
NSABB	National Science Advisory Board for Biosecurity
NSF	National Science Foundation
ONR	Office of Naval Research (US)
OPCW	Organisation for the Prevention of Chemical Weapons
PCSBI	Presidential Commission for the Study of Bioethical Issues (US)
PHAC	Public Health Agency, Canada
PI	principal investigator
PNAS	*Proceedings of the National Academy of Sciences*
PP	precautionary principle
PTSD	post-traumatic stress disorder
QALYs	quality-adjusted life-years
QTL	quantitative trait locus
RAC	Recombinant DNA Advisory Committee (US)
R&D	research and development
RCR	responsible conduct of research
rEVP	restricted expected-value procedure
RNJMS	Rutgers New Jersey Medical School
SAP	science affects people
S&T	science and technology
SEU	subjective expected utility
sPP	strong versions of the precautionary principle
TB	tuberculosis
TMS	transcranial magnetic stimulation
UK	United Kingdom
UN	United Nations
UNESCO	United Nations Educational, Scientific and Cultural Organisation
UNGA	UN General Assembly
UNMOVIC	United Nations Monitoring, Verification and Inspection Commission
UNODA	UN Office for Disarmament Affairs
UNSC	United Nations Security Council
UNSCOM	United Nations Special Commission

US	United States of America
USDA	US Department of Agriculture
VBM	valuable biological materials
VERTIC	Verification Research, Training and Information Centre
WHA	World Health Assembly
WHO	World Health Organisation
WMA	World Medical Association
WMD	weapons of mass destruction

Introduction

.

1. Ethics and Dual-Use Research

Michael J. Selgelid

[W]ho is most capable of treating friends well and enemies badly in matters of disease and health?

A doctor.

...

Isn't the person most able to land a blow, whether in boxing or any other kind of fight, also most able to guard against ... And the one who is most able to guard against disease is also most able to produce it unnoticed?[1]

Nuclear physics and contemporary genetics

One of the most dramatic—and ethically problematic—episodes in the history of science involved the making and use of the first atomic weapons. When key discoveries of things like atomic fission and the chain reaction were made during the revolution in physics under way during the first half of the twentieth century, the scientists involved realised that the new knowledge gained might be used for both good purposes (for example, energy production) and bad purposes (for example, weapons making).[2] The first atomic bombs were developed soon after the discovery of the chain reaction in particular. The atomic bombs dropped on Hiroshima and Nagasaki during World War II caused hundreds of thousands of deaths.[3] Since that time vast resources (which might have been used for other purposes) have been devoted to nuclear weapons production, and humanity has lived under the threat of nuclear holocaust for decades.

An implication of the contemporary scientific revolution in genetics and biotechnology is that the life sciences, at present, are in a situation analogous to that of physics when the key discoveries mentioned above were made. The very same biological discoveries that might be used to benefit humanity (for example, via advances in medicine) can sometimes also be used to cause harm (for example, via biological weapons). In some cases, the harms in question could be catastrophic. This is thus a key moment in the history of biology—and a crucial time for ethical decision-making and policymaking in the life sciences.

---

1   Plato 1992, *Republic*, G. M. A. Grube (trans.), C. D. C. Reeve (rev.), Hackett, Indianapolis, pp. 5–9.

2   Schweber, S. S. 2000, *In the Shadow of the Bomb*, Princeton University Press, Princeton, NJ.

3   Rhodes, R. 1986, *The Making of the Atomic Bomb*, Simon & Schuster, New York.

Research with potential to be used for both good and bad purposes is now commonly referred to as 'dual-use research'. While almost any knowledge and technology can be used for both kinds of purposes, the expression 'dual-use research of concern'[4] is used to refer to research that can be used for especially harmful purposes—that is, where the consequences of malevolent use would be potentially catastrophic. Of particular concern are advances in genetics that might enable development of a new generation of biological weapons of mass destruction. (In the remainder of this chapter, I will use the expression 'dual-use research' as shorthand for the expression 'dual-use research of concern'.)

Controversial cases

Such danger is illustrated by the following controversial studies that have been published during the new millennium.

The mousepox experiment

Australian scientists used genetic-engineering techniques to insert an interleukin (IL-4) gene into the mousepox virus. Their aim was to develop a strain of mousepox that would make mice infertile—and thus provide a potentially powerful new means of pest control. They unexpectedly discovered, however, that the altered virus killed both mice that were naturally resistant to and mice that had been vaccinated against ordinary mousepox. They published these findings, along with description of materials and methods, in the *Journal of Virology* in 2001.[5] A danger is that the same techniques might enable production of vaccine-resistant smallpox. Smallpox is one of the most feared biological weapons agents—and vaccine is our only defence against it.

Synthetic polio

Following the map of the polio genome (published on the Internet), American scientists (via mail order) bought and strung together corresponding strands of DNA. Addition of the synthesised genome to 'cell juice' (a solution containing cellular ingredients, but no live cells) led to production of 'live' polio virus that paralysed and killed mice. The scientists' aims were, inter alia, to show that

---

4   National Science Advisory Board for Biosecurity (NSABB) 2007, *Proposed Framework for the Oversight of Dual Use Life Sciences Research: Strategies for Minimizing the Potential Misuse of Research Information*, National Science Advisory Board for Biosecurity, Bethesda, Md, <http://oba.od.nih.gov/biosecurity/pdf/ Framework%20for%20transmittal%200807_Sept07.pdf> (viewed 7 April 2013).
5   Jackson, R. J., Christensen, C. D., Beaton, S., Hall, D. F. and Ramshaw, I. A. 2001, 'Expression of mouse interleukin-4 by a recombinant ectromelia virus overcomes genetic resistance to mousepox', *Journal of Virology*, vol. 75, no. 3, pp. 1205–10.

such a feat would be technically possible and to demonstrate that viruses are ultimately just chemicals.[6] They published their findings, along with description of materials and methods, in *Science* in 2002.[7] A danger is that similar techniques might be used to create biological weapons agents, such as smallpox, which bioterrorists and/or state-sponsored biological weapons programs might not otherwise be able to access easily.

Reconstruction of 1918 Spanish flu

American researchers used similar techniques (to those used with polio) to reconstruct the 1918 Spanish flu virus. The virus they created was more deadly to chickens than any flu virus that had previously been studied. The purpose of this research was to yield knowledge that would facilitate protection against possible future influenza pandemics—for example, vaccine production. They published their findings, along with description of materials and methods, in *Science* in 2005.[8] A danger is that malevolent actors (for example, bioterrorists) might use the published information to create and unleash the virus in question—which was responsible for one of the worst epidemics in human history, killing 20 to 100 million people over the course of a year or two.

Transmissible H5N1

The most recent and, to date, most controversial dual-use life-science research involved the study of H5N1 (avian) influenza transmissibility among ferrets, which provide the best model for influenza in humans. While it is estimated that H5N1 kills 60 per cent of humans infected, it is not (currently) transmissible between humans. Researchers in the Netherlands and the United States thus conducted experiments that aimed to determine whether H5N1 might develop into a human-to-human transmissible strain. Genetic engineering of the virus and 'passaging' of the altered virus between ferrets led to creation of strains that were airborne and easily transmissible among ferrets—thus indicating that natural evolution of a human-transmissible strain of H5N1 might be possible. Much debate surrounded the question of whether or not this research should be published in detail. On the one hand, it was argued that publishing these studies was important because this would facilitate vaccine development and/ or surveillance of relevant changes to H5N1 occurring in nature. The hope with regard to surveillance is that this would enable earlier detection of emerging

---

6   Selgelid, M. J. and Weir, L. 2010, 'Reflections on the synthetic production of poliovirus', *Bulletin of the Atomic Scientists*, vol. 66, no. 3, pp. 1–9.

7   Cello, J., Paul, A. V. and Wimmer, E. 2002, 'Chemical synthesis of poliovirus cDNA: generation of infectious virus in the absence of natural template', *Science*, vol. 297, pp. 1016–18.

8   Tumpey, T. M. et al. 2005, 'Characterization of the reconstructed 1918 Spanish influenza pandemic virus', *Science*, vol. 310, no. 5745, pp. 77–80.

pandemic strains and thus earlier implementation of protective measures. On the other hand, others argued that a human-transmissible strain of H5N1 could kill millions or perhaps even billions of people if produced and unleashed by bioterrorists or other malevolent actors (or in the event of accidental release from research laboratories). The US National Science Advisory Board for Biosecurity (NSABB) in December 2011 recommended that the findings of these studies be published, but that detailed description of materials and methods be omitted from the published articles.[9] After a highly publicised World Health Organisation (WHO) meeting reached the opposite conclusion in February 2012—recommending that the studies eventually be published in full[10]—the NSABB reversed its initial decision in March 2012.[11] Revised (fully detailed) versions of the papers in question were finally published in *Science*[12] and *Nature*[13] in June 2012. Part of the reason for the NSABB's reversal of decision apparently involved misunderstanding regarding whether or not the airborne transmissible strains were deadly to ferrets. While much of the initial public discussion indicated that the strains produced in the Netherlands were just as deadly as ordinary H5N1, it was later revealed that airborne strains created in the lab were not deadly to ferrets. Another cited reason for the NSABB's reversed decision is that the revised manuscripts they eventually approved provided additional data regarding ways in which the studies might have public health benefits—for example, via surveillance.

Levels of governance

The dual-use phenomenon requires ethical decision-making by various actors at different levels of the science governance hierarchy. Individual scientists (insofar as they are at liberty) must decide what research to conduct and/or publish. Research institutions (insofar as they are at liberty) must decide how to regulate potentially dangerous research within their confines; how to educate

---

9   US National Institutes of Health 2011, 'Press statement on the NSABB review of H5N1 research', <http://www.nih.gov/news/health/dec2011/od-20.htm> (viewed 7 April 2013).

10   World Health Organisation (WHO) 2012, 'Public health, influenza experts agree H5N1 research critical, but extend delay', <http://www.who.int/mediacentre/news/releases/2012/h5n1_research_20120217/en/> (viewed 7 April 2013).

11   National Science Advisory Board for Biosecurity (NSABB) 2012, 'March 29–30, 2012 meeting of the National Science Advisory Board for Biosecurity to review revised manuscripts on transmissibility of A/H5N1 influenza virus', <http://oba.od.nih.gov/oba/biosecurity/PDF/NSABB_Statement_March_2012_Meeting.pdf> (viewed 7 April 2013).

12   Herfst, S., Schrauwen, E. J. A., Linster, M., Chutinimitkul, S., de Wit, E., Munster, V. J., Sorrell, E. M., Bestebroer, T. B., Burke, D. F., Smith, D. J., Rimmelzwaan, G. F., Osterhaus, A. D. M. E. and Fouchier, R. A. M. 2012, 'Airborne transmission of influenza A/H5N1 virus between ferrets', *Science*, vol. 22, pp. 1534–41.

13   Imai, M., Watanabe, T., Hatta, M., Das, S. C., Ozawa, M., Shinya, K., Zhong, G., Hanson, A., Katsura, H., Watanabe, S., Li, C., Kawakami, E., Yamada, S., Kiso, M., Suzuki, Y., Maher, E. A., Neumann, G. and Kawaoka, Y. 2012, 'Experimental adaptation of an influenza H5 HA confers respiratory droplet transmission to a reassortant H5 HA/H1N1 virus in ferrets', *Nature*, vol. 486, pp. 420–8.

researchers working there (regarding dual use and/or ethics); what laboratory security measures to put into place, and so on.[14] Professional societies must make decisions about the development, promulgation and/or enforcement of ethical codes of conduct for scientists—and/or decisions about relevant (ethical) education of their members. Publishers must make decisions regarding processes of review of papers posing dual-use dangers—and they must ultimately decide which papers to publish. National governments must decide what research to fund, and the extent to (or manner in) which things like research review, publication review and/or relevant education of scientists will be mandated—and they must make decisions about the extent to which controls should be placed on access to potentially dangerous materials. International governance bodies (such as the WHO), finally, must make relevant decisions concerning global policy—for example, whether or not there should be international guidelines regarding dual-use research, or oversight thereof, and/or what the content of such guidelines should be.

Ethical dilemmas

In all cases the decisions will be difficult. On the one hand, responsible actors will want to take actions that will promote the development and use of beneficial science. On the other hand, they will want to take actions that will prevent the malevolent use of science (which might sometimes require avoidance of generation and/or publication of potentially dangerous information). An implication of the dual-use phenomenon, however, is that it is inherently difficult to achieve both goals at the same time in the cases where the very same research that is likely to be beneficial might also be used to cause harm. In the case of governmental decision-making, for example, a laissez-faire approach to scientific governance might facilitate scientific advance and the benefits thereby enabled—but it might also lead to especially dangerous research getting done and/or published. A more restrictive approach, on the other hand, might prevent generation and/or publication of dangerous information—but it might also stifle beneficial scientific advance at the same time. Hence the expression 'dual-use dilemma'.

In any case, it is important to recognise that key decisions posed by dual-use research are inherently *ethical* in nature. The decisions faced by the various actors enumerated above largely concern: 1) the responsibilities of the actors in question (for example, to what extent would a scientist be responsible if her research is used to cause harm?); 2) issues of how one should go about promoting

---

14 Research institutions, of course, already do such things to varying degrees—and numerous relevant measures (for example, regarding biosafety) are required by law. The point here, however, is that additional new measures are required to address dual-use research in particular.

benefits while avoiding harms or reducing risks (for example, should a paper be published if this has a good chance of promoting a significant amount of human wellbeing but a small chance of causing disaster?); and/or 3) questions about values and value conflict (for example, how should governments strike a balance between the goal to promote scientific freedom/progress and the goal to promote security?). Issues regarding responsibilities, harms, benefits and values—and, ultimately, what ought to be done—are exactly the kinds of things that ethics is about.

In the meantime, however, much of the debate about dual-use research has involved scientists and security experts rather than ethicists in particular.[15] While much bioethical discussion has focused on research ethics and the ethical implications of genetics/biotechnology, it is ironic that relatively little bioethical discussion has, to date, focused on dual-use research in particular. This is unfortunate because, given the potential for catastrophic consequences, dual-use research is surely one of the most important ethical issues regarding research and genetics/biotechnology.

This volume

The chapters in Part I of this volume, 'Dual Use in Context', map the terrain of dual-use issues emerging in various areas of life-science research—that is: nanotechnology (Chapter 2, Jim Whitman); neuroscience (Chapter 3, Valentina Bartolucci and Malcolm Dando); synthetic biology (Chapter 4, Alexander Kelle); agriculture (Chapter 5, Simon Whitby); and tuberculosis (Chapter 6, Nancy Connell).

The chapters in Part II, 'Ethical Frameworks and Principles', explore the relevance of various existing philosophical frameworks and/or tools of ethical analysis to the dual-use problem (in cases where their relevance to dual-use problems has received relatively little previous exploration). In particular, they examine the impact of environments and institutions on moral development and ethical reasoning (Chapter 7, Judi Sture); the ethics of weapons research in general (and the relevance thereof to dual-use research in particular) (Chapter 8, John Forge); application of 'rational decision theory' to the dual-use problem (Chapter 9, Thomas Douglas); the relevance of the 'doctrine of double effect' to the dual-use problem (Chapter 10, Suzanne Uniacke); implications of uncertainty for dual-use decision-making (Chapter 11, Michael Smithson); collective-action problems associated with dual-use research (Chapter 12, Seumas Miller); the

---

15 Selgelid, M. J. 2010, 'Ethics engagement of the dual use dilemma: progress and potential', in B. Rappert (ed.), *Education and Ethics in the Life Sciences: Strengthening the Prohibition of Biological Weapons*, ANU E Press, Canberra, pp. 23–34.

relevance of just-war theory to the dual-use problem (Chapter 13, Koos van der Bruggen); and the relevance of the 'precautionary principle' to dual-use decision-making and policymaking (Chapter 14, Steve Clarke).

Parts III and IV, 'Ethical Practices' and 'Ethical Futures', consider existing, developing and future practices and policies regarding ethics and dual-use life-science research—that is, the prospect of self-regulation by scientists (Chapter 15, David Resnik); lessons learnt from the history of nuclear physics (Chapter 16, Nicholas Evans); dual-use governance in developing countries (Chapter 17, Louise Bezuidenhout); the responsibilities of individual scientists in the context of incapacitating chemical and toxin agents (Chapter 18, Michael Crowley); and the WHO's project on 'Responsible Life Sciences Research' (Chapter 19, Emmanuelle Tuerlings and Andreas Reis). The Conclusion (Chapter 20) offers a wide-ranging and agenda-setting summary by Brian Rappert.

This volume is a product of a workshop (funded by the Wellcome Trust) on 'Promoting Dual Use Ethics' (organised by Michael Selgelid and Brian Rappert) held at the Centre for Applied Philosophy and Public Ethics (CAPPE) at The Australian National University in Canberra in January 2010.

Part I: Dual Use in Context

2. Nanotechnology and Dual-Use Dilemmas

Jim Whitman

Nanotechnology development and the accommodation of risk

The abundant dual-use potential of nanotechnology arises from the fact it is not materials-specific in either the organic or the inorganic realms. At the nanometre scale, familiar materials can display altered, unexpected and/or amplified qualities including tensile strength, viscosity, conductivity and antimicrobial properties. Our ability to manipulate matter at the molecular and even atomic levels opens up a breathtaking range of possibilities in engineering, medicine, materials science, energy and computing, to list but a few. At the nanometre level (a nanometre is one-billionth of a metre, approximately the diameter of a strand of DNA), disciplinary boundaries between biology and chemistry have become indistinct; and novel techniques notwithstanding, nanobiotechnology is biotechnology conducted at the nano level.[1] Technological convergence—the project to create a 'single engineering paradigm' combining nanotechnology, information technology, biotechnology and cognitive science—is being actively pursued in both the United States and the European Union.[2] These developments have already extended our understanding of what science and technology can achieve.

Although the proponents of nanotechnology and technological convergence exalt the possibilities for human betterment, a great many malfeasant objectives and adaptations are not merely a logical possibility, but also a risk integral to their development and dissemination.

---

1   Many issues in nanoscience and nanotechnology turn on the practical aspects of disciplinarity, for proponents of nanotechnology, science educators and practising scientists themselves. See Wienroth, M. 2002, 'Disciplinarity and research identity in nanoscale science and technologies', in J. S. Ach and C. Weidemann (eds), *Size Matters: Ethical, Legal and Social Aspects of Nanobiotechnology and Nano-Medicine*, Lit Verlag, Berlin, pp. 157–77.
2   Roco, M. C. and Bainbridge, W. S. (eds) 2002, 'Converging technologies for improving human performance: nanotechnology, biotechnology, information technology and cognitive science', *National Science Foundation/ Department of Commerce-sponsored Report*, <http://www.wtec.org/ConvergingTechnologies>; Nordmann, A. 2004, *Converging Technologies—Shaping the Future of European Societies*, <http://www.ntnu.no/2020/pdf/final_report_en.pdf>.

Nanotechnology: Pace, promise and peril

The promise of nanotechnology might well be overstated in some quarters, but there can be no doubting the political importance that has accompanied and furthered its establishment. It is now sufficiently well embedded in research and industrial settings in more than 60 countries that the appearance of nanomaterials in a range of mundane consumer products is unremarkable.[3] Indeed, such is the rate of research breakthroughs and the development of new nanotechnology products and processes that our deliberative systems and established means of regulating them are already being outpaced.[4] The following, from a report on nanomaterials by the Royal Commission on Environmental Pollution, under its depiction of 'risk governance', can also be read with a view to our capacities to apply dual-use bioethics across the full span of nanotechnology and nanobiotechnology:

> [Currently], regulators face a Sisyphean task. Innovation is, or soon will be, driving new products onto the market at rates that are orders of magnitude faster than they can currently hope to manage with the resources at their disposal. We heard from one regulatory body that it was not even considering how to address third and fourth generation nanomaterials because they were fully occupied with those currently at the commercial stage. The magnitude of the task combined with constraints on resources tends to create an attitude of regulatory fatalism ... One expert likened the challenge of risk governance in this field to that of shouting a warning to the driver of an express train as it thunders past.[5]

And, in the commission's view, what is true for regulatory deliberation also holds for public consideration more generally:

> There remains in our view a real question about whether the capacity for deliberation and (perhaps even more so) for public engagement in modern democracies is sufficient to sustain an approach that seeks to interrogate scientific and technological developments. The more specific the focus, the more numerous the cases will be, but as the focus becomes more generalized (looking at 'nanotechnologies' as a whole for example), the range of possible applications and implications threatens to make dialogue unmanageable. Informed and inclusive deliberation (however

---

3 See the inventory of nanotechnology-based consumer products at The Project on Emerging Nanotechnologies, Woodrow Wilson Center, <http://www.nanotechproject.org/inventories/consumer/>.
4 Whitman, J. 2007, 'The challenge to deliberative systems of technological systems convergence', *Innovation: The European Journal of Social Sciences*, vol. 20, no. 4, pp. 329–42.
5 Royal Commission on Environmental Pollution 2008, *Novel Materials in the Environment: The Case of Nanotechnology*, <http://www.official-documents.gov.uk/document/cm74/7468/7468.pdf>, p. 66.

conducted) on a huge range of potential developments seems as distant a prospect as the resolution of many of the technical uncertainties identified elsewhere.[6]

The commission's concerns appear to be validated by the difficulties that have met the many efforts to engage interested publics in serious ethical as well as practical deliberations about the likely impacts and implications of nanotechnology-driven change:

> This commitment to 'upstream' public engagement raises many unresolved questions. At what stages in scientific research is it realistic to raise issue of public accountability and social concern? How and on whose terms should such issues be debated? Are dominant frameworks of risk, ethics and regulation adequate? Can citizens exercise any meaningful influence over the pace, direction and interactions between technological and social change? How can engagement be reconciled with the need to maintain the independence of science, and the economic dynamism of its applications?[7]

Even as these morally weighty yet essentially abstract considerations are being aired, however, the development of nanotechnology is forging ahead, with practical and moral implications that can only be guessed at, as the royal commission describes with respect to whether nanomaterials might pose serious risks to human health and ecosystem integrity.

> Determining the fate of novel materials is vital when assessing the toxicological threat they pose. Nanomaterials are illustrative of the challenge. Techniques for their routine measurement in environmental samples are not widely available, nor are we currently able to determine their persistence in the environment or their transformation into other forms. Laboratory assessments of toxicity suggest that some nanomaterials could give rise to biological damage. But to date, adverse effects on populations or communities of organisms *in situ* have not been investigated and potential effects on ecosystem structure and processes have not been addressed. Our ignorance of these matters brings into question the level of confidence that we can place in current regulatory arrangements.[8]

---

6   Ibid., pp. 74–5.

7   Kearnes, M., Macnaghten, P. and Wilsdon, J. 2006, *Governing at the Nanoscale: People, Policies and Emerging Technologies*, Demos, London.

8   Royal Commission on Environmental Pollution, op. cit., p. 6. See also Morris, J., de Martinis, D., Hansen, B., Sintes, J. R., Kearns, P. and Gonzalez, M. 2011, 'Science policy considerations for responsible nanotechnology decisions', *Nature Nanotechnology*, vol. 6, pp. 73–7.

When set against the products and promise of nanotechnology, the very considerable risks and unknowns of the diffusion of nanomaterials might be expected to generate a great many dilemmas both at policymaking and at laboratory levels. Since these are not much in evidence either in public debate or in the momentum of nanoscientific advance, it is clear that the powerful interests involved and the expected gains and competitive advantages carry more weight than either ethical disquiet or practical concerns. Although this is hardly a novel position (of 30 000 bulk chemicals in use in the European Union, only 3000 or so have been formally assessed for health and environmental effects),[9] the moral significance and practical risks involved in nanotechnologies on present and planned scales cannot be regarded much as one might an incremental addition to an already chemicalised environment. After all, the 'transformative' qualities of nanotechnology have been much touted by its proponents, which accounts for the acknowledgment of serious ethical and societal implications that have found expression in the academic as well as policy and policy-related literatures.[10]

The acknowledged possibility of deleterious but unwanted outcomes of nanotechnology, however, in the form of kinds and degrees of toxicity—that is, a biochemical risk of a familiar sort, albeit perhaps on an exceptional scale—is not an end to the matter. It is in the realm of the many uses and adaptations of nanotechnology in practically every field of the physical and medical sciences and engineering that the larger dangers reside—and as part of them, the prospect of an explosion of dual-use issues. These are routinely gathered under the generic term 'societal implications', with 'ethical issues' often given freestanding consideration, by which it is unclear where military nanotechnology is (or should be) sited. After all, the research and development of military uses and adaptations of nanotechnology are by no means either merely prospective or peripheral to the establishment of nanotechnology. Although many of these are at least nominally defensive,[11] active research and already articulated war-fighting possibilities and other means of inflicting serious harm have not been

---

9   Royal Commission on Environmental Pollution, op. cit., p. 50.

10   See, for example, Roco, M. C. and Bainbridge, W. S. (eds) 2001, *Societal Implications of Nanoscience and Nanotechnology*, Kluwer Academic Publishers, Dordrecht; Roco, M. C. and Bainbridge, W. S. (eds) 2007, *Nanotechnology: Societal Implications II: Individual Perspectives*, Springer, Dordrecht; 'The coevolution of human potential and converging technologies', *Annals of the New York Academy of Sciences*, vol. 1013 (May 2004); 'Progress in convergence: technologies for human wellbeing', *Annals of the New York Academy of Sciences*, vol. 1093 (2006); Banse, G., Grunwald, A., Hronszky, I. and Nelson, G. (eds) 2007, *Assessing Societal Implications of Converging Technological Development*, Edition Sigma, Berlin; Allhoff, F., Lin, P., Moor, J. and Wekert, J. (eds) 2007, *Nanoethics: The Ethical and Social Implications of Nanotechnology*, Wiley-Interscience, Hoboken, NJ.

11   Kosal, M. E. 2009, *Nanotechnology for Chemical and Biological Defense*, Springer, New York; see also the MIT Institute for Soldier Nanotechnologies, <http://web.mit.edu/isn/>.

precluded.[12] Of the 13 US federal agencies which shared $1.6 billion under the US National Nanotechnology Initiative for 2009, the Department of Defense was allocated 28 per cent.[13]

Military uses and adaptations of nano-enabled products and processes are as yet not concentrated around dedicated means of offensive war-fighting or new weapons systems. Instead, the foreseeable nanotechnology developments most likely to be adopted for military purposes have a very wide range of applications: smart materials, micro-electromechanical systems, nanocomputing, robotics, micro-sensors and many more. Jurgen Altmann has noted that if 'the production facilities for raw material, feedstock, energy, and final products as well as the transport systems are themselves produced by' molecular nanotechnology, 'a very fast increase of the production and distribution of military goods is possible', and one possible outcome of this is that molecular nanotechnology 'production of nearly unlimited numbers of armaments at little cost would contradict the very idea of quantitative arms control and would culminate in a technological arms race beyond control'.[14] In any event, since the worldwide growth of nanotechnology research and development has been propelled by the expectation that it will usher in a 'new industrial revolution', an 'age of transitions' and other epoch-defining transformations, government sanction of nanotechnology developments has always had a strongly competitive edge— made all the more sharp now that it has been linked to conceptions/aspects of national security.[15] Realist fears inform the development of security dilemmas; and the dangers are already apparent: 'By taking cues from Moscow's centralized and opaque institutions, Washington risks misperceiving Russia's intentions and calculations … as well as prematurely locking into strategic competition. The result, if unaltered, could convert the promise of nanotechnology into a new realm of commercial rivalry and arms racing.'[16]

The perils of nanotechnology, however, are not limited to arms races of the sort that can, at least in principle, be addressed by dedicated arms-control initiatives.[17] There has already been concern over the comprehensiveness of the

---

12   Altman, J. 2006, *Military Nanotechnology*, Routledge, London. See also Shipbaugh, C. 2006, 'Offense-defense aspects of nanotechnologies: a forecast of military applications', *Nanotechnology*, pp. 741–7; Kharat, D. K., Muthurajan, H. and Praveenkumar, B. 2006 'Present and future military applications of nanaodevides', *Synthesis and Reactivity in Inorganic, Metal-Organic and Nano-Metal Chemistry*, vol. 36, pp. 231–5.

13   US National Nanotechnology Initiative, *Funding*, <http://www.nano.gov/html/about/funding.html> .

14   Altman, op. cit.

15   Vandermolen, LCDR T. D. 2006, 'Molecular nanotechnology and national security', *Air & Space Power Journal*, pp. 96–106.

16   Stulberg, A. N. 2009, '"Flying blind" into a new military epoch: the nanotechnology revolution, emerging security dilemmas, and Russia's double-bind', Paper presented at the International Studies Association Conference, February 2009.

17   Whitman, J. 2011, 'The arms control challenges of nanotechnology', *Contemporary Security Policy*, vol. 32, no. 1, pp. 99–115.

Biological Weapons Convention (BWC) and the Chemical Weapons Convention (CWC) relative to the malfeasant purposes to which nanomaterials and products can be put to use:

> Article 1 of the BWC prohibits the development, production, and possession of 'microbial or other biological agents, or toxins whatever their origin or method of production' ... Assuming microbial or biological agents should be considered living things, can a toxin be something inorganic or artificially created? Since the phrase 'whatever their origin or method of production' relates to toxins because of the comma placement, it seems to include so-called mechanical devices that could result from mature nanotechnology. In this sense, one can argue that nanotechnology—particularly nanorobots—can be treated as a toxin if it causes harm similar to already known toxins. However, because the BWC seems to deal only with biological organisms or products thereof, a strong argument can be made that the artificially assembled products that nanotechnology would produce cannot possibly fit under the BWC. Perhaps the only way nanotechnology can fall under the BWC's prohibitions without a doubt is if it were used to artificially create replicas of known biological weapons or toxins. Only then would it clearly be covered since no difference would exist between the natural product and the 'artificial' version. However, nanotechnology as a field is much broader than these narrow 'biological replica' applications.[18]

Nanorobots are not first and foremost an arms-control issue, but a dual-use one, with the main lines of current research in the field dedicated to bio, industrial and medical nanorobots.[19] The wide applicability and adaptability of nano-enabled products and processes mean that their dual-use potential is more often than not clearly evident. For example, where a nano-enabled medical advance also holds the promise of more effective delivery of a chemical or biological agent for offensive purposes, one might expect this potential to find expression as a dilemma of considerable weight and intractability. Measured against the many considerable benefits that can arise from their many uses, however, are the research and development of any dual-use nano-development likely to present practising scientists, technologists or policymakers with a *dilemma*?

---

18   Pinson, R. D. 2004, 'Is nanotechnology prohibited by the Biological and Chemical Weapons Conventions?' *Berkeley Journal of International Law*, vol. 22, no. 2, p. 298.
19   See the Special Issue of *The International Journal of Robotics Research*, vol. 28, no. 4 (April 2009): 'Current State of the Art and Future Challenges in Nanorobotics'.

Ethics, applied ethics and dual-use dilemmas

Implicit in the idea of a dual-use dilemma is that the beneficent purposes and/ or uses of a technology are roughly equal to the potential harm that could be brought about by pernicious applications. If harmful and destructive uses for a developing scientific or technological advance can be foreseen but are not expressed, or not expressed as a *dilemma*, the dual-use potential will not be sufficient on its own to necessitate the kinds of deliberation (scientific, managerial, legal, political) that will regulate, slow, modify or halt the development in question. Subsequent concerns about risks and dangers then become post-facto legislative and law-enforcement matters. Similarly, regulatory initiatives such as export bans to dangerous regimes or to untrustworthy end users do not address dilemmas, but arise from cost and risk/benefit analyses.

Within established scientific and technological arenas, real and potential dual-use dilemmas are not abstract: their 'visibility' in every sense is conditioned by some combination of political priorities; inter-business and international competitive impulses; patterns of public and private financing; and by the narrowness of specialist research that can contribute to, rather than exhibit, dual-use potential. Acknowledging this does not oblige one to subscribe to one or another form of technological determinism, but it takes into account that the proclaimed benefits on the plus side of dual use can serve promotional purposes (of which, beware nano-hype);[20] and that they are rarely if ever wholly disinterested. As discussed further below, these and related considerations narrow the range of actors likely to perceive or to voice dual-use dilemmas, as well as the circumstances when they will be able or likely to do so. At the same time, although many kinds of moral significance and forms of ethical predicament can follow in the wake of scientific and technological advances, these do not all take the form of dilemmas; there are, for example, detailed, ethical arguments both for and against various forms of human enhancement through genetic engineering.[21]

As a general consideration, dual-use dilemmas leave open the question of the affected and/or responsible agent—who or what experiences the dilemma—and indeed, whether such dilemmas are framed in ethical rather than, or in addition to, prudential considerations. For the purposes of this argument, it can be assumed that any dilemma arising from the possible misapplication of biological knowledge will entail practical dangers of considerable moral significance. We

---

20   Toumey, C. 2004, 'Nano hyperbole and lessons from earlier technologies', *Nanotechnology Law and Business*, vol. 1, no. 4, pp. 397–405; Berube, D. M. 2006, *Nano-Hype: The Truth Behind the Nanotechology Buzz*, Prometheus Books, Amherst, Mass.

21   Harris, J. 2007, *Enhancing Evolution: The Ethical Case for Making Better People*, Princeton University Press, Princeton, NJ; Sandel, M. J. 2007, *The Case against Perfection: Ethics in An Age of Genetic Engineering*, Belknap Press, Harvard, Mass.

can read as much into the definition of 'dual-use research of concern' proposed by the US National Science Advisory Board for Biosecurity: 'research that, based on current understanding, can be reasonably anticipated to provide knowledge, products, or technologies that could be directly misapplied by others to pose a threat to public health, agriculture, plants, animals, the environment, or materiel.'[22] Moral unease, however, is never automatically congruent with estimates of the practical significance of possible dangers; and it is in the nature of a dilemma that countervailing considerations are of roughly equal weight. In addition, countervailing considerations in life-science dual-use conundrums— that is, those that bolster the case for either the positive or the negative aspects of the dual use to be determined—most often entail structural considerations. This is because of the range of powerful interests involved and because the encompassing arenas are international and/or global.

The application of bioethics or ethical reasoning more generally is not a matter of informed individuals unfailingly applying knowledge to circumstance; and this applies to scientists no less than to any other moral agents.[23] In his comprehensive study of scientists whose research was furthered by collaboration with the Nazi regime, John Cornwell depicts the striking similarity of pressures facing scientists then and now: 'The greatest pressures on the integrity of scientists are exerted at the interface between the professional practice of science and the demands of award-giving patrons.' The narrative of scientific collaboration with the Nazi regime reveals 'the pressures of hubris, loyalty, competition and dependence leading to compromise. In the final analysis the temptation was a preparedness to do a deal with the Devil in order to continue doing science.'[24]

Even when there is no obvious devil in the form of a psychopathological regime, scientific research can but rarely be understood as 'pure' in the sense that it can wholly be abstracted from the potential for pernicious uses; and ethical concerns are embedded in risk/benefit calculations that do not present in clear, dichotomous form. So it is that particular cases of the dissemination of biological knowledge through open, peer-reviewed publication can be seen as posing grave risks of misuse;[25] but at the same time, there is no enthusiasm for general prohibitions on scientific exchanges, on which so much beneficent advance depends, and for which there is no commanding legal reach.[26] It follows that in

---

22   National Science Advisory Board for Biosecurity (NSABB) 2008, 'Frequently asked questions', <http://www.biosecurityboard.gov/faq.asp#1>.

23   Whitman, J. 2010, 'When dual use issues are so abundant, why are dual use dilemmas so rare?' Research report for the Wellcome Trust project 'Building A Sustainable Capacity in Dual Use Bioethics', <http://www.dual-usebioethics.net/>.

24   Cornwell, J. 2004, Hitler's Scientists: Science, War and the Devil's Pact, Penguin, London, p. 462.

25   Indicative examples are listed in Institute of Medicine and National Research Council 2006, Globalization, Biosecurity and the Future of the Life Sciences, The National Academies Press, Washington, DC, pp. 53–4.

26   Marchant, G. E. and Pope, L. L. 2009, 'The problems of forbidding science', Science and Engineering Ethics, vol. 15, pp. 375–94.

such instances, the moral and/or prudential reasoning of individual researchers, publishers and concerned institutions will be informed and conditioned by larger interests and concerns, many of which provide the basis on which particular dual-use dilemmas are generated—in this case, the maintenance of an open epistemic community that will support the range of biological sciences integral to twenty-first-century life. The cases of nanotechnology and technological convergence have been conceived and established not only in ways that highlight the beneficent potential (subtitles of key works include 'technologies for human wellbeing' and 'for improving human performance'), but that also depict 'ethical concerns' and 'societal implications' more as challenges to be dealt with than as outcomes to be avoided.

It is against this background that we must set the potential of bioethics to identify, articulate and address dual-use issues in the life sciences. What needs to be borne in mind is that, in common with other specialist forms of applied ethics within the life sciences (neuro-ethics, nano-ethics, and so on), dual-use bioethics operates *within* fields of endeavour that are engines for generating dual-use issues; it does not bring ethical scrutiny to bear on its own operating environments. In one sense, this is as it should be: subfields of ethics are delimited and specialist—that is their strength, particularly as the issues they engage increase in number and complexity. Yet the growth of specialist, applied ethics in the life sciences also has the effect of signalling the ethical validity of fields they are established to scrutinise, however many dual-use issues they seem set to produce. As the number and kind of ethical quandaries arising from far-reaching scientific and technological advances proliferate, attention naturally focuses on actual or impending difficulties, rather than on the conditions that produce them. Of course, the relationship between medicine and medical ethics is largely unproblematic in this regard, the much larger compass of modern biosciences and bioethics less clearly so, since enterprises such as genetic engineering remain contested in themselves and not merely in respect of some of their particulars or applications. Nanotechnology is still more vexed.

Yet at the same time, there is a great deal of developed-world government anxiety that, 'ethical and societal implications' notwithstanding, nothing should seriously impede the entrenchment and furtherance of nanotechnology and envisioned forms of technological convergence for which it is the foundation.[27] Hence the assertion in the path-breaking US National Science Foundation/ Department of Commerce (NSF/DOC) report on converging technologies that '[p]rogress can become self-catalyzing if we press ahead aggressively; but if we hesitate, the barriers to progress may crystalize and become harder to

---

27   Fisher, E. and Mahajan, R. L. 2006, 'Contradictory intent? US federal legislation on integrating societal concerns into nanotechnology research and development', *Science and Public Policy*, pp. 5–16.

surmount'.[28] Similarly, the principal EU report on nano-led convergence set out the plan for forging ahead despite acknowledging that each of the likely characteristics of converging technology applications 'presents an opportunity to solve societal problems, to benefit individuals, and to generate wealth. Each of these also poses threats to culture and tradition, to human integrity and autonomy, perhaps to political and economic stability.'[29]

Still more remarkable is the following assumption contained in the NSF/DOC report: 'The ability to control the genetics of humans, animals and agricultural plants will greatly benefit human welfare; widespread consensus about ethical, legal and moral issues will be *built into* the process.'[30]

Such a starkly instrumentalist understanding of ethical deliberation was characterised by Paul Ramsey more than 30 years ago:

> We need to raise … ethical questions with a serious and not a frivolous conscience. The man of frivolous conscience announces that there are ethical quandaries ahead that we must urgently consider before the future catches up with us. By this he often means that we need to devise a new ethics that will provide the rationalization for doing in the future what men are bound to do because of new actions and interventions science will have made possible. In contrast a man of serious conscience means to say in raising urgent ethical questions that there may be some things that men should never do. Good things that men do can be made complete only by the things they refuse to do.[31]

The reassurance in the EU report that '[e]nlightened exploitation of discoveries in' nanotechnology, biotechnology, information technology and cognitive science 'will humanize technology rather than dehumanize society'[32] also has the effect of making ethical deliberation appear capable of precluding dilemmas. Moreover, it appears that the task of practitioners in the social sciences and humanities is to act as guides across the space between anxious publics and prepared arenas, if not fixed ends. The following (also from the EU document cited above) on nano-facilitated technological convergence outlines their roles:

> The broadly transformative potential of [technological convergence] sets limits to [its] public acceptance. The pace of the diffusion of new technologies is constrained by the pace [at which we accept] and, if so, accommodate them. Here the social sciences and the humanities are

---

28   Roco and Bainbridge, 2002, op. cit., p. 3.
29   Nordmann, op. cit., p. 4.
30   Roco and Bainbridge, 2002, op. cit., p. 5 (emphasis added).
31   Ramsey, P. 1977, *Fabricated Man: The Ethics of Genetic Control*, Yale University Press, New Haven, Conn., pp. 122–3.
32   Nordmann, op. cit., p. 99.

needed to inform and accompany [technology convergence] research and to serve as intermediaries. They should create settings within which science and technology researchers on the one hand and the various publics on the other, can learn from each other.[33]

Such a position does not invalidate the many genuine and methodologically scrupulous efforts that have been made to engage both working scientists and the general public in dialogue about the risks and uncertainties of nanotechnology,[34] whether or not one agrees that the metaphor 'moving the debate upstream' adequately captures the degree to which debate is constrained by political, research, industrial and military developments already well advanced—and the degree to which advocates of nanotechnology foresee the compass and role of ethical deliberation. As depicted in the NSF/DOC report, the nanotechnology 'effort should have many stakeholders in education, healthcare, pharmaceuticals, social science, the military, the economy and the business sector, to name a few. No less than a comprehensive national effort will be required.' In addition, M. C. Roco, widely regarded as the founding father of the US National Nanotechnology Initiative, has recounted the efforts made in the 1990s to ensure widespread, interlocking social and institutional momentum behind the establishment and acceptance of nanotechnology:

> The US National Nanotechnology Initiative was conceived as an *inclusive process* where various stakeholders would be involved. In 1999 we envisioned a 'grand coalition' of academic, industry, governments, states, local organizations, and the public that would advance nanotechnology … Creating a chorus of approval for nanotechnology, from 1990 to March 1999 was an important preliminary step.[35]

Two considerations arise from the history of how nanotechnology has been introduced, institutionalised and consolidated. First is the difficulty any individual is likely to face in apprehending, acknowledging and voicing ethical concerns over dual use that run counter to a prevailing ethos of tightly interlocked institutions and interests. One need not subscribe to Daniel S. Greenberg's judgment that '[a]n infinity of researchable topics renders science insatiable for money and increasingly indiscriminate in ways to get it',[36] which depicts a corroded ethos, but leaves open the possibility for the ethical

---

33 Ibid., p. 18.
34 For example, see Rogers-Hayden, T. 2007, 'Moving engagement "upstream"? Nanotechnologies and the Royal Society and Royal Academy of Engineering's inquiry', *Public Understanding of Science*, vol. 16, no. 3, pp. 345–64; Kearnes, M., Macnaghten P. and Wilsdon, J. 2006, 'Governing at the nanoscale: people, policies and emerging technologies', Demos, London; and Nanologue.net, 2006, <http://www.nanologue.net/>.
35 Roco, M. C. 2006, *National Nanotechnology Initiative—Past, Present and Future*, <http://www.nano.gov/NNI_Past_Present_Future.pdf> (italics in original).
36 Greenberg, D. S. 2001, *Science, Money and Politics: Political Triumph and Ethical Erosion*, University of Chicago Press, Chicago, p. 463.

deliberation and moral courage of individuals. As psychologist Philip Zimbardo has observed, however, '[m]ost of us have a tendency both to overestimate the importance of dispositional qualities and to underestimate the importance of situational qualities when trying to understand the causes of other people's behavior'.[37] Zimbardo distinguishes between dispositional, situational and systemic causes of behaviour; and although his interest is in trying to account for cruel, inhumane and violent behaviours, his professional experiences, ranging from the Stanford Prison Experiment to Abu Ghraib trial testimony, illustrate how difficult it is for individuals to maintain and enact an oppositional ethical position against a powerful, prevailing expectation. If, as quoted above, the proponents of nano-led technological convergence foresee that 'widespread consensus about ethical, legal and moral issues will be built into the process', the practical prospects for dual-use bioethics will rest on levels of moral courage unsupported by the records of social psychology or history.

Second, it might reasonably be asked whether, if ethicists and bioethicists are stakeholders in a nano-led technological convergence initiative, they can also bring ethical scrutiny to the enterprise itself. There is a strong case for nonconformity with the functional roles that some advocates of nanotechnology foresee for the social sciences and humanities:

> We object to the narrow apprehension that the function of the humanities and the social sciences consists only of achieving public trust concerning nanotechnology. We believe that philosophy and ethics have a critical function regarding the implementation of new technologies, which for instance encompasses asking fundamental questions such as, What impact will this new technology have on humanity? What is a good life? Will this new technology affect the realization of a good life? What kind of society do we want? and How does this new technology relate to that kind of society? (And the list could easily be extended.)[38]

This argument has purchase as a matter of professional orientation, but still greater meaning within contexts that contain such as the following, which concerns the prospect of a nanotechnology-enabled battlefield in which soldiers 'could be accompanied by machines capable of making their own decisions' (again, from the NSF/DOC report):

> [C]ognitive scientists can do research on how a cyborg system makes decisions about what constitutes a legitimate target under varying conditions, including [the] amount of information, how the information is presented, processing time, and the quality of the connection to

37   Zimbardo, P. 2007, *The Lucifer Effect: How Good People Turn Evil*, Rider, London, p. 8.
38   Ebbsen, M., Andersen, S. and Besenbacher, F. 2006, 'Ethics in nanotechnology: starting from scratch?' *Bulletin of Science, Technology and Society*, vol. 26, no. 6, p. 453.

higher levels of command. Practical ethicists can then work … with cognitive scientists to determine where moral decisions, such as when to kill, should reside in this chain of command … Practical ethicists and social scientists need to act as stand-ins for other global stakeholders in debates over the future of military nanotechnology.[39]

None of the above diminishes the importance of awareness and ethical engagement of dual-use potentials by practising life scientists and by applied ethicists, but it does contextualise it. Where, one might ask, are the ethicists and their awareness of the dual-use potential of nanotechnology? We surely know enough in outline about the practical potential and hoped-for applications of nanotechnology (and the considerable efforts being expended to secure them) to recognise them as morally weighty and ethically challenging matters. In respect of dual-use bioethics as it applies to nanotechnology, moral engagement with the project might best be thought of as pre-emptive rather than speculative. Contrary to this, Alfred Nordmann has argued:

> Ethical reflection of science and technology typically reacts to issues that present themselves in the form of classical dilemmas, actual and current predicaments, or hypothetical cases. In the case of reproductive technologies, for example, ethical discussion has proven its relevance by being very close on the heels always of novel techniques. In contrast, nanotechnologies develop a tool-box for technological development. As such they prepare the ground for a technical convergence at the nanoscale. By enabling such a convergence, nanotechnologies create a methodological challenge in that ethical engagement with presenting issues becomes displaced by a perceived need to proactively engage emerging issues. Lay and professional ethicists are only beginning to meet this challenge.[40]

To characterise nanotechnological developments as a 'toolbox', however, and to assert that the ethical challenges presented at this stage are 'methodological' are to abstract them from their social, political and economic contexts, which shape their purposes and drive their momentum. In short, this line of argument insulates nanotechnological developments from ethical scrutiny. But the political sanction and investments that have directed the nanotechnology enterprise have brought about a proliferation of new dual-use issues and have extended or hybridised familiar ones. The weight and competitive orientation of the interests involved have also established it on foundations shot through

---

39   Roco and Bainbridge, 2002, op. cit., p. 370.
40   Nordmann, A. 2007, 'If and then: a critique of speculative nanoethics', *Nanoethics*, vol. 1, no. 1, p. 34.

with moral and practical risks both profound and pervasive. These are hardly propitious operating conditions for the scope and efficacy of either nano-ethics or dual-use bioethics as applied to nanotechnology.

Dual use and dual-use dilemmas in nanotechnology

As we have seen, even at this early stage of development, the thoroughgoing international commitment to nanotechnology research and practical applications has brought about a condition in which new scientific breakthroughs, novel materials and processes and technological adaptations of nanotechnology are coming on stream at speeds and with complex interrelations that challenge professional comprehension at the community level. These have already begun to outpace our wider deliberative systems—including our means of ethical scrutiny. To this we can add the studied reluctance of nanotechnology's many advocates to engage ethical deliberation outside a framework in which ethical considerations are but one of a set of 'issues' in a scientific and technological progression that must not lose its inertia. This has been the case from the start, as the following two examples illustrate. The first, a report by the US National Research Council in 2002, explained why there was then a paucity of social science research on the 'societal implications' of nanotechnology: 'There appear to be a number of reasons for the lack of activity' in social science research on societal implications.

> First and foremost, while a portion of NNI [National Nanotechnology Initiative] support was allocated to various traditional disciplinary directorates, no funding was allocated directly to the Directorate of Social and Behavioral and Economic Sciences, the most capable and logical directorate to lead these efforts. As a consequence, social science work on societal implications could be funded in one of two ways: (1) it could compete directly for funding with physical science and engineering projects through a solicitation that was primarily targeted at that audience or (2) it could be integrated within a nanoscience and engineering center.[41]

More recently, the US National Nanotechnology Initiative's funding opportunities for 'societal implications' in fiscal year 2010 are typical (and, it should be added, only slightly above 1 per cent of the total disbursement):

---

41 National Research Council (NRC) 2002, *Small Wonders, Endless Frontiers: A Review of the National Nanotechnology Initiative*, Committee for the Review of the National Nanotechnology Initiative, Division on Engineering and Physical Sciences, National Research Council, National Academies Press, Washington, DC, p. 34.

Research directed at identifying and quantifying the broad implications of nanotechnology for society, including social, economic, workforce, educational, ethical, and legal implications ($5.78 million). The application of nanoscale technologies will stimulate far-reaching changes in the design, production, and use of many goods and services. Factors that stimulate scientific discovery at the nanoscale will be investigated, effective approaches to ensure the safe and responsible development of nanotechnology will be explored and developed, and the potential for converging technologies to improve human performance will be addressed. The Nanotechnology in Society Network will extend its national and international network.[42]

A second difficulty is that the range of uses of nanotechnology is so extensive and the promise so considerable that 'dual use' will not accurately or fully capture the likely relationship between promise and peril. This is particularly compelling in the case of nano-facilitated medical advances. For example, in an experiment that shows great promise for treating brain cancer in humans, University of Washington researchers have been able to cross the blood–brain barrier in mice. 'Until now, no nanoparticle used for imaging has been able to cross the blood–brain barrier and specifically bind to brain-tumor cells. With current techniques doctors inject dyes into the body and use drugs to temporarily open the blood–brain barrier, risking infection of the brain.'[43] But the possibilities of nano-facilitated drug-delivery systems[44] extend greatly beyond the precise imaging of brain tumours, particularly with respect to malign applications of neurobiology.[45]

Third, the very adaptability of nanotechnology makes a determination of dual use at early stages of development very difficult, or introduces the possibility at a threshold that is below that required for a genuine dilemma. For example, in the following, it is very unlikely that a scientist engaged in the development of nanowires would regard the full span of their possible applications as posing an ethical dilemma:

University of Arkansas researchers have created assemblies of nanowires that show potential in applications such as armor, flame-retardant fabric, bacteria filters, oil cracking, controlled drug release, decomposition of

---

42   National Nanotechnology Initiative (NNI), *Funding Opportunities, at NSF in FY 2010,* <http://www.nsf.gov/crssprgm/nano/core10_nsf_wta_nni.doc>.
43   'Nanoparticles cross blood–brain barrier to enable "brain tumor painting"', *University of Washington News*, 3 August 2009, <http://uwnews.org/article.asp?articleid=51245>
44   Suril, S. S., Fenniri, H. and Singh, B. 2007, 'Nanotechnology-based drug delivery systems', *Journal of Occupational Medicine and Toxicology*, vol. 2, no. 16, <http://www.occup-med.com/content/pdf/1745-6673-2-16.pdf>.
45   Wheelis, M. and Dando, M. 2005, 'Neurobiology: a case study of the imminent militarization of biology', *International Review of the Red Cross*, vol. 87, no. 859, pp. 1–16.

pollutants and chemical warfare agents. This two-dimensional 'paper' can be shaped into three-dimensional devices. It can be folded, bent and cut, or used as a filter, yet it is chemically inert, remains robust and can be heated up to 700 degrees Celsius.[46]

In all probability, nanotechnology dual use—that is, malign possibilities and pernicious applications as readily achievable as beneficent ones—will continue to feature in research and development programs ranging across the physical and medical sciences, but will rarely appear as dual-use *dilemmas*. The way in which the nanotechnology and technological convergence enterprises have been structured indicates faith at the policymaking level that the worrying developments or applications of nanotechnology can be dealt with on a case-by-case basis, which essentially relegates dual-use conundrums to considerations of risk. There is nothing in the way that nationally supported nanotechnology programs have been initiated and sustained that acknowledges ethical quandaries of a structural kind; and similarly, nor is it the case, as the Fink Report of 2004 asserts, that '[b]iotechnology represents a "dual use" dilemma'.[47] Genuine dilemmas are incapacitating—and there is little to suggest that our individual and collective ethical unease over the biotechnology and nanobiotechnology prospects will inhibit these enterprises, despite the number and range of dual-use possibilities they will continue to generate.

In order for dual-use potentials to be recognised and publicised as matters of social concern and public policy, dual-use bioethics in nanotechnology will need to be sited as far as possible 'upstream', towards current political initiatives and research and development commitments. But the malign possibilities are too wide-ranging in kind and too numerous to be dealt with as and when a genuine dual-use dilemma appears. Indeed, bioethics and nano-ethics will find little argumentative purchase in highlighting dual use if they concentrate on the apprehension of dilemmas. After all, when set against the already extant dual-use potential of nanotechnology, the number of expressed dilemmas is exceedingly rare.[48] What are required are ethical studies that regard the nano-enterprise from 'outside' or 'above', which will provide a context for, and inform, applied ethical studies, such as bioethics and nano-ethics can provide.

---

46 US National Nanotechnology Initiative, 2006, <http://www.nano.gov/html/research/home_research.html>.
47 Committee on Research Standards and Practices to Prevent the Destructive Application of Biotechnology, National Research Council of the National Academies 2004, *Biotechnology Research in An Age of Terrorism* [Fink Report], The National Academies Press, Washington, DC, p. 15.
48 Whitman, 2010, op. cit.

3. What Does Neuroethics Have to Say about the Problem of Dual Use?

Valentina Bartolucci and Malcolm Dando

Introduction

It is clear that in the past advances in neuroscience were used for hostile as well as peaceful purposes. Lethal chemical nerve agents, after all, interfered with the acetylcholine neurotransmitter system[1] and, during the twentieth century's East–West Cold War, both sides clearly also made efforts to develop 'non-lethal' chemical agents for various purposes.[2] The use of some form of fentanyl derivative(s) by Russian security forces to break the 2002 Moscow theatre siege shows that today at least one major state has deployed such a weapons system. Many commentators fear that Russia would be far from alone in having an interest in developing novel incapacitating capabilities if the advances in neuroscience provide suitable opportunities.[3]

This chapter sketches the areas of interest and methods used by neuroethicists to ask what they have had to say about the problem of dual use: the fact that advances in benignly intended civil neuroscience could produce materials, knowledge and technologies that might then be used for hostile purposes by others. Of course, it should be understood from the start that this is no small problem, as has been made abundantly clear, for example, in the Lemon-Relman report of the US National Academies in 2006, which, in its second recommendation, stated that it was necessary to '[a]dopt a broadened awareness of threats beyond the classical "select agents" and other pathogenic organisms and toxins, so as to include, for example, approaches for disrupting host homeostatic and defence systems and for creating synthetic organisms'.[4]

---

1   Dando, M. R. 1996, *A New Form of Warfare: The Rise of Non-Lethal Weapons*, Brassey's, London.
2   Dando, M. R. and Furmanski, M. 2006, 'Midspectrum incapacitant programs', in M. L. Wheelis, L. Rózsa and M. R. Dando (eds), *Deadly Cultures: Biological Weapons Since 1945*, Harvard University Press, Cambridge, Mass., pp. 236–51.
3   Pearson, A. M., Chevrier, M. and Wheelis, M. L. 2007, *Incapacitating Biochemical Weapons: Promise or Peril?* Lexington Books, Lanham, Md.
4   Committee on Advances in Technology and the Prevention of their Application to Next Generation Biowarfare Threats 2006, *Globalization, Biosecurity and the Future of the Life Sciences*, The National Academies Press, Washington, DC.

Host homeostatic and defence systems obviously would include hormones of the endocrine system, neurotransmitters of the nervous system and cytokines of the immune system—understanding all of which is critical to our continuing advances in neuroscience.

The rise of neuroethics

As Robert Blank argued before the turn of the century, whilst there may be some issues unique to the societal impacts of neuroscience, the issues raised by advances in this area are basically similar to those raised in other areas of medical advances.[5] He suggested that there are, in fact, three clear levels of policy dimensions in regard to all of these areas of technology:

> 1. ... [D]ecisions must be made concerning the research and development of the technologies ...

> 2. The second policy dimension relates to the individual use of technologies once they are available ...

> 3. The third dimension of biomedical policy centers on the aggregate societal consequences of widespread application of a technology.[6]

These roots within the general growth of bioethical concerns are generally acknowledged by neuroethicists. As Raymond de Vries noted in the 2007 special issue of *EMBO Reports* on 'The Biology of Behaviour: Scientific and Ethical Implications':[7] 'Most histories of neuroethics are varieties of the technology story. Illes and Bird (2006) place the history of neuroethics squarely in the standard account of bioethics that runs from the Nuremberg Code in 1947, to the 1964 Declaration of Helsinki ... to the Belmont Report in 1979.'

The technology story, de Vries explains, is one in which new technologies bring ethical questions that are too difficult for ordinary practitioners to answer and they therefore need expert guidance from ethicists. Being a sociologist, de Vries does not think this is the only possible history, but it is one that will be here accepted for the moment.

Despite the acknowledgment of these long roots within bioethics, it appears to be widely accepted that, as a particular field of study, 'neuroethics' originated

---

5   Blank, R. H. 1999, *Brain Policy: How the New Neuroscience Will Change Our Lives and Politics*, Georgetown University Press, Washington, DC.
6   Ibid., pp. 11 and 12.
7   de Vries, R. 2007, 'Who will guard the guardians of neuroscience? Firing the neuroethical imagination', *EMBO Reports*, vol. 8 (S1), pp. S65–9.

after the turn of the century. Martha Farah,[8] in a thoughtful attempt to delineate the field, argued, for example, that '[b]eginning in 2002, neuroscientists began to address these issues in the scientific literature ... and the field gained a name "neuroethics". At the same time, key meetings brought together large numbers of experts and specific neuroethics membership organisations began to be founded.'[9]

The comparatively recent delineation of neuroethics as a special field it is not too surprising; however, given its acknowledged deep roots within bioethics, as Parens and Johnston carefully pointed out, it is important to understand that bioethicists have made major errors in the past. If neuroethicists forget past mistakes they could end up doing the same in the near future.[10] Parens and Johnston point to three particular problems that could easily arise in the new field of neuroethics:

1. ... [T]he problem of reinventing the bioethical wheel ...

2. ... [T]he problem of exaggerating what the science can teach us about who we are ...

3. ... [T]he problem of exaggerating what bioethics research can deliver.[11]

Furthermore, given the recent delineation of the specific field of neuroethics, we should not be surprised to find that there is as yet no comprehensive, widely accepted view of the scope of the field. For example, one might argue that finding out what happens in the brain when ethical decisions are being made should be central to any conception of neuroethics, but that does not appear to be what practitioners have decided to do. Rather, the field seems to deal more with what Farah called 'the practical and the philosophical': the implications of advances in neuroscience for practical social issues and the implications of advances in neuroscience for our understanding of ourselves (which, of course, overlaps to some extent with the question of how we make ethical decisions). The sections of Farah's paper provide illustrations of these different aspects (Table 3.1).

8 Farah, M. J. 2005, 'Neuroethics: the practical and the philosophical', *Trends in Cognitive Sciences*, vol. 9, no. 1, pp. 34–40, see Figure 3.1: 'Milestones in the history of ethics in neuroscience'.
9 See, for example, 'The History of Neuroethics' section of the entry on 'Neuroethics' in *Wikipedia*.
10 Parens, E. and Johnston, J. 2007, 'Does it make sense to speak of neuroethics?' *EMBO Reports*, vol. 8 (S1), pp. S61–4.
11 From *Wikipedia*, op. cit., p. 64.

Table 3.1 Neuroethics: Practical and Philosophical

Practical: Brain imagining and brain privacy

• 'Among the neuroscience technologies that present new ethical challenges of a practical nature is functional brain imaging.'

• 'For example, in "neuromarketing" brain imaging is used to measure limbic system response to a product that may indicate consumer's desire for it.'

Philosophical: Science and the soul

• 'Recent neuroimaging research has shown a characteristic pattern of brain activation associated with states of religious transcendence, which is common to Buddhist meditation and Christian prayer.'

• 'The idea that there is somehow more to a person than their physical instantiation runs deep in the human psyche and is a central element in virtually all the world's religions.'

• 'Neuroscience has begun to challenge this view, by showing that not only perception and motor control, but also character, consciousness and sense of spirituality may all be features of the machine.'

Source: de Vries, R. 2007, 'Who will guard the guardians of neuroscience? Firing the neuroethical imagination', *EMBO Reports*, vol. 8 (S1), pp. S65–9.

Neil Levy, in his introduction to the new journal *Neuroethics*, accepted this twofold division of the subject but stressed that, to the extent that neuroscience shows that we are less than rational and autonomous in our decision-making, that must impact on our understanding of the practical impact of the advances in neuroscience.[12]

In an attempt to advance the field of neuroethics, Georg Northoff argued that both of Farah's two subdivisions of neuroethics (which might be simply characterised as what we can do and what we know) should be termed 'empirical neuroethics', and we should accept that there is no sharp distinction between them (Table 3.2). So, in his view: 'Empirical neuroethics deals with the empirical and practical aspects of the linkage between neuroscientific and ethical concepts.'[13]

---

12   Levy, N. 2008, 'Introducing neuroethics', *Neuroethics*, vol. 1, pp. 1–8.
13   Northoff, G. 2009, 'What is neuroethics: empirical and theoretical neuroethics', *Current Opinion in Psychiatry*, vol. 22, pp. 565–9.

Table 3.2 Aspects of Empirical Neuroethics

Technology	Questions	
	Old technology	New questions
Brain imaging	Safety	Intentional deception
	Researchers' obligations	Neuromarketing
	Validity	Personal characteristics
Pharmacological enhancement	Safety	Attention
	Validity	Memory
		Mood
		Equity
BMI and N-P enhancement		Military research on cyborgs

Source: Northoff, G. 2009, 'What is neuroethics: empirical and theoretical neuroethics', *Current Opinion in Psychiatry*, vol. 22, pp. 565–9.

He writes, however, that '[a]lthough there has been much discussion of various issues in empirical neuroethics, the discussion of methodological and conceptual issues and thus theoretical neuroethics has remained rather sparse so far'.

Thus, Northoff argues for the need for theoretical neuroethics to focus 'on the methodological and conceptual aspects' of the linkage between neuroscientific facts and ethical concepts and for a theoretical neuroscience that is able to give proper weight to both norms and facts in a 'norm–fact circularity'.

Towards the end of his paper, Northoff notes that his ideas may strike some as a mere playground for theoreticians—a criticism that might be particularly applied to stressing theoretical neuroscience in this paper on the severely practical issue of dual use. We think, however, he has a general point to make on neuroethics methodology that is significant: 'If neuroethics wants to establish itself as a separate discipline that is different from its neighboring disciplines like philosophy, ethics and neuroscience, it must develop a special methodology.'

In this view, a discipline or field of study cannot claim to be distinct just because it studies certain things; it must have developed its own distinctive methodology as well. It would certainly be wrong to consider that this will be a simple task given that there are still great philosophical differences about how we should go about understanding our brains and behaviour.[14]

---

14 Evers, K. 2007, 'Towards a philosophy for neuroethics', *EMBO Reports*, vol. 8, pp. 848–51.

Although the recent conceptualisation of neuroethics originated in the United States and Europe, research and publications on neuroethics are increasingly international. Lombera and Illes suggested that despite the broad scope of the field, 'neuroethics has attempted to frame its efforts in terms of four "pillars": brain science and the self, brain science and social policy, ethics and the practice of brain science, and brain science and public discourse'.[15]

On this basis, they carried out a wide-ranging literature survey and concluded that their results demonstrated 'a steady increase in global participation in neuroethics from 1989 to 2005, characterized by an increase in numbers of articles published specifically on neuroethics, journals publishing these articles, and countries contributing to the literature'—and they clearly expect this trend to continue as neuroscience advances and the associated technologies spread around the world.

Neuroscientists and neuroethicists have been commendably interested in the problem of communication with the public about their work[16] and about how neuroscientists might be better enabled to carry out such tasks. Yet there are also reasons to believe that neuroscientists do not readily take to ethical analyses and the communication of their ethical positions,[17] and that they are often not trained to do so.[18] This, of course, is cause for considerable concern because, as the sociologist Raymond de Vries pointed out, there are other possible accounts of the rise of bioethics besides the technology story. For example, de Vries cites one account in which '[u]nlike the conventional technology story, in which bioethicists are cast as the guardians who oversee and regulate doctors and scientists ... bioethicists are less-than-critical allies of medicine and medical science'.[19]

Raymond de Vries comments that neuroethicists would do well to understand these different possibilities because if they fail to do so, 'they are less inclined to appreciate the way in which funding sources, and the structure of industry and academic research, shape bioethics and neuroethics'. Against that brief background context, the rest of this chapter is concerned with some examples of what neuroethicists do—particularly in regard to the problem of dual use.

---

15 Lombera, S. and Illes, J. 2009, 'The international dimension of neuroethics', *Developing World Bioethics*, vol. 9, no. 2, pp. 57–64.

16 Illes, J. et al. 2005, 'International perspectives on engaging the public in neuroethics', *National Review of Neuroscience*, vol. 6, no. 12, pp. 977–82; Illes, J. et al. 2010, 'Neurotalk: improving the communication of neuroscience', *National Review of Neuroscience*, vol. 11, no. 1, pp. 1–20.

17 Wolpe, P. R. 2006, 'Reasons scientists avoid thinking about ethics', *Cell*, vol. 125, no. 6, pp. 1023–5.

18 Sahakian, B. J. and Morein-Zemir, S. 2009, 'Neuroscientists need neuroethics teaching', *Science*, vol. 325, p. 147.

19 Blank, op. cit.

Case studies

Whilst the intention here is to concentrate on the ethical issues involved in the *practical* consequences of the advances in neuroscience, it should be made clear that this is not because we underestimate the importance of the philosophical issues raised by modern neuroscience. This point has been made regularly in reviews of the emerging field: an inadequate model of the brain and mind is likely to cause great misunderstandings and practical difficulties.[20] For example, if someone is in a persistent vegetative state but neuroimaging shows that he or she is able to respond to some questions,[21] can informed consent be achieved for treatment? Clearly, the brain is not without response, but is there an ability to decide on complex questions?

If such questions at the more philosophical end of the spectrum of the empirical issues are set aside, it is clear that neuroethicists have considered questions arising from three types of new technologies: neuroimaging; pharmacological enhancement; and non-pharmacological enhancement (such as brain–machine interfaces, transcranial magnetic stimulation, and direct current stimulation). There is also now some consideration being given to how the combination of these technologies with genomics[22] and information technology[23] may affect the ethical questions.

Some of the ethical questions that arise with these new technologies are far from new. In this regard neuroethics has carefully noted the issues relating to safety of the technologies, the validity of results drawn from complex analyses derived from neuro-'imaging' techniques, and how these images are understood by the media and general public. Neuroethicists have also discussed the obligations that researchers have for their findings—for example, what to do about findings that suggest that the person involved could have an increased likelihood of illness in the future.

Despite acknowledging these issues that have long-running parallels in bioethics, neuroethicists' writing makes frequent reference to the fact that some of the issues that arise from the application of these new technologies are novel. As

---

20   Glannon, W. 2009, 'Our brains are not us', *Bioethics*, vol. 23, no. 6, pp. 321–9.
21   Owen, A. M. et al. 2006, 'Detecting awareness in the vegetative state', *Science*, vol. 313, p. 1402.
22   Tairyan, K. and Illes, J. 2009, 'Imaging genetics and the power of combined technologies: a perspective from neuroethics', *Neuroscience*, vol. 164, pp. 7–15.
23   Amari, S.-I. 2002, 'Neuroinformatics: the integration of shared databases and tools towards integrative neuroscience', *Journal of Integrative Neuroscience*, vol. 1, no. 2, pp. 117–28.

Farah and Wolpe express it: 'The brain is the organ of mind and consciousness, the locus of our sense of selfhood. Interventions in the brain therefore have different ethical implications than interventions in other organs.'[24]

With regard to neuroimaging, there are numerous discussions of the possibility, and thus implications, of being able to detect when people are intentionally carrying out a deception. There has been wide discussion of the implications of being able to detect, through neuroimaging, people's desire for certain products and the consequences of the growth of 'neuromarketing'. Concerns have also been expressed about the dangers to privacy if such personal characteristics can be elucidated by neuroimaging.

With respect to pharmacological enhancement, there are again issues of safety, validity and communication of findings to a non-expert audience that are not specific to the concerns of neuroscientists and neuroethicists. Again, however, the neuroethicists' writing clearly indicates that they believe the possibilities of enhancing attention, memory and mood and the reverse possibility of damping down memories (to help people suffering from post-traumatic stress disorder) do raise *new* ethical issues. What happens, for example, to those who choose not to be enhanced, or who cannot afford to be enhanced, in an enhancement-ridden society? And, again touching on the philosophical, what happens to our sense of self and worth if we can have a better attention span, memory or mood not by work to achieve such developments but by 'popping a pill'?

A similar set of novel and not so novel issues arises in regard to non-pharmaceutical enhancement, but one unusual point can be noted in the concerns expressed about military funding of work on brain–mind interfaces and of 'new breeds of cyborgs'. Nevertheless, it is clear that this worry about the military implications of the advances in neuroscience and related technologies is very limited in the neuroethics literature. The limited nature of the neuroethical debate was noted by Joelle Abi-Rached in her discussion of the launch of the European Neuroscience and Society Network. In her view, this has to change and the debate 'must also include the controversies surrounding the potential application of various technologies in "neurosecurity" and "counter-terrorism"'.[25]

She correctly references Jonathan Moreno's book *Mind Wars: Brain Research and National Defense*[26] when making her point. Even there, however—in the only extended discussion of the possible misapplications of various technologies—there is very little discussion of the problem of dual use. Indeed, although in

---

24   Farah, M. J. and Wolpe, P. R. 2004, 'Monitoring and manipulating the human brain: new neuroscience technologies and their ethical implications', *Hastings Center Reports*, vol. 34, no. 3, pp. 35–45.
25   Abi-Rached, J. M. 2008, 'The implications of the new brain sciences', *EMBO Reports*, vol. 9, no. 12, pp. 1158–62.
26   Moreno, J. D. 2006, *Mind Wars: Brain Research and National Defense*, Dana Press, Washington, DC.

a 2007 editorial in *Science*, Henry Greely's reflections on neuroethics included the view that '[o]ther studies may have military implications: Suppose brain stimulation created an indefinitely awake and alert soldier or pilot? *Will neuroscience be a new source of dual-use technologies such as those we worry about for biological or chemical warfare?*' (emphasis added).[27]

It is very difficult to find even a reference to, let alone a discussion of, the problem of dual use in the neuroethics literature, even when it deals with national security issues.[28]

For those trained in natural science, a further striking feature of the neuroethics literature is the generality of the discussion around the ethical implications of neuroscience.[29] There is very good reason to believe that the ongoing advances in civil neuroscience could produce materials, technologies and knowledge that could later be applied, for example, to develop new forms of incapacitating chemicals for use in the 'War on Terror'. As well, they could lead to the erosion and eventual demise of the prohibition on the use of the modern life sciences for non-peaceful purposes that is embodied in the Biological and Toxin Weapons Convention and the Chemical Weapons Convention.[30] So what are civil neuroscientists to do to help protect their work from such misuse? Does it perhaps conform to the paradigm described by Lawrence Schmidt and Scott Marratto:

> The technological imperative has opened every aspect of human life to relentless transformation. The assumption is that it is always ethically acceptable to experiment to find out whether we can do something, and if we can, we ought to. But the adoption of the technological imperative has meant the liberation of means from ends.[31]

That, of course, is a huge question, but one way to approach it, perhaps, is to look more closely at one of the technologies that neuroethicists have focused upon: cognitive enhancement.

---

27   Greely, H. 2007, 'Editorial: neuroethics', *Science*, vol. 318, p. 533.

28   Conli, T. et al. 2007, 'Neuroethics and national security', *The American Journal of Bioethics*, vol. 7, no. 5, pp. 3–13; Resnik, D. B. 2007, 'Neuroethics, national security and secrecy', *The American Journal of Bioethics*, vol. 7, no. 5, pp. 14–26.

29   See, for example, Illes, J. 2007, 'Empirical neuroethics', *EMBO Reports*, vol. 8, pp. S57–60, Figures 2 and 3; and Glannon, op. cit., Figure 2.

30   See <www.opbw.org> and <www.opcw.org>.

31   Schmidt, L. E. and Maratlo, S. 2008, *The End of Ethics in a Technological Society*, McGill-Queen's University Press, Montreal and Kingston.

Cognitive enhancement

In December 2008, *Nature* carried a commentary by Henry Greely and five colleagues titled 'Towards responsible use of cognitive-enhancing drugs by the healthy'. The stated aim of this piece was to 'propose actions that will help society accept the benefits of enhancement, given appropriate research and evolved regulation'.[32]

Arguing that these drugs are just another means devised by our innovative species to improve itself, they suggested cognitive-enhancing drugs should 'be viewed in the same general category as education, good health habits and information technology'. They listed what they viewed as standard arguments against the use of these drugs—cheating, unnaturalness and drug abuse—and dismissed them. They did accept, however, that questions of safety, freedom (from coercion to enhance) and fairness (towards those who could not afford enhancement) would need to be addressed. They also suggested a program of research, professional guidance, public understanding and regulation development be undertaken so that responsible use was facilitated. In a direct critique of this commentary, two Canadian neuroethicists struck a much more cautious note, arguing that the question of safety was far from settled: 'it is important to stress that sizeable gaps exist in our current understanding of the effects, both positive and negative, of neuropharmaceuticals on healthy individuals.'[33]

And they pointed out that the simple tasks studied so far in laboratories 'do not reflect the complexity and diversity of activities in learning and thinking'. They thus implied that there was a degree of overstatement amongst those favouring cognitive enhancement in the same way that gene therapy had been oversold by supporters. A more neutral term than cognitive enhancement, they suggested, might be 'non-medical use of prescription drugs'. These critics also had concerns about the impact on health resources of interest in cognitive enhancement—for example, about the safety research program suggested by advocates eating up resources that were more urgently needed elsewhere, for example, in treating people who are ill. In short, there does not yet appear to be a settled view amongst neuroethicists about cognitive enhancement.[34]

Yet there is a considerable literature on this subject and certainly enough to ask two questions relevant to dual use: 1) do neuroethicists dealing with cognitive

---

32 Greely, H. et al. 2008, 'Towards responsible use of cognitive-enhancing drugs by the healthy', *Nature*, vol. 456, pp. 702–5.

33 Racine, E. and Forlini, C. 2009, 'Expectations regarding cognitive enhancement create substantial challenges', *Journal of Medical Ethics*, vol. 35, pp. 469–70.

34 Harris, J. and Chatterjee, A. 2009, 'Head to head: is it acceptable for people to take methylphenidate to enhance performance?' *BMJ*, vol. 338, pp. 1532–3.

enhancement consider the issue of dual use, and 2) do the investigations of the mechanism of enhancement by neuroscientists indicate any awareness of the possibility that their work could be used for the very opposite manner by those with malign intent?

Most discussions of cognitive enhancement are concerned with the use of drugs such as: methylphenidate (used medically to help people suffering from attention deficit hyperactivity disorder) to improve attention; modafinil (used medically to help people suffering from sleep problems to improve *alertness*; and SSRIs— selective serotonin reuptake inhibitors (used medically to help people suffering from problems of mood)—to improve the mood of people who are unwell. A particularly interesting subject in regard to this chapter is the use of the drug propranolol, not to improve memory but to help people to *forget* emotionally laden traumatic events—memories that recur in post-traumatic stress disorder (PTSD).[35]

In such discussions it is certainly possible to find what appears to be approval of military-funded research that could fundamentally change war-fighting and force employment because of, for example, the elimination of the need for sleep and the maintenance of a high level of cognitive performance.[36] On the other hand, it is rare indeed to find a clear recognition in the neuroethics literature of the problem of dual use in regard to cognitive enhancement. As Kathinka Evers pointed out:

> It has been suggested that therapeutic forgetting is interesting for military purposes, for example, to provide soldiers with propranolol before a battle. A problem here is that if it helps them forget what they have been subjected to it also helps them forget what they have done to others.[37]

This comment is clearly a special case because Evers is one of the few bioethicists who has addressed the general issue of dual-use bioethics.[38]

If we turn to the neuroscience, it is clear that safety issues involved in cognitive enhancement are sometimes well understood. As Cakic recently pointed out: 'For ... psychostimulants such as methylphenidate, the dangers are real and

---

35   Glannon, W. 2008, 'Psychopharmaceutical enhancement', *Neuroethics*, vol. 1, pp. 45–54.

36   Sahakian, B. J. and Morein-Zamir, S. 2010, 'Neuroethical issues in cognitive enhancement', *Journal of Psychopharmacology* [epub ahead of print, doi: 10.1177].

37   Evers, K. 2007, 'Perspectives on memory manipulation: using beta-blockers to cure post-traumatic stress disorder', *Cambridge Quarterly of Healthcare Ethics*, vol. 16, pp. 138–46.

38   *Research Ethics and Bioethics*, University of Uppsala, <www.crb.uu.se>.

relatively well known. Aside from its abuse potential, methylphenidate may aggravate mental illness, produce sleep disturbances and is associated with cerebrovascular complications.'[39]

It is also clear that we know a great deal about how emotion-laden memories are laid down in mammals and the way in which propranolol can be used to interfere with the role of noradrenaline in the consolidation and reconsolidation of such memories.[40]

Whilst accepting that some modest improvements in capabilities may be achieved, recent detailed reviews of the science and ethics of cognitive enhancement have emphasised caveats: 'first ... doses most effective in facilitating one behavior could at the same time exert null or even detrimental effects on other cognitive domains. Second, individuals with "low memory span" might benefit from cognitive-enhancing drugs, whereas "high span subjects" are "overdosed".'[41]

And, finally: 'evidence suggests that a number of trade-offs could occur. For example, increases of cognitive stability might come at the cost of a decreased capacity to flexibly alter behaviour.'

This particular review also discusses six ethical issues found in the literature: safety; societal pressure; fairness and equality; enhancement versus therapy; authenticity and personal identity; and happiness and human flourishing. In the last of these it does refer to the possibility that the blunting of memory could involve violation of 'a duty to remember and bear witness of crimes and atrocities'. Thus it touches on the question of dual use, but again there is no indication that the general point—that all of the work in cognitive enhancement could be dual use—has been understood.

Another recent review that notes the problem of such misuse in the blunting of memories without drawing the general conclusion about dual use does make the crucial point that our knowledge of the brain remains limited: 'the fundamental question is: are we technically ready and do we have sufficient basic knowledge to develop such drugs without risking a deadly brain doping?'[42]

A possible response is to say that cognitive enhancement has been much overhyped and that major brain modifications are years away and therefore

39 Cakic, V. 2009, 'Smart drugs for cognitive enhancement: ethical and pragmatic considerations in the era of cosmetic neurology', *Journal of Medical Ethics*, vol. 35, pp. 611–15.
40 Tully, K. and Bolshakov, V. Y. 2010, 'Emotional enhancement of memory: how norepinephrine enables synaptic plasticity', *Molecular Brain*, vol. 3, pp. 15–24; Dando, M. R. 2007, 'Scientific outlook for the development of incapacitants', in Pearson et al., op. cit., pp. 123–48.
41 de Jongh, R. et al. 2008, 'Botox for the brain: enhancement of cognition, mood and pro-social behavior and blunting of unwanted memories', *Neuroscience and Biobehavioral Reviews*, vol. 32, pp. 760–76.
42 Lanni, C. et al. 2008, 'Cognitive enhancers between treating and doping the mind', *Pharmacological Research*, vol. 57, pp. 196–213.

we should not worry too much about the state of neuroethics and its lack of coverage of the problem of dual use. That is indeed what one neuroscientist stated recently in a review of a book of essays on neuroethics: 'the participants in this discussion often claim that their speculative approach provides us with the unique opportunity to discuss the ethical consequences of new technologies before they are fully developed … However, do we really need a debate on a technology that will probably never materialise?'[43]

In this view, we do not really need to be concerned about neuroethics because the debate amongst neuroethicists is not of great importance. Many people took amphetamines during and after World War II so the present uses of drugs for cognitive enhancement is nothing new and, given the limited benefits that are likely to be available for the foreseeable future, the present phase will surely pass quietly away.

Back to dual use

The problem with the benign viewpoint voiced above (which does not take into account malign misappropriation of neuroscience advancements) is that whilst it may be appropriate for such civil uses of cognitive enhancement, it surely cannot be said to apply to the problem of military dual use. Here, in addition to a long history of misuse of the ongoing advances in neuroscience, we clearly have evidence of state-level development and use of novel chemical incapacitants. Furthermore, there is an obvious concern that continued interest in such developments could lead to the erosion and, potentially, the destruction of the prohibition of chemical and biological weapons embodied in the Chemical Weapons Convention and the Biological and Toxin Weapons Convention.

It is not as if it is difficult to find evidence of real concern amongst experts about this issue. For example, when the International Committee of the Red Cross launched its appeal on 'Biotechnology, Weapons and Humanity', the eminent neuroscientist Professor Tamas Bartfai worried about the dangers of the misuse of advances in neuroscience. He noted that the acetylcholine system had been the target of successive generations of lethal nerve gases and for drugs designed to help people suffering from Alzheimer's disease, and went on to point out a number of clear-cut ways in which current developments in the neurosciences might be subject to misuse.[44] Similarly, Robert McCreight, a senior US civil servant, gave a lecture on the implications of the advances in neuroscience for

---

43 Quednow, B. B. 2010, 'Ethics of neuroenhancement: a phantom debate', [Books Forum], *BioSocieties*, vol. 5, no. 1, pp. 153–6.
44 Bartfai, T. and Sellstem, A., 'Neurobiology—weapons and humanity', Presentation at the launch of the appeal on Biotechnology, Weapons and Humanity, International Committee of the Red Cross, Montreux.

national security and future strategic weapons in January 2007. As he noted: 'Scientific research, concept development, examination of bioethical issues related to enhanced mental health and considerable long term funding support has been ongoing for several years in the broad area of neuroscience.'[45]

Then he asked: 'Have we adequately analysed and discussed the dual-use implications of neuroscience, particularly its various military applications, and the extent to which operational safeguards and societal controls are needed to manage or control its most destructive weapons outcomes or debilitating systems?'

McCreight did not think that there had been adequate analysis and discussion and suggested that unless measures were taken 'we may face new categories of weapons before 2010 held by several nations both friendly and hostile'.

Certainly, some practising neuroscientists have raised concerns about the hostile misuse of advances in neuroscience,[46] and there are a few bioethicists who have tried to contribute to a better understanding of the problem of dual use.[47] Given this relatively early stage in the development of neuroethics, it is not clear whether neuroethicists will have anything special to add to the work of bioethicists in general, but if they do aspire to help as 'architects of moral space' by 'fostering open and constructive dialogue, discussion and debate' about critical issues,[48] in order to develop 'smarter regulation' where that is necessary[49] then surely there is no better time than now to begin to develop their view of the neuroethics of dual use.

Conclusion

The field of neuroscience has grown considerably in the past decade. Advances in neuroscience already have raised important questions on a wide range of policy issues, such as those affecting neurotoxins and the environment, mental health, child development, cognitive enhancement, criminal behaviour, the safety and efficacy of pharmaceuticals and medical devices, and the ethics and regulation of emerging discoveries. To an ever-increasing understanding of the

---

45   McCreight, R. 2007, 'Protecting our national neuroscience infrastructure: implications for homeland security', Presentation to National Security and the Future of Strategic Weapons, George Washington University, Institute of Crisis, Disaster and Risk Management, <http://www.chds.us/?fs:file&mode=dl&drm= ..%2F..%2Fresources%2Fsummit%2F%2Fsummit07&f=McCreight-GeorgeWashUniv.ppt&altf=McCreight-GeorgeWashUniv.ppt>.
46   Bell, C. 2010, 'Neurons for peace: take the pledge, brain scientists', New Scientist, vol. 2746 (8 February).
47   Atlas, R. M. 2009, 'Responsible conduct by life scientists in an age of terrorism', Science and Engineering Ethics, vol. 15, pp. 293–301.
48   Robert, J. S. 2009, 'Toward a better bioethics', Science and Engineering Ethics, vol. 15, pp. 283–91.
49   Sutton, V. 2009, 'Smarter regulations', Science and Engineering Ethics, vol. 15, pp. 303–9.

brain mechanisms associated with core human attributes and values should also correspond an increasing interest in the possible dual-use implications of such advancements, giving the various ways, not always benign, in which the new knowledge could be used. Neuroethics addresses the various philosophical issues around the relationship between brain and mind as well as practical issues about the impact upon society of our ability to understand and manipulate the brain. The problem of dual use, consisting in a malign appropriation of knowledge initially designed for benign purposes, should be an important focus of neuroethicists' analyses. In particular, the long-term applications and impacts of neuroscience are likely to be powerful and profound. As pointed out by Marchant and Gulley: 'Military and intelligence agencies, with the most at stake from such applications in terms of both benefits and risks, recognize the potential of neuroscience to revolutionize intelligence gathering and warfare.'[50]

Furthermore, as pointed out by Jonathan Moreno, the 11 September 2001 attacks have resulted in increased efforts to exploit all technical possibilities for enhancing security. The Pentagon's Defense Advanced Research Projects Agency (DARPA) is supporting work at Lockheed Martin on remote brain prints and the scientist in charge already claims to be able to tell if a person is thinking of a certain number. In the words of Moreno, 'a striking aspect of much of this and other national security work being done in the field of neuroscience is that it is "dual use"—potentially applicable to medical therapy or other peaceful purposes as well as combat, riot control, hostage situations, or other security problems'.[51]

Worryingly, from a preliminary review of the literature on neuroethics, it clearly emerges that, while publications abound on issues such as lie detection, informed consent for certain patients and around the implications of neuroimaging, the problem of dual use is very marginally addressed.

In the majority of research centres and institutions, no mention is made of the problem of dual use, and the overwhelming assumption is that neuroscience works for the betterment of humanity worldwide (see Appendix A). If this is what neuroethicists are working for, more attention should surely be devoted to the way advances in neuroscience could be misused. Leading neuroethicists also very rarely address the issue of dual use in neuroscience, with few notable exceptions.[52]

---

50   Marchant, G. and Gulley, L. 2010, 'National security neuroscience and the reverse dual-use dilemma', *AJOB Neuroscience*, vol. 1, no. 2, pp. 20–2.

51   Ibid.

52   Moreno, J. 2005, 'Dual use and the "moral taint" problem', *The American Journal of Bioethics*, vol. 5, no. 2, pp. 52–3; Moreno, J. 2008, 'Using neuro-pharmacology to improve interrogation techniques', *Bulletin of the Atomic Scientists*; Huang, J. Y. and Kosal, M. E. 2008, 'The security impact of the neuroscience', *Bulletin of the Atomic Scientists*.

Nevertheless, if neuroethicists have paid almost no attention to the problem of dual use, the issue should be of great concern to both neuroscientists and neuroethicists, who should critically approach dual use in neuroscience in order to save neuroscience from dreadful distortions of its intended purpose. In the current environment, a similar remarkable omission, if not addressed urgently, will be soon deplored.

4. Synthetic Biology as a Field of Dual-Use Bioethical Concern

Alexander Kelle

Introduction

Over the past decade synthetic biology has emerged as one of the most dynamic subfields of the post-genomic life sciences. According to a European high-level expert group, synthetic biology comprises 'the synthesis of complex, biologically based (or inspired) systems which display functions that do not exist in nature ... [and] is a field with enormous scope and potential'.[1] Some of the areas where this expert group argues that synthetic biology could have a major impact include biomedicine, a sustainable chemical industry, environment and energy, and biomaterials. *If* the emerging discipline of synthetic biology can deliver on the promises of some of its leaders and become as pervasive as computing has become in the past few decades, we might very well be witnessing a fundamental shift similar to the one that happened to chemistry with the introduction of the periodic table. *If* synthetic biologists live up to some of the more far-reaching expectations, biology ultimately may become a mechanistic science.

On the one hand, synthetic biology developments show promise of leading to beneficial applications in a number of areas, such as drug development,[2] biodegradation[3] and biofuels.[4] At the same time, the dual-use character of this new technoscience carries with it the possibility of synthesised biological parts, modules and systems being malignly misused. This dual-use potential has—at a rather abstract level and with a focus on one particular subfield of synthetic biology, that is, DNA synthesis—been recognised by practitioners in the field as well as analysts. While this is a positive development, however, these mostly technical governance measures that are addressing DNA synthesis capabilities need to be broadened so as to cover all aspects of synthetic biology and to allow for a comprehensive bioethical analysis of the field's dual-use implications. In addition, part of the discourse on the broader societal implication of synthetic biology can be traced back to the debates on ethical, legal and social implications

---

1   *Synthetic Biology. Applying Engineering to Biology*, Report of a New and Emerging Science and Technology (NEST) High-Level Expert Group, European Commission, Brussels, 2005, p. 5.
2   See Neumann, H. and Neumann-Staubnitz, P. 2010, 'Synthetic biology approaches in drug discovery and pharmaceutical biotechnology', *Applied Microbiology and Biotechnology*, vol. 87, pp. 75–86.
3   See Kirby, J. R. 2010, 'Designer bacteria degrades toxin', *Nature Chemical Biology*, vol. 6, pp. 398–9.
4   See Dellmonaco, C. et al. 2010, 'The path to next generation biofuels: successes and challenges in the era of synthetic biology', *Microbial Cell Factories*, vol. 9.

(ELSI) of genetic engineering. Yet, given their limited nature, these debates do not provide a solid foundation for a comprehensive discussion and assessment of synthetic biology's dual-use potential.

This chapter will first outline the scope of synthetic biology as a new subfield in the life sciences in which different science and engineering disciplines converge. This will be followed by a brief discussion of synthetic biology's potential for malign misuse as well as some of the proposals for governance of this new technoscience. Thus far, the mostly technical character of these proposals has resulted in a rather limited appreciation of the wider governance issues related to the full breadth of approaches usually subsumed under the synthetic biology label. The final section will discuss both academic and institutional contributions to a bioethically informed discourse on the misuse potential of synthetic biology.

Scope of synthetic biology as a subfield in the life sciences

Attempts at defining synthetic biology

Not surprisingly for a discipline that is still in its formative stages, several definitions exist for synthetic biology. One that has received the most attention describes synthetic biology as 'the design and construction of new biological parts, devices, and systems, and the re-design of existing, natural biological systems for useful purposes'.[5] This definition clearly reflects the approach to synthetic biology pioneered by scientists at the Massachusetts Institute of Technology (MIT), and puts centre stage the idea to develop a registry of standardised biological parts that can be assembled in devices and systems with predefined functions. Although the MIT framing of the issue area has certainly sparked the development of the whole field (for example, by organising the first international Synthetic Biology Conference in Boston in 2004, by supporting the Biobricks Foundation, and by organising the annual iGEM student competitions), an exclusive focus on the parts-based approach to synthetic biology tends to overlook other important subfields in synthetic biology (see Table 4.1 for an overview). For ease of reference, four different substrands of synthetic biology that have developed since the early years of the twenty-first century are distinguished here[6]

---

5 See <http://syntheticbiology.org/Who_we_are.html> (viewed 6 November 2008).
6 This subdivision follows Schmidt, M. 2009, 'Do I understand what I can create?' in M. Schmidt, A. Kelle, A. Ganguli-Mitra and H. de Vriend (eds), *Synthetic Biology. The Technoscience and Its Societal Consequences*, Springer, Dordrecht, pp. 81–100.

- engineering DNA-based biological circuits, by using standardised biological parts
- identifying the minimal genome
- constructing protocells—in other words, living cells from base chemicals
- creating orthogonal biological systems in the laboratory through chemical synthetic biology.

Two enabling technologies also tend to be usually subsumed under the heading of synthetic biology, although they have a more supportive role to the four fields mentioned above. This applies first and foremost to the increasingly more affordable large-scale DNA-synthesis capabilities that some companies are providing,[7] but also to the more generic bioinformatics capabilities utilised by those attempting to identify a minimal genome.

Thus, synthetic biology at the very least aims at merging molecular biology with engineering, by designing and producing new biological parts, devices and systems. To achieve this goal, synthetic biology utilises high-throughput commercial DNA-synthesis capabilities to provide the actual biological material for the assembly of genetic circuits. In addition, synthetic biology is relying on increasingly powerful information technology tools that allow for the modelling of certain desired biological functions.

A report by the Royal Academy of Engineering in the United Kingdom has detailed the design cycle that informs the engineering approach to synthetic biology.[8] Accordingly, the design cycle for biological systems starts with an initial specification stage for bioparts, which is 'followed by a detailed design step', drawing on 'the ability to undertake detailed computer modeling'.[9] In the subsequent implementation stage, synthesised DNA is usually inserted into *E. coli*, yeast or some other suitable chassis. In the final validation stage, the functionality of the original specification is verified. Standardised bioparts that have undergone such quality controls can then be utilised to build standard devices, which, in turn, can be assembled into biological systems. As O'Malley has recently pointed out, however, research and development (R&D) in synthetic biology have so far generally not been that straightforward. Instead, strictly hypothesis-driven work is regularly complemented by '[e]xploration, iterativity and kludging'.[10] In other words, realisation of the engineering-driven and standards-based 'plug and play' ideal of many synthetic biologists still awaits realisation and will for some time.

---

7   See, for example, the web pages of GENEART (<www.geneart.com>) or Blue Heron (<www.blueheronbio.com>). Another prominent company in this area, Codon Devices, recently closed. See Hayden, E. C. and Ledford, H. 2009, 'A synthetic biology reality check', *Nature*, vol. 458, p. 818.

8   The Royal Academy of Engineering 2009, *Synthetic Biology: Scope, Applications and Implications*, The Royal Academy of Engineering, London, esp. pp. 18–21.

9   Ibid., p. 19.

10   O'Malley, M. 2011, 'Exploration, iterativity and kludging in synthetic biology', *C. R. Chimie*, vol. 14, pp. 406–12.

Table 4.1 Breakdown of Subfields of Synthetic Biology

	DNA-based bio-circuits	Minimal genome	Protocells	Chemical synthetic biology
Aims	Designing genetic circuits, for example, from standardised biological parts, devices and systems	Finding the smallest possible genome that can 'run' a cell, to be used as a chassis; reduced complexity	To construct viable approximations of cells; to understand biology and the origin of life	Using atypical biochemical systems for biological processes, creating a parallel world
Method	Design and fabrication; applying engineering principles using standardised parts and abstraction hierarchies	Bioinformatics-based engineering	Theoretical modelling and experimental construction	Changing structurally conservative molecules such as DNA
Techniques	Design of genetic circuits on the blackboard, inserting the circuits in living cells	Deletion of genes and/or synthesis of entire genome and transplanting the genome in a cytoplasm	Chemical production of cellular containers, insertion of metabolic components	Searching for alternative chemical systems with similar biological functions
Examples	'AND' gate, 'OR' gate; genetic oscillator repressilator; Artemisinin metabolism, 'Bactoblood'	DNA-synthesis and transplantation of Mycoplasma genitalium	Containers such as micelles and vesicles are filled with genetic and metabolic components	DNA with different sets of base pairs, nucleotides with different sugar molecules

Source: Schmidt, M. 2009, 'Do I understand what I can create?' in M. Schmidt, A. Kelle, A. Ganguli-Mitra and H. de Vriend (eds), *Synthetic Biology. The Technoscience and Its Societal Consequences*, Springer, Dordrecht, p. 84.

According to a widely accepted historiography of the term 'synthetic biology', it entered the scientific vocabulary in 1912 with the publication of the French chemist Stephane Leduc's monograph of the same title.[11] As Campos outlines, the desire among scientists to redesign life can be observed in a number of different approaches throughout the twentieth century.[12]

Synthetic biology in its contemporary manifestation was initially promoted under labels such as 'open source biology' and 'intentional biology', which were proposed for the renewed attempt by some to convert biology into a predictive science by incorporating elements of the engineering design cycle. As Rob Carlson, for example, stated in 2001: 'When we can successfully predict the behavior of designed biological systems, then an intentional biology will exist. With an explicit engineering component, intentional biology is the opposite of the current, very nearly random applications of biology as technology.'[13] Besides the Molecular Sciences Institute of the University of California at Berkeley at which Rob Carlson was envisaging a distributed open-source biological manufacturing system driving future industry, another early institutional hub of contemporary synthetic biology was located at MIT, where in the Computer Science and Artificial Intelligence Laboratory (CSAIL) Tom Knight had set up a biology laboratory and started developing the Biobricks standard for biological parts with support from a Defense Advanced Research Projects Agency/Office of Naval Research (DARPA/ONR) contract on 'Computing with Synthetic Biology'.[14]

The two main vehicles for developing the field of synthetic biology since then have been the annual student iGEM competition as well as the four SBx.0 conferences. While the first two of these conferences were held in the United States (Cambridge and Berkeley), SB3.0 took place in Zurich in the summer of 2007 and SB4.0 in Hong Kong in October 2008. Over the course of these SB conferences it has become obvious that synthetic biology encompasses more than only the parts-based approach emphasised here. While in this context clearly some relabelling of more traditional biotechnology and molecular biology approaches can be observed in order to participate in a newly established, 'cool' and potentially well-funded discipline, there clearly are different facets to contemporary synthetic biology that go beyond the engineering of biological parts, devices and systems.[15]

---

11   See, for example, de Lorenzo, V. and Danchin, A. 2008, 'Synthetic biology: discovering new worlds and new words', *EMBO Reports*, vol. 9, pp. 822–7.

12   Campos, L. 2009, 'That was the synthetic biology that was', in Schmidt et al., op. cit., pp. 5–21.

13   Carlson, R. 2001, 'Open source biology and its impact on industry', *IEEE Spectrum*, pp. 15–17, as quoted by Campos, op. cit., p. 17.

14   Knight, T. F. 2002, *DARPA BioComp Plasmid Distribution 1.00 of Standard Biobrick Components*, <http://dspace.mit.edu/handle/1721.1/21167>.

15   For an even more detailed subdivision of the field than the one used here, see Lam, C. M. C., Godinho, M. and dos Santos, V. 2009, 'An introduction to synthetic biology', in Schmidt et al., op. cit., pp. 23–48.

Beneficial uses of synthetic biology

Optimistic assessments of synthetic biology's potential, such as the one contained in the abovementioned report by an EU high-level expert group, expect it to 'drive industry, research, education and employment in the life sciences in a way that might rival the computer industry's development during the 1970s to the 1990s'.[16] The same group of experts envisages synthetic biology to have such a dramatic effect by 're-organizing biotechnological development' in a way so that 'research & development are likely to proceed in a much faster and much more organized way'.[17] The report identifies six areas that could benefit from such a streamlining of R&D processes: biomedicine, synthesis of biopharmaceuticals, sustainable chemical industry, environment and energy, production of smart materials and biomaterials, and counter-bioterrorism measures.[18]

Probably the most often quoted example for the imminent breakthrough of a high-value synthetic biology application is related to the insertion of an engineered metabolic pathway into a yeast strain to produce artemisinic acid.[19] This in turn can be converted into artemisinin, which forms the basis of antimalarial drugs. The key goal of the attempt to synthesise this artemisinin precursor is to reduce production costs for the therapy and thus increase its availability in developing regions of the world. As Chang and Keasling have pointed out in relation to 'this approach, the genes related to the biosynthetic pathway for a target natural product are transplanted from the natural host into a genetically tractable host system such as *E. coli* or *S. cerevisiae*'.[20] Similarly, synthetic biologists have made some progress in utilising genetically modified microbes to produce biofuels.[21]

---

16 *Synthetic Biology*, op. cit., p. 13.
17 Ibid.
18 Ibid., pp. 13–17.
19 Keasling, J. et al. 2006, 'Production of the antimalarial drug precursor artemisinic acid in engineered yeast', *Nature*, vol. 440, pp. 940–3, <http://www.nature.com/nature/journal/v440/n7086/abs/nature04640.html>.
20 Chang, M. C. Y. and Keasling, J. D. 2006, 'Production of isopronoid pharmaceuticals by engineered microbes', *Nature Chemical Biology*, <doi:10.1038/nchembio836>.
21 See, for example, Clomburg, J. M. and Gonzales, R. 2010, 'Biofuel production in *Escherichia coli*: the role of metabolic engineering and synthetic biology', *Applied Microbiology and Biotechnology*, vol. 86, pp. 419–34; Keasling, J. D. and Chou, H. 2008, 'Metabolic engineering delivers next-generation biofuels', *Nature Biotechnology*, vol. 26, pp. 298–9.

Synthetic biology's potential for malign misuse

Potential misapplication(s)

The ability to understand, modify and ultimately create new life forms at the molecular level clearly represents a scientific paradigm shift with a substantial misuse potential. In general terms, synthetic biology can be misused to engineer biological parts or modules that increase the efficiency/stability/usability of known warfare agents or to synthesise new ones.

While currently there are still formidable challenges to overcome—for example, in the creation of synthetic pathogens[22]—there is a general acknowledgment that these hurdles will be lowered by scientific and technological advances over the next few years. In the words of a report jointly published by MIT, the Centre for Strategic and International Studies (CSIS) and the J. Craig Venter Institute: 'In the near future, however, the risk of nefarious use will rise because of the increasing speed and capability of the technology and its widening accessibility.'[23] While this statement refers to the risks related to synthetic genomics, it is also applicable to the wider area of synthetic biology.

A brief look into two specific applications of major impacts expected from synthetic biology as described in the abovementioned New and Emerging Science and Technology (NEST) report—smart drugs and vectors for therapy— should suffice to illustrate the quantum leap in biological warfare or bioterrorist capabilities that may result from advances in synthetic biology. According to the NEST report, a 'smart drug includes a diagnostic module that … is capable of directly sensing of molecular disease indicators … it will only become active in cells affected by disease'.[24] The misuse potential of such smart-drug technology is obvious: if the sensing mechanism were programmed to detect other, not disease-related indicators and/or the activated chemical compound were to harm, not cure or would simply be administered in the wrong dosage, considerable harm could be done with such a device. Similarly, it is conceivable that newly designed or modified viral vectors that can 'deliver healthy genes to the target tissue' or that 'can recognize specific cells and target them for

---

22   See Epstein, G. L. 2008, 'The challenges of developing synthetic pathogens', *Bulletin of the Atomic Scientists*, vol. 64, pp. 46–7.

23   Garfinkel, M. S., Endy, D., Epstein, G. L. and Friedmann, R. M. 2007, *Synthetic Genomics: Options for Governance*, <http://www.jcvi.org/cms/fileadmin/site/research/projects/synthetic-genomics-report/synthetic-genomics-report.pdf>, p. 12.

24   *Synthetic Biology*, op. cit., p. 14.

destruction'[25] could be easily diverted from their intended benign use to malign applications that would, for example, aim at delivering pathogenic genes or target not cancer, but nerve or other essential cells.

Recognition of dual-use concerns

In parallel with the focus of early activities in synthetic biology in general, which have been concentrated in the United States, the origins of the discourse on preventing the misuse of this new field of scientific inquiry also can be traced back to the US scientific and policymaking communities. In 2004, for example, George Church put forward 'A Synthetic Biohazard Non-Proliferation Proposal' to address some of the biosecurity concerns of synthetic biology. The underlying rationale for this proposal was his identification of the misuse of synthetic biology as a low-probability, high-consequence event.[26] At around the same time, the first big synthetic biology gathering, SB1.0 at the MIT, also included a couple of presentations on safety, security and ethical issues related to synthetic biology. Echoing Church's assessment of the misuse of synthetic biology as a low-probability, high-consequence event, George Poste of the Biodesign Institute at Arizona State University also made reference to the move away from the 'bug' towards 'biological circuit disruptors' as the object of concern.[27] At the Second International Conference on Synthetic Biology, the third day of the conference was dedicated to four key societal issues associated with synthetic biology, one of which was 'biosecurity and risk'.[28] Discussions during the corresponding session were informed by the preparatory work undertaken by the Berkeley SynBio Policy Group, which had produced a white paper on 'Community-Based Options for Improving Safety and Security in Synthetic Biology'.[29]

Also in 2006 the National Research Council tasked a Committee on Advances in Technology and the Prevention of their Application to Next Generation Biowarfare Threats to analyse the impact of the revolution in the life sciences on the evolving biosecurity threat spectrum.[30] In dispensing of its task, this committee rejected a list-based approach, for which it felt that, because of 'the pace of research discovery in the life sciences', the 'useful lifespan of any

25 Ibid.
26 Church, G. 2004, *A Synthetic Biohazard Nonproliferation Proposal*, <http://arep.med.harvard.edu/SBP/Church_Biohazard04c.htm>.
27 Poste, G. 2004, 'Synthetic biology: charting rational public policies for the oversight and regulation of vanguard technologies', 11 June, <http://openwetware.org/images/3/3a/SB1.0_George.Poste.pdf>.
28 See <http://syntheticbiology.org/SB2.0/Biosecurity_and_Biosafety.html>.
29 See Maurer, S. M., Lucas, K. V. and Terrell, S. 2006, *From Understanding to Action: Community-Based Options for Improving Safety and Security in Synthetic Biology*, Goldman School of Public Policy, University of California, Berkeley, Draft 1.1, <http://gspp.berkeley.edu/iths/UC White Paper.pdf>.
30 National Research Council 2006, *Globalization, Biosecurity, and the Future of the Life Sciences. Committee on Advances in Technology and the Prevention of their Application to Next Generation Biowarfare Threats*, The National Academies Press, Washington, DC, <www.nap.edu/catalog.php?record_id=11567>.

such list would be measured in months, not years'.[31] To emphasise this point, the report pointed out that '[n]ew, unexpected discoveries and applications in RNAi and synthetic biology arose even during the course of deliberations by this Committee'.[32] Instead, the committee developed a classification scheme for science and technology (S&T) advances with four different groups. These four groups are

1. technologies that seek to acquire novel biological or molecular diversity

2. technologies that seek to generate novel but predetermined and specific biological or molecular entities through directed design

3. technologies that seek to understand and manipulate biological systems in a more comprehensive and effective manner

4. technologies that seek to enhance production, delivery and 'packaging' of biologically active materials.[33]

Synthetic biology is explicitly mentioned by the committee in relation to the first two of these categories. A concise discussion of the future applications of synthetic biology in the report acknowledges that 'DNA synthesis technology could allow for the efficient, rapid synthesis of viral and other pathogen genomes—either for the purposes of vaccine or therapeutic research and development, or for malevolent purposes or with unintended consequences'.[34] It is thus fair to conclude that the biosecurity community during the deliberations of the Lemon-Relman Committee had clearly identified synthetic biology, albeit with an emphasis on DNA synthesis and not the four subfields of synthetic biology outlined above, as one of the technologies that will have a major impact on the future biothreat spectrum.

This emphasis on DNA synthesis is also reflected in the approach of the National Science Advisory Board on Biosecurity (NSABB).[35] In order to conduct its work, NSABB can set up working groups to address specific issues including one in the field of synthetic biology. In the first phase of its work, the NSABB synthetic biology working group has sought to address biosecurity implications of the *de novo* synthesis of select agents.[36] A report of the synthetic biology working group on this issue was discussed during a NSABB meeting in October 2006

---

31   Ibid., p. 3.
32   Ibid., p. 103.
33   Ibid., p. 3.
34   Ibid., p. 109.
35   National Science Advisory Board for Biosecurity (NSABB) 2006, *Addressing Biosecurity Concerns Related to the Synthesis of Select Agents*, National Science Advisory Board for Biosecurity, Washington, DC, <http://oba.od.nih.gov/biosecurity/pdf/Final_NSABB_Report_on_Synthetic_Genomics.pdf>.
36   Select agents are those biological agents and toxins that can pose a severe threat to public, animal or plant health, or to animal or plant products. For the current list of select agents, see <http://www.cdc.gov/od/sap/docs/salist.pdf>.

and has subsequently been submitted to the US Government. Clearly, here as well the focus of the threat assessment had not been on engineered bioparts and modules, but on improved DNA-synthesis capabilities as they relate to select agents. Thus, although it is reasonable to assume that since the end of 2006 US policymakers have been aware of the potential biosecurity risks of synthetic biology, discussion of these risks has up to now been limited to only a part of the range of approaches subsumed under the synthetic biology label.

Evolving dual-use potential

Although the main focus of synthetic biologists is on the design and engineering aspects of this new field of scientific inquiry, some also emphasise the field's contribution to 'achieving a better understanding of life processes'.[37] Clearly, this better understanding is sought in order to improve the human condition via improved diagnostics, therapeutics and other beneficial applications; however, as one study has pointed out, a better understanding of life processes in relation to regulatory systems in the human body also opens new doors for potential misuse of biologically active chemical compounds that can target the human nervous, immune or endocrine systems with a higher degree of specificity.[38] To the extent that synthesised bioparts, devices and modules are utilised for such misuse, synthetic biology's dual-use potential will represent an incremental change in the misuse potential of the life sciences in general. In principle, however, synthetic biology has a much more profound dual-use potential. If the standardisation of parts and modules progresses to a point where these are truly compatible and, in addition, can be inserted in a robust chassis for application/ dissemination, a threshold in dual-use potential will be crossed. Given the early stage of development at which parts-based synthetic biology finds itself, and the issues related to standardisation referred to above, and more generally the fact that most of synthetic biology lies at the cutting edge of today's life-science research, the progress made by synthetic biologists is most likely to be misused first in an offensive state-level biological weapons (BW) program.

This will change, however, as the dissemination of the knowledge and hands-on experience with synthetic bioparts and modules continue. The already mentioned deskilling efforts by some leading synthetic biologists go hand in hand with the dissemination of do-it-yourself synthetic biology kits and 'how to' protocols. In addition to these, the iGEM competition also supports this trend. Its 'broader goals include: to enable the systematic engineering of biology; to promote the open and transparent development of tools for engineering biology and to help

---

37 *Synthetic Biology in Europe*, TESSY Information leaflet, <www.tessy-europe.eu/public-docs/ SyntheticBiology_TESSY-Infomation-Leaflet.pdf>.
38 See Kelle, A., Nixdorff, K. and Dando, M. 2006, *Controlling Biochemical Weapons: Adapting Multilateral Arms Control for the 21st Century*, Palgrave, Basingstoke, UK, esp. chs 4–6.

construct a society that can productively apply biological technology'.[39] Thus, over time, the amount of specialist knowledge that is limited to (relatively) few 'experts' will become increasingly smaller, thereby increasing the likelihood that sub-state actors and individuals will utilise synthetic bioparts and modules for nefarious purposes.

Governance proposals for synthetic biology

From the formative days of contemporary synthetic biology, addressing the question of governance mechanisms and their relative utility has been a feature of the unfolding discourse on the discipline's societal implications. This early engagement of the scientific community with the risks associated with synthetic biology can in part be traced back to earlier debates about the ethical, legal and social implications of genetic engineering.

As briefly mentioned above, one of the earliest proposals to address synthetic biology's dual-use implications was put forward by George Church in his 'Synthetic Biohazard Non-Proliferation Proposal'. In it he advocated developing a system for both 'instrument and reagent licensing', and to screen for select agents, including a 'DNA agent clearinghouse'.[40]

While Church's proposal foresaw some form of government involvement or oversight in both these systems, subsequent governance proposals placed more emphasis on the self-governance efforts of the scientific community and commercial DNA-synthesis providers.

Such proposals were discussed at the SB2.0 conference, which took place in Berkeley, in May 2006, and during which a full day was devoted to discussion of societal issues surrounding synthetic biology.[41] The subsequently formulated declaration of the conference contains four resolutions that aim at addressing some of the dual-use implications of synthetic biology—in particular, DNA synthesis that may give rise to safety or security concerns.[42] The focus on DNA synthesis is also reflected in two of the four resolutions contained in the final declaration. These resolutions support the

> development of improved software tools that can be used to check DNA synthesis orders for DNA sequences encoding hazardous biological systems …

---

39 <http://parts2.mit.edu/wiki/index.php/About_iGEM>.
40 The proposal is available at <http://arep.med.harvard.edu/SBP/Church_Biohazard04c.htm>.
41 Much of the debate was informed by the white paper produced by Maurer and colleagues mentioned above.
42 See the revised public draft of the SB2.0 declaration at <https://dspace.mit.edu/handle/1721.1/18185>.

[A]doption of best-practice sequence checking technology, including customer and order validation, by all commercial DNA synthesis companies …

[O]ngoing and future discussions within international science and engineering research communities for the purpose of developing creative solutions and frameworks that directly address challenges arising from the ongoing advances in biological technology, in particular, challenges to biological security and biological justice …

[O]ngoing and future discussions with all stakeholders for the purpose of developing and analyzing governance options … such that the development and application of biological technology remains overwhelmingly constructive.[43]

In terms of practical next steps to be pursued, the declaration announces the formation of an open working group in support of the improvement of existing software tools for checking DNA sequences, as well as the completion of a study to 'develop policy options that might be used to govern DNA synthesis technology'.[44]

This study, which was conducted jointly by the MIT, the J. Craig Venter Institute and the CSIS, supported the trend to focus governance options in relation to synthetic biology on DNA-synthesis technology. The report's authors identify DNA synthesis itself, as conducted by gene-synthesis firms and oligonucleotide manufacturers, and with the help of DNA synthesisers, as the most effective intervention point for preventing the misuse of synthetic genomics. The authors of the report concluded that for both gene foundries and oligo manufacturers, a combination of screening orders by companies and the certification of orders by a biosafety/biosecurity officer provide the greatest benefits in terms of preventing incidents. The storage of order information by firms was regarded as the most useful tool for responding after an incident had occurred. Lastly, concerning equipment such as DNA synthesisers, the report concluded that the licensing of both equipment and reagents was most likely to enhance biosecurity by preventing misuse.[45]

Ideas developed in this report were subsequently taken up in proposals by the (then existing) two industry associations in the area of synthetic biology. The first of these groups, the International Consortium for Polynucleotide Synthesis (ICPS), put forward a 'tiered DNA synthesis order screening process'.[46] This

---

43 Ibid., p. 3.
44 Ibid.
45 Garfinkel, M. S., Endy, D., Epstein, G. L. and Friedman, R. M. 2007, *Synthetic Genomics. Options for Governance*, <http://www.csis.org/media/csis/pubs/071017_synthetic_genomics_options.pdf>.
46 Bügl, H. et al. 2007, 'DNA synthesis and biological security', *Nature Biotechnology*, vol. 25, pp. 627–9.

proposal would put DNA-synthesis companies and their industry association at the centre of a governance structure that would, however, not be a self-contained system of oversight, but rather rely on 'agreed-upon guidelines'. Such guidelines would be operationalised, inter alia, through lists of 'select agents or sequences' that would determine whether and how to process DNA-synthesis orders on the part of those companies that follow the guidelines.

Members of the Industry Association Synthetic Biology (IASB) have focused on a number of interrelated issues that also revolve around the screening of DNA orders by synthesis companies. These were formulated during a workshop that was held in Munich in April 2008 on 'Technical Solutions for Biosecurity in Synthetic Biology'.[47] Motivated by 'our responsibility for the scientific field to which we provide services and products',[48] workshop participants agreed on the adoption of five distinct work packages

1. harmonisation of screening strategies for DNA-synthesis orders

2. creation of a central virulence-factor database

3. publication of an article on the status quo of synthetic biology

4. establishment of a technical biosecurity working group with members from both organisations in order to 'discuss improvements and next steps for bio-security measures'

5. formulation of a code of conduct.[49]

With this last work package, the IASB started to expand its activities away from the technically focused measures into the political arena of setting standards and promoting best practices.

The focus of proposals put forward has been on the bottleneck of long-strand DNA synthesis, which has evolved into one of the key enabling technologies for parts and module-based synthetic biology. Some of these proposals place a higher emphasis on government involvement than others, but all try to minimise the negative impact governance might have on scientific progress and economic benefits. By focusing on DNA synthesis these governance measures have in a sense targeted the most developed part of the supply chain for the synthesis of bioparts or modules, whose ancestors can be traced back to the advent of recombinant DNA technology three decades ago, and which is relatively easily targetable by traditional supply-side controls.

---

47 Industry Association Synthetic Biology n.d., *Report on the Workshop 'Technical Solutions for Biosecurity in Synthetic Biology'*, <http://www.ia-sb.eu>.
48 Ibid., p. 2.
49 Ibid., p. 16.

In contrast, the notion of making a library of bioparts and modules the object of dual-use informed governance measures has yet to receive substantial attention. It seems that in this context more thought has been devoted to issues of open source versus intellectual property rights (IPR), with large biotechnology companies increasingly discovering synthetic biology for commercial purposes.[50] Unless a systematic discourse on dual-use governance structures for parts-based and the other synthetic biology subfields identified above will commence soon, it could thus be pre-empted by the IPR-driven attempts to formulate governance solutions. This could also complicate the realisation of any bioethically informed dual-use governance approach for synthetic biology that might be developed out of the approaches and deliberations discussed in the next section.

Bioethical analyses to address the dual-use aspects of synthetic biology

One of the earliest ethical commentaries on synthetic biology was published in 1999 in *Science*[51] and has been regularly quoted by practitioners in the field and observers alike. It acknowledged that in order to 'ensure responsible use of knowledge that could be applied to the construction of biological weapons, we need to give serious thought to monitoring and regulation at the level of national and international public policy'.[52]

Yet, as Yearley has concluded a decade later, while Cho and her colleagues 'highlight things that people may have ethical concerns about, the paper does not set out or determine what the ethical analysis might conclude'.[53] More fundamentally, Yearley's review of bioethics as a template to conduct an ethical review of synthetic biology leads him to caution that this might actually be 'counterproductive since the apparatus constructed to conduct the social and ethical review will come to look like a mere legitimatory cloak for synthetic biology's advance'.[54] Yet, Yearley remains somewhat elusive in specifying what else might serve as a suitable foundation for an ethical review of synthetic biology.

50   Oye, K. A. and Wellhausen, R. 2009, 'The intellectual commons and property in synthetic biology', in Schmidt et al., op. cit., pp. 121–40.
51   Cho, M. K., Magnus, D., Caplan, A. L., McGee, D. and the Ethics of Genomics Group 1999, 'Ethical considerations in synthesizing a minimal genome', *Science*, vol. 286, pp. 2087, 2089–90.
52   Ibid., p. 2089.
53   Yearley, S. 2007, 'Review: the ethical landscape: identifying the right way to think about the ethical and societal aspects of synthetic biology research and products', *Journal of the Royal Society Interface*, <doi:10.1098/rsif.2009.0055.focus>, p. 3.
54   Ibid., p. 6.

Miller and Selgelid, in contrast, provide a much more detailed and in-depth 'ethical and philosophical consideration' of dual-use issues in the life sciences.[55] Although not specifically targeted at synthetic biology, their analysis illuminates many aspects of bioethical reasoning that are of relevance to synthetic biology too. Based on 'a particularly morally problematic species of the dual-use dilemma',[56] in the form of a number of experiments of concern, they discuss the permissibility of certain kinds of research, debate dissemination of dual-use research results, and analyse different ethically informed governance models with which to tackle the dual-use issues presented by the biological sciences. With regard to this last aspect, Miller and Selgelid discard both the laissez-faire option of giving the individual scientist complete autonomy over their research with dual-use potential and the rather draconian option of complete governmental control. Instead they argue for either a mixed system of institutional and governmental controls or a governance approach that would rely on an independent authority being set up.[57] Thus, while providing a detailed discussion of ethical issues in relation to dual-use life-sciences research, and a narrowing of—in their view—suitable governance options, the analysis remains in this latter dimension somewhat inconclusive.

In contrast with Miller and Selgelid, Ehni's discussion of the ethical responsibilities of scientists engaged in dual-use research is more limited in scope.[58] He approaches the issue by discussing the 'basic conflict between the freedom of science and the duty to avoid causing harm' from two perspectives—that of 'moral skepticism and the ethics of responsibility by Hans Jonas'.[59] On this basis, Ehni evaluates

> four basic duties … to define the prospective responsibility of scientists: 1) stopping research in some cases, 2) systematically exploring dangers of dual use in some cases, 3) informing public authorities about possible dangers resulting from research and the application of its results, and 4) not publishing results and descriptions of research results and possible dual-use applications.[60]

Given the nature of the dual-use issues at hand and the way science is organised, Ehni concludes along the lines of Miller and Selgelid that '[i]t is no solution to

---

55   Miller, S. and Selgelid, M. J. 2007, 'Ethical and philosophical considerations in the dual-use dilemma in the biological sciences', *Science and Engineering Ethics*, vol. 13, pp. 523–80.

56   Ibid., p. 531.

57   Ibid., p. 573.

58   Ehni, H.-J. 2008, 'Dual use and the ethical responsibility of scientists', *Archivum Immunologiae Et Therapiae Experimentalis*, vol. 56, pp. 147–52.

59   Ibid., p. 147.

60   Ibid., p. 151.

the "dual-use" problem to transfer total responsibility to individuals'.[61] Instead he advocates a 'mixed authority' for the governance of dual-use issues, the details of which he also leaves unspecified.

While the recommendation of not assigning sole responsibility to individual scientists is shared by Kuhlau et al., they point out that the moral duty to prevent harm on the part of an individual researcher does exist and includes both intentional and unintentional harms.[62] Furthermore, such moral duty linked to the professional role of the researcher carries with it a requirement for an 'awareness of relevant regulation and potential dangers'.[63] This awareness in turn 'entails a continuous process of reviewing one's work in a wider context'.[64] Based on this reasoning, they identify five criteria for the obligation to prevent harm.

> In order to take social responsibility and due care, life scientists should strive to prevent harm that is: Within their professional responsibility … Within their professional capacity and ability … Reasonably foreseeable … Proportionally greater than the benefits … [and] Not more easily achieved by other means.[65]

Kuhlau et al. subsequently apply these criteria to a number of proposed obligations for life scientists in relation to dual-use issues. They conclude that scientific responsibility 'does not involve preventing the *act* of misuse but rather involves obligations concerned with preventing foreseeable and highly probable harm' (emphasis in original).[66] Such reasonable obligations include, in their view, the duties to consider negative research implications, to protect sensitive material, technology and knowledge from unauthorised access, and to report suspicious activities.[67] While this again represents a useful clarification of the obligation of life scientists including synthetic biologists in relation to dual-use issues they face in their work, it leaves both the institutional and the wider governance contexts unexplored within which these duties need to be considered.

The European Group on Ethics (EGE) in Science and New Technologies to the European Commission in its Opinion No. 25 on the 'Ethics of synthetic biology' usefully points out in this respect that '[g]overnance is an overarching concept

---

61 Ibid.
62 Kuhlau, F., Erikson, S., Evers, K. and Höglund, A. T. 2008, 'Taking due care: moral obligations in dual use research', *Bioethics*, vol. 22, pp. 477–87.
63 Ibid., p. 481.
64 Ibid.
65 Ibid., p. 481 ff.
66 Ibid., p. 487.
67 Ibid.

including legal, political and ethical considerations. Since synthetic biology may result in major changes of traditional biology, governance needs to be reflected on all these levels, finally entering the legal sphere.'[68]

With respect to ethical consideration of synthetic biology, the EGE distinguishes between conceptual and specific issues, and addresses both biosafety and biosecurity under the latter heading.[69] In its discussion of potential steps to be taken, the EGE opinion states under the heading of 'Biosecurity, prevention of bioterrorism and dual uses' that '[e]thical analysis must assess the balance between security and transparency',[70] and moves on to recommend that 'ethical issues that arise because of the potential for dual use should be dealt with at the educational level. Fostering individual and institutional responsibility through ethics discussion on synthetic biology is a key issue'.[71]

In addition, the EGE opinion contains three formal recommendations: 1) linking dual-use bioethics to the Biological Weapons Convention (BWC) by recommending that this international treaty 'should incorporate provisions on the limitation or prohibition of research in synthetic biology'; 2) requesting the European Commission to define a 'comprehensive security and ethics framework for synthetic biology'; and 3) requesting the establishment of DNA-sequence databases with supporting, legally based rules and procedures.[72]

A similarly wide-ranging attempt to chart the ethical issues surrounding synthetic biology was undertaken by the US Presidential Commission for the Study of Bioethical Issues (PCSBI), which in December 2010 produced its first report, entitled *New Directions: The Ethics of Synthetic Biology and Emerging Technologies*.[73] Guided by five ethical principles—that is, '(1) public beneficence, (2) responsible stewardship, (3) intellectual freedom and responsibility, (4) democratic deliberation, and (5) justice and fairness'[74]—the report arrives at 18 recommendations, some of which are informed by the dual-use character of synthetic biology or seek to address its implications. Of particular relevance in this context are recommendations 12 and 13. Acknowledging the dynamic character of the field and the resulting changes in dual-use issues of relevance, the committee recommends periodic assessments of safety and security risks be undertaken. It states:

---

68 European Group on Ethics 2009, *Ethics of Synthetic Biology*, Opinion No. 25, European Union, Brussels, p. 36.
69 Ibid., pp. 42–4.
70 Ibid., p. 51.
71 Ibid., p. 52.
72 Ibid.
73 Presidential Commission for the Study of Bioethical Issues (PCSBI) 2010, *New Directions: The Ethics of Synthetic Biology and Emerging Technologies*, Presidential Commission for the Study of Bioethical Issues, Washington, DC.
74 Ibid., p. 5.

Risks to security and safety can vary depending on the setting in which research occurs. Activities in institutional settings, may, though certainly do not always, pose lower risks than those in non-institutional settings. At this time, the risks posed by synthetic biology activities in both settings appear to be appropriately managed. As the field progresses, however, the government should continue to assess specific security and safety risks of synthetic biology research activities in both institutional and non-institutional settings including, but not limited to, the 'do-it-yourself' community ... An initial review should be completed within 18 months and the results made public to the extent permitted by law.[75]

In case this review identifies 'significant unmanaged security or safety concerns', recommendation 13 foresees changes to existing oversight and control mechanisms with a view to 'making compliance with certain oversight or reporting measures mandatory for all researchers ... regardless of funding sources'.[76] This last point would lead to a significant tightening of existing oversight mechanisms as it would expand their reach beyond publicly funded life-science research and oblige commercial research activities to abide by the same regulatory framework.

Conclusions

This chapter set out to first illustrate that synthetic biology is one of the most dynamic new subfields of the life sciences. It offers the potential to live up to the promise that the discipline behind the label of genetic engineering has long aspired to: the engineering of biological parts, devices and systems, either to modify existing or to create new ones. By applying the toolbox of engineering disciplines and information technology to biology, a wide range of potential applications becomes possible, ranging across scientific and engineering disciplines. Some of the anticipated benefits of synthetic biology, such as the development of low-cost drugs or the production of chemicals and energy by engineered bacteria, are potentially very significant. There are, however, also significant risks due to deliberate or accidental damage. In a way, synthetic biology can be described as the prototypical emerging dual-use technoscience.

Although first attempts at formulating governance mechanisms can be identified, these are for the most part focusing only on a subfield, or, as some would say, enabling technology, of synthetic biology—that is, large-scale commercial DNA synthesis. The conceptualisation of dual-use governance processes and structures for the bioparts and modules-based approach within synthetic

75 Ibid., p. 13.
76 Ibid., p. 14.

biology is, by contrast, still in its infancy (as is the case for the other subfields outlined above). Academic work on the characteristics of dual-use issues from an (bio-)ethical perspective has increased in numbers over recent years, but stops short of considering embedding their analysis and recommendations into the wider institutional or political context that synthetic biologists find themselves in. Similarly, the opinions and recommendations of advisory bodies and committees briefly discussed are quite generic and deal with dual-use issues among many other ethical questions raised by synthetic biology.

What are required are thus more detailed analyses of the dual-use implications of the whole of synthetic biology and systematic dual-use bioethics awareness-raising efforts that reach all practising synthetic biologists and that are supplemented by education and training efforts as well as the formulation of codes and other governance tools that go well beyond the rather technically orientated order screening by DNA-synthesis companies.

5. Crops Agents, Phytopathology and Ethical Review

Simon Whitby

Introduction

This chapter explores the dual-use quality of scientific research and technological development in the field of phytopathology (plant science). It offers a brief survey of naturally occurring pathogens that have been developed for use in weapons and considers areas of convergence and overlap between the hostile use of disease organisms as a form of *warfare* and the *peaceful* deployment of bio-control and plant inoculants. The relevance of bio-control agent and plant inoculant production to the Biological Weapons Convention (BWC) is then considered. Included in this chapter also is a snapshot of some significant developments in civil plant science, alluding to the scale and speed of progress in plant science and technology.

I argue that since they can be used for both peaceful and hostile purposes, plant science and technology raise issues of dual-use biosecurity concern that are thus worthy of ethical consideration. In this connection, this chapter argues that ethical review processes could usefully be located alongside deliberative processes that facilitate consideration of the legal and social implications of plant-science research. Deliberation regarding its potential as dual-use research of concern may therefore be best located within the context of a comprehensive system for oversight of scientific research such as that recommended by the US National Science Advisory Board for Biosecurity (NSABB).[1] Therefore, a brief survey of the contours of the latter is included in the concluding section of the chapter, which focuses specifically on the role of the principal investigator (PI)— 'the most critical element in the oversight of dual-use life-sciences research'[2]— and the requirement to seek to ensure through improved awareness and training that PIs are sufficiently aware of dual-use research issues and concerns.

---

1 National Science Advisory Board for Biosecurity (NSABB) 2007, *Proposed Framework for the Oversight of Dual Use Life Sciences Research: Strategies for Minimizing the Potential Misuse of Research Information*, National Science Advisory Board for Biosecurity, Bethesda, Md.

2 Ibid., p. 11.

Deliberate disease for hostile purposes: Biological warfare

As part of its offensive biological warfare program that originated in the early 1940s, the United States developed an extensive anticrop program that resulted in the production, development, stockpiling and assimilation of anticrop chemical and biological weapons agents. As is well documented,[3] chemical anticrop agents and defoliants were utilised extensively during the course of offensive operations in Vietnam in the 1960s and 1970s.

A range of biological warfare anticrop agents was the focus of extensive research and development during the course of the offensive US anticrop program between the mid 1940s and 1969, and research on production, scale-up, storage, dissemination and effectiveness saw the standardisation of agents for crop destruction with targets including both staple food crops and narcotics production. Since the United States provides an example of a systematic offensive program from research to assimilation, attention will turn briefly to list the agents standardised for the conduct of anticrop warfare.

Agents in the US stockpile as documented[4] in 1970, including for the destruction of food crops, are listed in Table 5.1. An early attempt to identify agents for the destruction of narcotics crops is also included.

Table 5.1 US Stockpiles in 1970 for Food Crop Destruction

Agents category	Type	Hosts
Category A agents	Stem rust of wheat	Wheat, barberry and certain grasses
Category B agents	Rice blast	Rice, possibly some other grasses
	Stripe rust of wheat	Wheat, barley, various grasses
Category C agents	Hoja blanca of rice	Rice, wheat, corn, barley, rye, sorghum and various other grasses
	Bacterial leaf blight of rice	Rice and various grasses
	Downy mildew of poppy	Species of Papaver and Argemone

Source: Stockholm International Peace Research Institute (SIPRI) 1973, *The Problem of Chemical and Biological Wafare*, vol II, Stockholm International Peace Research Institute, Stockholm.

---

3   Cecil, F. 1986, *Herbicidal Warfare: The RANCH HAND Project in Vietnam*, Praeger, New York. See also, Karnow, S. 1997, *Vietnam: A History*, 2nd edn, Penguin, New York.
4   Stockholm International Peace Research Institute (SIPRI) 1973, *The Problem of Chemical and Biological Wafare*, vol II, Stockholm International Peace Research Institute, Stockholm.

Following President Richard Nixon's unilateral renouncement of offensive biological warfare in 1969 the stockpile scheduled for destruction[5] in 1970 included 158 684 lb (71 t) of the causal agent of wheat rust, and 1865 lb (846 kg) of the causal agent of rice blast (as well as munitions for their deployment).

In a 1980s study based on publicly available secondary sources, Geissler noted that prior to 1969—at a time of an emerging international consensus towards the agreement of a complete ban on biological warfare—the majority of military work on pathogens had involved bacteria and fungi. The last included those agents that were developed to attack some of the world's most economically and socially significant food and cash crops. In a program of research and development that appeared to parallel the US biological weapons (BW) program, Iraq's modest late-1980s attempts to conduct research and development into crop warfare also focused on investigations into the effectiveness of fungal plant pathogens.[6] According to Geissler, though, by 1983, the majority of military work on pathogens had switched from bacteria and fungi to instead focus on investigations into the effectiveness of viruses.

The latter featured as agents of choice in what might be regarded as 'second-generation' programs, not least as they appear to have featured in work in the former Soviet Union.[7] With the advent of genetic-engineering techniques, Geissler noted a renewed military interest in biological warfare post 1981, with Dr Kenneth Alibek[8] (a former Soviet biological weapons scientist) subsequently alluding to activities in the former Soviet Union in the early 1980s that included investigations into the production of genetically engineered antibiotic strains of a number of zoonotic, and antipersonnel agents, including anthrax and glanders. According to Geissler, these events were characteristic of the emergence of a 'third generation' of scientific and technological applications in offensive biological warfare. Although there is no publicly available information relating to third-generation research and development in offensive anticrop military programs, it would be unwise to rule out the possibility of improvements in destructive effectiveness. Indeed, outside military programs, ample evidence suggests that routine genetic manipulation is being deployed for peaceful purposes in the burgeoning area of research, development and deployment of bio-control agents and plant inoculants in agriculture.

---

5   Ibid.
6   United Nations Special Commission (UNSCOM) 1995, *Report to the Secretary-General*, 11 October.
7   Tucker, J. 1999, 'Biological weapons in the former Soviet Union: an interview with Dr. Kenneth Alibek', *The Nonproliferation Review*, (Spring–Summer), p. 2.
8   Ibid.

Peaceful bio-control that could serve biological warfare

The focus of research in this area was in the development and deployment of naturally occurring pathogens and insects in the protection of crops from disease caused by pathogens and disease caused or transmitted by insect vectors. The deployment of bio-control agents and plant inoculants marks this area out as distinct since their use for peaceful purposes is not prohibited by the 1972 BWC. Nevertheless, during the course of talks to negotiate a legally binding protocol to strengthen the effectiveness and improve the implementation of the BWC in the 1990s and early 2000s, discussion focused on initiatives intended to ensure that scientific and technological developments in this area were in compliance with the objectives and scope of the convention.

In a statement[9] by the South African delegation to the Fifth Review Conference, on 19 November 2001, Peter Goosen, Department of Foreign Affairs, reminded states parties that in all of their work related to the BWC the threat against plants was usually considered to be of a lower priority than the threat against humans. This had occurred, Goosen pointed out, in spite of the widespread appreciation amongst states parties that major elements of biological weapons programs since the 1920s had been directed against crops and that numerous plant pathogens had been researched, developed and produced together with weapons as part of offensive BW programs for the purpose of the widespread dissemination of anticrop agents.[10]

No clear distinctions, however, separate pathogens in this area from those deployed in offensive biological warfare programs. Bio-control agents are living organisms, such as bacteria, fungi, insects, mites or weeds, or microorganisms that are used in the control of microbes or other organisms. A large number of bio-control agents are currently available—for example, in the United States, where they are marketed as bio-pesticides and include bacteria such as Agrobacterium, the widely used *Bacillus thuringiensis* that produces a protein toxic to species of insect pests belonging to the orders Lepidoptera (caterpillars), Diptera (flies) and Coleoptera (beetles and weevils), Pseudomonas and Streptomyces. Further bio-pesticides include fungi such as Ampelomyces, Candida, Coniothyrium

---

9 Goosen, P. 2001, Statement by Chief Director: Peace and Security, Department of Foreign Affairs, Pretoria, South African Delegation to the Fifth Review Conference of the Convention on the Prohibition of the Development, Production, and Stockpiling of Bacteriological (Biological) and Toxin Weapons and on their Destruction, Geneva, 19 November, <http://www.opbw.org/rev_cons/5rc/docs/statements/5RC-OS-SAFRICA.pdf>.
10 For a systematic study of such state offensive anticrop biological warfare programs, see: Whitby, S. 2001, *Biological Warfare against Crops*, Palgrave, London.

and Trichoderma.[11] Plant inoculants are formulations containing living micro-organisms, used in the treatment and propagation of seeds and plant propagation matériel for enhancing growth and disease resistance in plants. They are also used for the restoration of the microflora of soil. Indeed the technologies associated with the dissemination of such agents appear to equate with those used in the dissemination of biological warfare agents.

Prior to the First Review Conference of the BWC, the Preparatory Committee requested that depositary governments prepare a background paper[12] on new scientific and technological developments relevant to the convention and invited states parties to submit their views on new scientific and technological developments relevant to the convention. Prepared by experts of the depositary governments, the review[13] focused on new scientific and technological developments relevant to the convention and looked inter alia at the microbial control of pests.[14] Since significant environmental and human health implications arose from the deployment of synthetic chemical pesticides that had seen extensive use in Vietnam, this section of the report noted environmental and human health concerns and questioned the efficacy of the use of agents against plants that might develop resistance to their use. The review noted, however, that there had been a remarkable increase in interest in this area. This was summarised as follows:

> Microbiological methods involve the large-scale production of certain live micro-organisms or their extractable toxins, the formulation of a liquid or powder product and dissemination of the product by vehicle or aircraft-borne sprays (or in rodent control, the use of ground bait) over crops or forests. With live microbial agents death of insect or rodent occurs through infection; with microbial toxins death is produced by toxic effects. *In some basic respects the whole sequence resembles biological warfare.* [Emphasis added]

Table 5.2 illustrates the methods of production and dissemination of viral, bacterial and fungal bio-control agents of relevance to the BWC.

---

11 McSpadden Gardener, B. B. and Fravel, D. R. 2002, 'Biological control of plant pathogens: research, commercialisation, and application in the USA', *Plant Health Progress*, [Online], <doi:10.1094/PHP-2002-0510-01-RV>.
12 *Report of the Preparatory Committee for the Review Conference of the Parties to the Convention on the Prohibition of the Development, Production, and Stockpiling of Bacteriological (Biological) and Toxin Weapons and on their Destruction*, BWC/CONF.I/5, 6 February 1980, <http://www.opbw.org>.
13 Not all states parties submitted information to the Secretary-General of the United Nations that referred either directly or indirectly to potential problems posed by the use of microbial agents against crops. For the purpose of this discussion, it has been necessary to refer selectively to the official documentation.
14 *Report of the Preparatory Committee*, op. cit., Appendix E.

The United Kingdom's contribution[15] to a subsequent BWC Review Conference focused also on the microbial control of pests. It included an assessment of increased interest in biological control and noted apparent changes to the methods of production:[16]

> Increases in use since 1980 have not been spectacular except possibly in the nations of Eastern Europe, where *Lepidoptera* pests are a greater agricultural problem than elsewhere. In such nations about 30 different microbial preparations or formulations have been developed and some are produced on often multi-tonne scales. GE [genetic engineering] is being applied in many nations to the development of improved and novel agents for pest control. Obviously large-scale industrial microbiology is a key aspect of production ... Insect viruses for pest control continue to be relatively expensive to produce but the possibility that viruses could be more cheaply and effectively produced through GE, rather than by bulk production in insects, is likely to result in widely-adopted production methods in nations where susceptible pests are a problem.

Table 5.2 Methods of Production and Dissemination of Viral, Bacterial and Fungal Bio-control Agents

Viruses	Nuclear polyhedrosis and granulosis viruses: produced on a large scale by a few nations, using mass rearing of insect hosts. The viruses extracted at a concentration of 12×10^6 infective units/ml are sprayed by aircraft. Viral insecticides are more expensive to produce than bacterial insecticides but they have the possible advantage of high target specificity.
Bacteria	*Bacillus thuringiensis*: produced by several nations on a multi-tonne basis in deep-aerated vessels. The final product contains about 3×10^{10} bacterial spores/g, and is stable for two to three years. Disseminated by aircraft spray as liquid or powder aerosol, the bacterium is highly valued for controlling a wide variety of insect pests.
	Bacillus popilliae: another agent produced and used in much the same way as *Bacillus thuringiensis* for controlling Japanese beetle larvae.
	Pseudomanas seruginosa (and *Pl. fluorescens*) and *Chromobacterium prodigiosum*: produced and used in a few countries for dissemination by aircraft spray on reservoirs (at 10×10^6 organisms/sq cm water surface) in mosquito larvae control. These agents are, however, facultative* pathogens for humans.
Fungi	Various species such as Trichoderma, Sporotrichum, Beauveria and Cuelomemvces are produced on a multi-tonne basis by several nations. They are disseminated by aircraft spray to infect insect pests and sometimes to attack other fungal diseases of crops. Additionally, a number of other microbial agents are currently being studied or evaluated in field trails.

* Facultative pathogens include those with mechanisms for infecting human body tissue.

Source: *Report of the Preparatory Committee for the Review Conference of the Parties to the Convention on the Prohibition of the Development, Production, and Stockpiling of Bacteriological (Biological) and Toxin Weapons and on their Destruction*, BWC/CONF.I/5, 6 February 1980, <http://www.opbw.org>.

---

15 *Background Document on New Scientific and Technological Developments Relevant to the Convention on the Prohibition of the Development, Production and Stockpiling of Bacteriological (Biological) and Toxin Weapons and on their Destruction*, BWC/CONF.11/4, '6. Microbial Control of Pests', p. 8, <http://www.opbw.org>.

16 Ibid., para. 6.2, p. 8.

Regarding methods of dissemination for microbial methods of pest control, the UK contribution[17] noted the following:

Methods of dissemination of microbial pest control agents continue to be the subject of increasing R&D and trials. Mobile jet-engined devices are capable of disseminating agent aerosols, notably insect viruses, over vast tracts of land. There has been continuing R&D on ultra-low volume spraying systems, methods of studying spray deposition, formulations, the problems of disseminating dusts and powders, micro-encapsulation and other relevant topics.

A section on developments of relevance to the BWC in regard to the microbial control of pests was also included in the UK contribution, which noted the following (Table 5.3).

Table 5.3 Microbial Control of Pests

a.	GE-derived bacteria with high toxin yields.
b.	The production through GE of toxins in species beyond those bacteria that produce them in nature.
c.	Development of formulations aimed at enhanced retention of microbial viability during storage and in aerosol.
d.	Protection of aerosolised micro-organisms by the incorporation of protective UV-light screening dyes.
e.	Improvements in the spray-drying and milling of micro-organisms and toxins.
f.	The formulation of synergistic combinations of live micro-organisms and toxic anticoagulants, together with drug-delivery systems.
g.	Development of automated production lines for insects, used in the production of some viruses.
h.	Vastly increased knowledge of aerobiological aspects of dissemination and the elucidation of the factors that control viability and stability in dissemination, aerosol and in respect of persistence.
i.	Computer-controlled continuous culture systems and improved purification systems.

Source: *Report of the Preparatory Committee for the Review Conference of the Parties to the Convention on the Prohibition of the Development, Production, and Stockpiling of Bacteriological (Biological) and Toxin Weapons and on their Destruction*, BWC/CONF.I/5, 6 February 1980, <http://www.opbw.org>.

In its concluding remarks, the UK contribution noted a greater potential for abuse across a range of civil capabilities than was present at the time of the First BWC Review Conference in 1980. In particular, with regard to microbial methods of pest control, it noted

the 'biotechnology explosion' in the civil sectors of many nations and the realization of the potentials of GE and industrial microbiology. We have drawn attention again to developments in microbial methods of pest control and to the increasing knowledge of nations in the

---

17   Ibid., para. 6.2, p. 8.

large-scale production and dissemination of micro-organisms and microbial products. Such developments in the civil sector are relevant to the BWC and could be abused to support offensive programmes.

The South African submission[18] commented extensively and in considerable detail in regard to developments relating to plant inoculants and biological control relevant to the convention. In regard to plant inoculants, separate detailed subsections of the South African submission gave an overview of: the history of their use and development; their purpose; their mode of action; types of inoculants; methods of inoculation; production methods; and the relevance of plant inoculants for the BWC. And in regard to bio-control agents, separate detailed subsections of the South African submission gave an overview of: differing approaches to biological control; the complexity of factors affecting their application; and biological agents against plants—the last including the controversial area of attacking drug crops.

According to the South African submission,[19] plant inoculants are relevant to the convention in terms of:

a. A growing industry and more sophisticated production facilities that have the potential to be diverted to BW producing facilities, as in the case of vaccine production facilities.

b. The genetic research and development that is conducted to improve the micro-organisms that form the active ingredients of inoculants.

c. The development of liquid inoculants that will make their application by spraying and aerosolisation a possibility.

In comparison, the control of plant pests, weeds and plants with biological control agents is relevant to the BWC in terms of:

a. The less clear distinction between the peaceful use of biocontrol agents and their use as BW due to the dual-use nature of these agents.

b. Undesired plants, exotic plants or even noxious plants in one country may be natural, essential and in many cases utilised for commercial purposes (crops) in other countries.

The failure of states parties at the Fifth Review Conference to produce a final declaration[20] meant that the usual 'extended understandings' were not produced during the BWC in 2001–02; South Africa urged co-depositaries and

---

18   Ibid., para. 6.2, p. 5.
19   Ibid., para. 6.2, p. 7.
20   Pearson, G. S. and Sims, N. A. 2005, *Preparing for the BTWC Sixth Review Conference in 2006*, Bradford Review Conference Paper No. 10 (February), <http://www.brad.ac.uk/acad/sbtwc/briefing/RCP_10.pdf>.

states parties to give careful consideration to the issues of biological control at the Sixth Review Conference of the Convention in 2006 and recommended that CBM declarations be extended to include animal and plant pathogen research and production facilities.

As noted by Kelle and colleagues[21]—and echoing some of the sentiments expressed by Geissler above—biological warfare capabilities have been determined by developments in the life sciences, but the genomics[22] revolution has changed dramatically the nature and scope of biological warfare.

Recognition of the challenge posed by rapid advances in science and technology was, for example, included in the Final Declaration[23] of the Second Review Conference of the Biological and Toxin Weapons Convention in 1986. Thus, such concern was embodied in additional understandings agreed on by states parties that:

> The Conference, conscious of apprehensions arising from relevant scientific and technological developments, *inter alia*, in the fields of microbiology, genetic engineering and biotechnology, and the possibilities of their use for purposes inconsistent with the objectives of the Convention, reaffirms that the undertaking given by States Parties in Article 1 applies to all such developments.

Subsequent reviews by states parties of scientific and technological developments of relevance to the BWC have been conducted (on a voluntary basis) by states parties, and since 1980 submissions to the five-year reviews of the convention have noted that significant advances of relevance to the convention have taken place in the fields of biotechnology, genetic modification and genomics.

---

21 Kelle, A., Nixdorff, K. and Dando, M. 2006, *Controlling Biochemical Weapons: Adapting Multilateral Arms Control for the 21st Century*, Palgrave, Basingstoke, UK, p. 35.

22 The sum total of genetic information of an individual, which is encoded in the structure of deoxyribonucleic acid (DNA), is called a genome. The study of the genome is termed 'genomics'. Recently, the order of most of the chemical building blocks, or bases, which constitute the DNA of the genomes of human beings (estimated to amount to three billion), several other animal species and a variety of human pathogens and plants has been determined. Over the next few years this remarkable achievement will be completed and augmented by research into functional genomics, which aims to characterise the many different genes that constitute these genomes and their variability of action. Such research will also determine how these genes are regulated and interact with each other and with the environment to control the complex biochemical functions of living organisms, both in health and in disease.

23 Second Review Conference of the Parties to the Convention on the Prohibition of the Development, Production and Stockpiling of Bacteriological (Biological) and Toxin Weapons and on their Destruction, Final Document PART II Final Declaration, BWC/CONF.II/13/II, <www.opbw.org>.

Science and technology: Phytopathology

In the field of phytopathology, it became apparent by 1996 that genomics would play an increasingly important role in plant biotechnology, and 2006 marked a decade of important scientific and technological developments. During the course of this period plant science provided a glimpse into the huge agronomic and social potential of plants.

In recognition of the fulfilment of objectives set by the Arabidopsis genome initiative in 1996, a news article in a December 2000 edition of *Nature*,[24] titled 'A green chapter in the book of life: the sequencing of an entire plant genome is now complete ...', and the genome analysis in the same edition, signalled a landmark in plant science. The sequencing of the remaining three of five chromosomes, and therefore of the complete gene of *Arabidopsis thaliana*, a small flowering plant that is a member of the mustard family commonly known as thale cress, represented a major scientific and technological breakthrough, with Arabidopsis representing a model organism in plant biology. This development opened up the possibility of investigating the genetic complexity of more economically and socially significant plant life. Citing Bonny, the World Health Organisation (WHO)[25] highlights the further potential benefits of the new science and technology thus: 'The potential uses of modern biotechnology in agriculture include: increasing yields while reducing inputs of fertilizers, herbicides and insecticides; conferring drought or salt tolerance on crop plants; increasing shelf-life; reducing postharvest losses; increasing the nutrient content of produce; and delivering vaccines.'

The cheap availability, short life cycle, small physical size and the small size of the Arabidopsis genome—118.7 million base pairs—relative to other more complex plants, meant that plant science was now able to undertake, under controlled conditions, the identification of the genes responsible for a wide range of physiological processes. In fundamental terms, finding the genes responsible for a plant's physiological response to, inter alia, light, soil and soil nutrients, bacterial, fungal and viral plant pathogens and insect pests, now promised to be much simpler.

Rapid progress in plant science was soon also made regarding the rice genome. The 10-nation International Rice Genome Sequencing Project announced the

---

24   Walbot, V. 2000, 'A green chapter in the book of life', *Nature*, vol. 408, p. 794. See also The Arabidopsis Genome Initiative 2000, 'Analysis of the genome sequence of the flowering plant *Arabidopsis thaliana*', *Nature*, vol. 408 (December), pp. 796–815.
25   World Health Organisation (WHO) 2005, *Modern Food Biotechnology, Human Health and Development: An Evidence-Based Study*, World Health Organisation, Geneva, p. 37, <http://www.who.int/foodsafety/publications/biotech/biotech_en.pdf>.

sequencing of the second complete genome,[26] that of rice, *Oryza sativa*, in 2005. The economic and social importance of rice is significant. Rice is the world's most important food crop, consumed by more than half of the world's population, and to meet projected demand over the next 20 years production will have to rise by an estimated 30 per cent.

In connection with progress in plant science outlined above, a first generation of genetically modified crop products that have emerged in the marketplace over the past decade has expressed a limited number of characteristics. In particular, since 1996, millions of acres have been used for the production of genetically modified crops with, in the case of the United States, large-scale production of modified varieties of corn, cotton and soya (soya beans) enhanced by gene-transfer techniques that confer herbicide tolerance and insect resistant in crops. According to the International Service for the Acquisition of Agri-Biotech Applications report *Global Status of Biotech Crops*,[27] as early as 2005 the United States had approximately 49.8 million ha planted with crops that were the product of genetic modification. The extent to which some of these crops have been adopted in US agriculture and the short time span over which this has taken place are perhaps indicative of the willingness with which agricultural enterprises in the United States have embraced such technologies in spite of concerns raised regarding human health and the environment. According to the US Department of Agriculture (USDA),[28] planting of herbicide-tolerant soya beans expanded from 17 per cent of US soya bean acreage in 1997 to more than 85 per cent in 2005. Planting with herbicide-tolerant cotton expanded from 10 per cent of US acreage in 1997 to more than 60 per cent in 2005. Planting with insect-resistant transgenic corn and cotton containing the *Bacillus thuringiensis* (*Bt*) toxin gene also increased significantly over this period in the United States. USDA figures[29] for 1995 reveal the '[a]doption of all GE cotton, taking into account the acreage with either or both HT and *Bt* traits, reached 79 percent in 2005, versus 87 percent for soybeans. In contrast, adoption of all biotech corn was 52 percent'.

According to the International Service for the Acquisition of Agri-biotech Applications (ISAAA),[30] global biotech planting had exceeded 1 billion ha by 2010.

---

26   For an analysis of some of the salient features of the rice genome, see: International Rice Genome Sequencing Project 2005, 'The map-based sequence of the rice genome', *Nature*, vol. 436 (August), p. 793.

27   Available at: <http://www.isaaa.org/>.

28   United States Department of Agriculture (USDA) n.d., *Adoption of Genetically Engineered Crops in the U.S.: Extent of Adoption*, Economic Research Service, United States Department of Agriculture, Washington, DC, <http://www.ers.usda.gov/Data/biotechcrops/adoption.htm>.

29   Ibid.

30   International Service for the Acquisition of Agri-biotech Applications (ISAAA) 2010, *Global Status of Commercialized Biotech/GM Crops: 2010*, ISAAA Brief, <http://www. ISAAA.org>.

In the past two decades, genome studies have facilitated manipulation of the genetic characteristics of food crops. Crops can now be produced with built-in defences against insect pathogens such as *Bacillus thuringiensis*. They can also be manipulated to delay ripening, as in the case of the slow-ripening Flavr Savr tomato, which was approved for sale in the United States in 1994. Infertility can be conferred on plant seeds, as in the case of the controversial Terminator gene.

Plant science is, however, now beginning to focus on a second generation of crops that have been genetically modified to express a broader and more complex range of plant traits. The challenge now also extends to assigning functions to genes, and in the case of Arabidopsis much work[31] had been done by 2007. According to a recent edition of *Current Opinion in Plant Biology*,[32] advances in understanding at the level of functional genomics will result, in the case of Arabidopsis, in:

> An understanding of the networks through which these genes interact to control plant development, metabolism, reproduction and other fundamental processes will accelerate the advent of a new generation of improved crop products to benefit growers, processors and consumers.

> Bringing together knowledge of the function of genes and gene networks, and of their regulation within the contexts of cell, organ, organism, and environment will be crucial for achieving the level of precision in crop engineering that will be required to fuel the development of the next-generation products.

Considerable progress has been made in regard to the re-annotation[33]—including both the structure and the functions of genes—of the gene sequences of Arabidopsis, and a similar re-annotation of rice is also under way.

The revolution in genomics signals a transition in plant biology from a descriptive to a predictive science. As Dixon[34] points out: 'Genomics (originally DNA and transcript based, but recently extended to integrate the proteome and metabolome) has revolutionized the speed of gene discovery for important plant traits.'

---

31   Lan, H., Carson, R., Provart, N. J. and Bonner, A. J. 2007, 'Combining classifiers to predict gene function in *Arabidopsis thaliana* using large-scale gene expression measurements', *BioMedCentral*, <http://www.biomedcentral.com/1471-2105/8/358>.
32   Salmeron, J. and Herrera-Estrella, L. 2006, 'Plant biotechnology: fast-forward genomics for improved crop production', *Current Opinion in Plant Biology*, vol. 9, pp. 177–9.
33   Rensink, W. A. and Buell, R. C. 2005, 'Microarray expression profiling resources for plant genomics', *Trends in Plant Science*, vol. 10, no. 12 (December).
34   Ibid.

Indeed a brief review of scientific and technological developments emerging since the turn of the twenty-first century offers a glimpse into the broad range of activities relating to a more sophisticated understanding of the function of plant genomes.

A feature article[35] published by the American Phytopathological Society (APS) in 2000 described four significant areas of research and possible future approaches involving genetic-engineering techniques that could confer plant resistance to pathogen invasion. Research focused on enhancing resistance with plant genes sought to identify the genes involved in defences against, and resistance to, plant pathogens (*resistance is against pathogens, rather than diseases*) to facilitate the conferring of disease resistance. Use of this approach increased the plant's ability to defend itself against pathogen invasion, and plant biologists using recombinant DNA biotechnology could now adopt a number of new strategies. As Fermin-Munoz[36] demonstrated, the insertion of a specific transgene conferred resistance not normally present in a host plant. The insertion of transgenes also could trigger a plant's intrinsic defence mechanism against both pathogen invasion and abiotic stress. Also described were similar techniques for the insertion of proteinaceous and non-proteinaceous compounds having antibacterial and antifungal properties. One such example has led to the discovery of a protein called harpin (produced by members of the plant pathogenic bacterial genus Erwinia, which is sometimes thought of as a toxin and sometimes as a defence chemical) that could be used prior to pathogen invasion to activate crop defences.

Another approach[37] was developed to protect plants from pathogen invasion through 'pathogen-derived resistance'. This strategy involved engineering genes into plants, and important recent work has included the insertion of viral transgenes as an alternative to the use of harmful pesticides against insect vectors. Pioneering research has shown that both protein-mediated and RNA-mediated pathogen-derived resistance can be conferred using viral transgenes, with some success being achieved using a number of plant viruses[38] affecting alfalfa, cucumber, tobacco, tomato and potato.

---

35   Fermin-Muñoz, G. A., Meng, B., Ko, K., Mazumdar-Leighton, S., Gubba, A. and Carroll, J. E. 2000, 'Biotechnology: a new era for plant pathology and plant protection', *APSnet Feature*, May, <http://www.apsnet.org/publications/apsnetfeatures/Pages/Biotechnology.aspx>.

36   Fermin-Munoz, G. A. 2000, 'Enhancing a plant's resistance with genes from the plant kingdom', *APSnet Feature*, May, <http://www.apsnet.org/publications/apsnetfeatures/Pages/EnhancingPlantResistance.aspx>.

37   Meng, B. and Gubba, A. 2000, 'Genetic engineering: a novel and powerful tool to combat plant virus diseases', *APSnet Feature*, May, <http://www.apsnet.org/publications/apsnetfeatures/Pages/GeneticEngineering.aspx>.

38   Ibid.

Another approach[39] for conferring plant disease resistance involves antimicrobial peptides and proteins that confer antimicrobial properties, thus strengthening immunity and resistance to fungal and bacterial plant pathogens. This review noted that proteins potentially useful in improving plant disease resistance could be found outside the plant kingdom in insects, animals, humans and fungi. Antibacterial peptides/proteins and enzymes have been shown to inhibit pathogen invasion in a variety of plants including some of considerable socioeconomic significance. Furthermore, immunity or resistance to pathogen invasion has been attempted at the molecular level by *in planta* expression of an antibody against a protein necessary for pathogenesis.

The past decade has seen continued progress in the development of plant disease-resistance mechanisms, and a wide array of new tools is being developed to produce plants expressing a broader range of such traits. Interesting areas of development were described in 2005 in *Trends in Plant Science*.[40]

In China, Wang et al.[41] focused on improving rice, applying molecular marker-assisted breeding, functional genomics and genetic modification techniques to the identification of gene function in elite rice cultivars having important socio-agronomic traits such as enhanced pest and stress resistance, good grain quality, and high and stable yield potential.

In the area of molecular marker-assisted breeding, which is the application of molecular biotechnologies to breeding, tools include marker-assisted selection (MAS), quantitative trait locus (QTL) analysis, and genetic transformation techniques. While plant scientists have enjoyed success in improving cultivars with important traits using genes with known desirable traits, the adoption of molecular marker-assisted breeding techniques may facilitate the identification of desirable plant traits that result from the expression of multiple genes.

Wang et al. also delved into functional genomics,[42] demonstrating how micro-arrays, reverse genetics and map-based cloning are being used for identifying important characteristics in rice genes, including '[e]xpression pattern, chromosomal position, perceived biological function, and behaviour of alleles under phenotypic selection'.

---

39   Ko, K. 2000, 'Using antimicrobial proteins to enhance plant resistance in biotechnology: a new era for plant pathology and plant protection', *APSnet Feature*, May, <http://www.apsnet.org/publications/apsnetfeatures/Pages/AntimicrobialProteins.aspx>.

40   Dixon, R. A. 2005, 'Plant biotechnology kicks off into the 21st century', *Trends in Plant Science*, vol. 10, no. 12; Neal-Stewart, C., jr, 2005, 'Plant functional genomics: beyond the parts list', *Trends in Plant Science*, vol. 10, no. 12 (December).

41   Wang, Y., Xue, Y. and Li, J. 2005, 'Towards molecular breeding and improvement of rice in China', *Trends in Plant Science*, vol. 10, no. 12 (December), pp. 610–14.

42   Ibid.

Although scientists have had some success in the development of transgenic crops with resistance to plant disease and plant pests, a number of hurdles, not least those relating to public health and the environment, have (in China) thus far prevented the widespread commercialisation and human consumption of rice varieties with improved traits.

In their article 'Microarray expression in profiling resources for plant genomics',[43] Rensink and Buell discuss various approaches to facilitate the identification of gene function and understanding of, inter alia, basic physiology, developmental processes and environmental stress responses, using information derived from micro-array platforms. They note, in particular, the significance for plant researchers of micro-array-derived bio-information and the importance of the worldwide availability, via the Internet, of bioinformatics data sets.

Likewise, in the article 'Genomics-assisted breeding for crop improvements',[44] Varshney et al. note the importance of information on molecular markers (functional markers), and the relevance of the rapidly advancing area of bioinformatics, which is providing a means for the integration and structured interrogation of data sets that will facilitate cross-fertilisation of disciplines in the evolution of future genomics-assisted breeding.

A 2006 review in *Current Opinion in Plant Biology*[45] provided a further glimpse into research in functional genomics. For example, Bohnert et al. describe a 'palette of tools' that facilitates a more detailed understanding of the spectrum of plant responses to developmental and environmental stimuli, including tolerance to drought, soil salinity and cold stresses. According to Salmeron and Herrera-Estrella,[46] tools such as metabolite and protein profiling, subcellular imaging, transcript clustering, comparative biology and reverse genetics reveal a range of valuable genes, alleles and promoters. Research by Valliyodan and Nguyen in the same edition shows understanding of abiotic stress tolerance, 'specific gene components' and transcriptional and cis-acting regulatory elements important in possible future engineering of specific plant traits has progressed significantly.

A review of work by Fernie et al.[47] describes progress in the development of genomics-based techniques that use molecular markers in the identification of desirable plant trait alleles. According to Salmeron and Herrera-Estrella:[48]

---

43   Rensink and Buell, op. cit.
44   Varshney, R. K., Grner, A. and Sorrells, M. E. 2005, 'Genomics-assisted breeding for crop improvement', *Trends in Plant Science*, vol. 10, no. 12 (December), pp. 621–30.
45   Salmeron and Herrer-Estrella, op. cit., pp. 177–9.
46   Ibid.
47   Ibid.
48   Ibid.

'Genomics will make possible a level of surgical precision in breeding that allows such traits to be efficiently extracted by using molecular markers to tag specific desirable alleles.'

Alleles of importance for plant traits such as those related to nutrition and yield have already been identified in some solanaceous and other important crops. Also reviewed by van Schie et al. in the same edition is research on plant fragrance mechanisms, and the possibility that it could be manipulated at the genome level.

A further article on functional genomics centres on emerging techniques for gene silencing using geminiviral vectors. Salmeron and Herrera-Estrella[49] outline some possible applications as follows:

> 1) [T]hey can be used for functional genomics in plants species for which the production of transgenic lines is difficult or time-consuming; 2) they provide the ability to work with genes whose knockout mutants are lethal; 3) inoculation with Gemini viral vectors is rather simple and phenotypes can be analysed a few days after the host has been infected; and 4) they can be adapted easily for high-throughput genomic studies.

Malign applications

The above developments offer a glimpse of the rapid progress in plant biology over the past 10 years. The genomics revolution opens up a range of new possibilities for improvements in the quality and quantity of crop yields and an increasing number of techniques and applications will become available for combating disease that is caused naturally and accidentally; however, the same developments also open up a range of possibilities for malign applications.

Analysts have for many years expressed concern regarding the ways in which naturally occurring plant pathogens might be deployed for malign purposes. This might involve the simple introduction into a crop species, for example, of a pathogen to which no natural immunity exists. Van der Plank's well-known observations of the seemingly explosive spread of some plant pathogens in the absence of immunity remain particularly salient in spite of great progress that has been made in phytopathology over the past 40 years.

Scenarios may also include the introduction into crops of pathogens that have mutated naturally in the environment—witness, in this connection, the near-future possibility of the re-emergence in regions such as the Middle East and in countries such as India of a new and virulent strain of wheat rust, Ug99

---

49   Ibid.

(Ug99 has been found in Egypt at least and perhaps other countries in the Middle East), and its predicted associated 'immense potential for social and human destruction' as it has been described by Borlaug[50] in the *New Scientist*. According to Mackenzie,[51] to combat this mutation effectively, the production of enough Ug99-resistant seed to plant our wheat fields might take up to eight years.

As in the case of bio-control agent production discussed above, there is also a possibility that crop pathogens could be genetically modified deliberately. The result could include increased toxicity or pathogenicity. Or, as suggested by Kelle et al.,[52] plants' innate immune systems are vulnerable to manipulation, possibly affecting the response to pathogen invasion. For example, a plant's response could be manipulated so as to trigger 'systemic', rather than localised, hypersensitive reactions to pathogen invasion. Nixdorff explains the mechanisms involved in systemic plant resistance mechanisms thus: 'The main systemic signals include salicylic acid, jasmonate and ethylene, which are produced in response to wounding and insect attack. $H2O2$ is ... the most important' response mechanism to pathogen invasion

> involved in downstream signalling, leading to the activation of signalling cascades in *Arabidopsis* as well as activation of genes controlling the production of proteins involved in HR [hypersensitive reaction]. [Therefore] ... plants may be attacked through their innate immune systems, for example, by targeting either the receptors of signalling cascades, or by inhibiting or producing an over-reaction in a signalling cascade with the use of inhibitors of key components in that cascade.

Kagan et al.[53] present concerns about the introduction of noxious DNA material in the form of, for example, a bio-regulator into a bio-control agent such as *Bacillus thuringiensis* in quantities sufficiently large to contaminate a food product. Indeed, further to this, Chofness et al.[54] also note in relation to transgenic plants that such plants

> could be malevolently engineered to produce large quantities of bioregulators or toxic proteins, which could either be purified from plant cells or used directly as biological agents. As with legitimate production, using transgenic plants as bioreactors would eliminate the

---

50   Mackenzie, D. 2007, 'Rusting defences in the battle for wheat', *New Scientist*, 3 April.
51   Mackenzie, D. 2007, 'Billions at risk from wheat super-blight', *New Scientist*, 3 April, pp. 6–7.
52   Kelle et al., op. cit., p. 76.
53   Kagan, E. 2006, 'Bioregulators as prototypic nontraditional threat agents', *Clinics in Laboratory Medicine*, vol. 26, no. 2 (June), pp. 421–43.
54   Choffnes, E. R., Lemon, S. M. and Relman, D. A. 2006, 'A brave new world in the life sciences, the breadth of biological threats is much broader than commonly thought and will continue to expand', *Bulletin of Atomic Scientists*, vol. 62, no. 5 (September–October), pp. 26–33.

need for mechanical equipment normally associated with the process. The technology would be limited to producing protein-based agents. But because transgenic plants would be largely indistinguishable from non-transgenic crops, biopharming could potentially provide a covert means for producing large amounts of product.

Conclusion

This survey of offensive biological warfare agents, bio-control agents and plant inoculants of relevance to the BWC, and recent trends in phytopathology and plant technology, has alluded to a broad range of scientific discovery and technological innovation. This highlights a number of areas where consideration—including ethical deliberation—regarding the dual-use nature of scientific discovery and technological application might usefully be applied. Indeed, where deliberation might take place, who might be involved, and what mechanisms and systems might be put in place to facilitate deliberative processes have been the subjects of considerable attention in the United States.

Since 2006, the NSABB has been tasked,[55] inter alia, with advancing thinking around proposals for the development of recommendations for scientific oversight measures and for recommending how such measures might be developed so as to minimise the risk of misuse of scientific information. One of the challenges inherent in pursuing this mandate is how oversight mechanisms might be created that are both efficient and effective but constructed and implemented in such a way as to mitigate the stifling of life-science innovation. This initiative was viewed as a preliminary step towards establishing a mechanism of oversight through the development and 'implementation of a comprehensive system for the responsible identification, review, conduct, and communication of dual use research'.[56] The NSABB proposed that a system of oversight might include the following seven 'key features':

- The development of Federal Guidelines for oversight of dual use life science research
- Enhanced levels of awareness of dual use research of concern amongst practicing life scientists
- Enhanced, ongoing, mandatory education that raises awareness of dual use research of concern and addresses the roles and responsibilities of life scientists

---

55   See NSABB, op. cit., Appendix A, p. 9.
56   Ibid., p. 7.

- With appropriately trained Principal Investigators, a system of local evaluation and review of research of dual use potential.

- Risk Assessment and Risk Management

- Mechanisms for Periodic Local and Federal Evaluation

- Local and Federal Mechanisms for ensuring Compliance.[57]

In regard to a system of 'local evaluation and review of research of dual use potential', the NSABB recommended that the initial evaluation of 'research for its potential as dual use research of concern' should fall to the principal investigator (PI). Indeed, as further noted by NSABB, appropriate expertise could thus be brought to bear in assessing the dual-use potential of scientific research, but significantly NSABB also noted a requirement in this area for 'appropriate training'.[58]

As part of its recommendations, NSABB identified a series of oversight roles and responsibilities and elaborated upon how both would relate to researchers, research institutions, institutional review entities, ongoing independent review and the Federal Government. In regard to the roles and responsibilities of researchers, NSABB noted a further potential deficit in the area of awareness and training. As it argued, PIs are the ones who will be best placed to discharge their responsibility to be able to assess the kinds of knowledge generated, its potential for misuse and how such threats might be mitigated. According to NSABB,[59] however, PIs would need both to be 'cognizant of the concept of dual use research of concern and aware of the risk that technologies or information produced by life sciences research may be misused'.[60] Indeed, NSABB[61] places considerable emphasis on improving levels of awareness in that it argues that an 'enhanced culture of awareness is essential to an effective system of oversight and is a critical step in scientists taking responsibility for the dual use potential of their work'.

This chapter argues that ethical discussion and deliberation, alongside consideration of the potential social and legal ramifications of scientific research,[62] could be usefully embedded in a broader system of scientific oversight such as that envisaged by NSABB, and could have an important deliberative function in assessing experimentation of dual-use concern in the area of plant-science research. Indeed, opportunities for ethical and other types of deliberation could

---

57   Ibid., p. 8.
58   Ibid., p. 9.
59   Ibid., p. 11.
60   Ibid., p. 11.
61   Ibid., p. 9.
62   Ibid. See also Committee on Research Standards and Practices to Prevent the Destructive Application of Biotechnology, Development, Security, and Cooperation, Policy and Global Affairs 2004, *Biotechnology in An Age of Terrorism*, National Research Council, The National Academies Press, Washington, DC, 'Recommendation 1: Educating the Scientific Community', p. 4.

be included at each respective step (PIs, institutional and independent review entities, governments, and so on) in the research oversight process as identified by NSABB.

Whilst the publication of the framework proposed by NSABB represented a step-change in thinking about how to address the oversight challenges posed by dual-use research of concern, as argued by NSABB,[63] deliberative processes will need to be informed by increased levels of awareness and education about dual-use research of concern, including 'all applicable policies as well as the provision of guidance and tools that facilitate compliance with the policies'.

Much work needs to be done by ethicists, however, in respect of how such review processes will be informed by ethical considerations. Indeed, awareness raising, education and training courses will need to be developed that facilitate ethical deliberation around dual-use research of concern but also that facilitate deliberation and assessment of both the legal and the social implications of such research. Such material must also be oriented to provide life scientists with guidance and information that facilitate compliance with relevant guidelines, policies and legislation.

Unfortunately, few accredited university courses or non-accredited short courses currently exist that seek to engage life scientists in deliberation about the ethical, legal and social implications of the scientific research they conduct. One notable exception is a combined hybrid biosafety (bio-risk management) and dual-use biosecurity training course[64] that is being designed and developed in a collaboration between the University of Bradford, UK, and the Public Health Agency, Canada (PHAC). Following the implementation of Canada's *Human Pathogens and Toxins Act*,[65] which sets out a compliance requirement that biosafety officers are trained and have requisite 'qualifications', this initiative seeks to develop training material (with a Canadian focus) that engages life scientists in deliberation about life-science research of potential dual-use concern by addressing a range of dual-use issues of relevance to laboratory biosafety and beyond the laboratory door. It is argued here that this combined biosafety/dual-use biosecurity approach possibly represents a model of best practice for the subsequent development of awareness raising, education and training. The last could be incorporated into the training of plant scientists in order that they are cognisant of dual-use research issues of concern and better able to assess the ethical, legal and social implications of their work.

---

63   NSABB, op. cit., p. 15.
64   The dual-use biosecurity element of this initiative has evolved from an existing online distance-learning course that focuses on applied dual-use biosecurity education that has been running at the University of Bradford since September 2010.
65   *Human Pathogens and Toxins Act*, 2009, Canada, <http://lois-laws.justice.gc.ca/eng/acts/H-5.67/index.html>.

Finally, building on research that was previously set out in the influential 2004 Fink Report,[66] NSABB also set out seven criteria for the identification of 'endeavours' and 'discoveries' that might trigger discussion and review. The following seven classes of experiments might provide a useful framework for considering 'the types of endeavours or discoveries' that, if proposed, might trigger review by PIs including, where appropriate, 'review and discussion by informed members of the scientific and medical community before they are undertaken or, if carried out, before they are published in full detail'.[67]

The experiments include those that:

1. Would demonstrate how to render a vaccine ineffective. This would apply to both human and animal vaccines. Creation of a vaccine-resistant smallpox virus would fall into this class of experiments.

2. Would confer resistance to therapeutically useful antibiotics or antiviral agents. This would apply to therapeutic agents that are used to control disease agents in humans, animals, or crops. Introduction of ciprofloxacin resistance in *Bacillus anthracis* would fall in this class.

3. *Would enhance the virulence of a pathogen or render a nonpathogen virulent. This would apply to plant, animal, and human pathogens. Introduction of cereolysin toxin gene into* Bacillus anthracis *would fall into this class.*

4. *Would increase transmissibility of a pathogen. This would include enhancing transmission within or between species. Altering vector competence to enhance disease transmission would also fall into this class.*

5. *Would alter the host range of a pathogen. This would include making nonzoonotics into zoonotics agents. Altering the tropism of viruses would fit into this class.*

6. *Would enable the evasion of diagnostic/detection modalities. This could include microencapsulation to avoid antibody-based detection and/or the alternation of gene sequences to avoid detection by established molecular methods.*

7. *Would enable the weaponization of a biological agent or toxin.* [Emphasis added]

Whilst a detailed analysis of NSABB's oversight system is beyond the purview of this chapter, the above criteria could be used to trigger interventions

---

66 Committee on Research Standards and Practices to Prevent the Destructive Application of Biotechnology, Development, Security, and Cooperation, Policy and Global Affairs, op. cit.
67 NSABB, op. cit., p. 18.

in plant-science research at the level of PI. A multilayered oversight system could include a system of checks and balances so as to ensure that appropriate review is applied, where appropriate to do so, at each stage of the research/ oversight process. Application of the criteria to developments in plant science and technology (as discussed above) would seem to suggest that discussion and review concerning the ethical, legal and social implications of such research would be triggered in no less than five out of seven classes of experiments identified in Fink and by NSABB (see emphases).

6. The Super TB Experiment: Evolution and Resolution of an Experiment with Dual-Use Concerns

Nancy Connell

Introduction

Tuberculosis (TB) is a devastating disease that leads to 10 million deaths per year worldwide.[1] Despite enormous advances in TB research in recent years, the surge of drug-resistant strains and deadly synergy with the HIV virus have threatened to destabilise gains made in its control. During my career as a microbial geneticist, I have studied the physiology of *Mycobacterium tuberculosis*, the causative agent of TB, and its interaction with the human macrophage, its primary host cell. The major function of the macrophage is to engulf and destroy invading organisms, but many pathogens, such as TB, have developed intricate and clever ways to avoid or subvert these host defences.

This chapter discusses how a classical approach to asking a very basic question about microbial pathogenesis can lead to a surprising result and a dual-use dilemma. The elimination of a single gene from *M. tuberculosis* created a strain with increased virulence when measured in a model system (in vitro) but not during animal infection (in vivo).

Experiments of concern

The concept of dual-use research—that certain kinds of experimental endeavours might be used for malignant purposes despite their original beneficent intent—has entered the biomedical lexicon.[2] My personal interest in the matter grew out

---

1 <http://www.who.int/tb/publications/global_report/2010/en/index.html>.
2 National Research Council and the American Association for the Advancement of Science (NRC/AAAS) 2009, *A Survey of Attitudes and Actions on Dual Use Research in the Life Sciences: A Collaborative Effort of the National Research Council and the American Association for the Advancement of Science*, Washington, DC; American Association for the Advancement of Science (AAAS) 2009, *Building the Biodefense Policy Workforce*, American Association for the Advancement of Science, Washington, DC; American Association for the Advancement of Science (AAAS) 2008, *Professional and Graduate-Level Programs on Dual Use Research and Biosecurity for Scientists Working in the Biological Sciences*, American Association for the Advancement of Science, Washington, DC; Atlas, R. and Dando, M. 2006, 'The dual-use dilemma for the life sciences: perspectives, conundrums, and global solutions', *Biosecurity and Bioterrorism*, vol. 4, no. 3, pp. 1–11.

of a longstanding interest in biological weapons and arms control beginning in the 1980s. At the time, many scientists were contributing to the development of a Verification Protocol for the Biological and Toxin Weapons Convention (BTWC).[3] Part of that commitment included the responsibility of scientists to be aware of the societal impact of their work and to communicate that impact to the public.

In response to concerns about dual-use research, a number of advocacy groups started up at this time such as Science for the People and the Committee for Responsible Genetics (now the Council for Responsible Genetics). Simultaneously, scientists and policy analysts realised that the potential for harm caused by naturally occurring organisms and toxins could be easily surpassed by the creative development of recombinant agents expressing altered or novel virulence characteristics. Commonly available genetic technologies in laboratories improved, allowing for a deeper understanding and dissection of the mechanisms of infection and virulence.

As the Verification Protocol fell by the wayside in the autumn of 2001, one month before the anthrax mailings in the United States, interest in strengthening the BTWC took on new directions. One of those was an interest in how knowledge and techniques being generated as part of peaceful work in the life sciences could facilitate the deliberate spread of disease—what is now known as 'dual use'. A list of the experiments considered to be of dual-use concern was published in several reports shortly after the events of 2001. This included the landmark Fink Report.[4] These lists included several approaches to creating novel mutants of existing pathogens, such as: 1) developing antibiotic-resistant strains, 2) expanding host range, 3) increasing transmissibility, and 4) altering virulence.

To understand a biological system, one of the first approaches is to perturb the system and observe the effects of that manipulation. In a genetic analysis, the approach is to alter or even eliminate the function of a gene and observe the effects of that alteration on the biological process. By integrating information from a number of different alterations of a single gene, its role or function in a system or pathway can be deduced. In the case of disease-causing organisms, each of the 'experiments of concern', listed above, fits into this classic approach to understanding biological function

- antibiotic-resistant strains are selected and characterised to understand the mechanism of action of an antibiotic, and the identification of the target will allow development of more effective antibiotics directed to that target

---

3   Zilinskas, R. A. 1998, 'Verifying compliance to the biological and toxin weapons convention', *Critical Reviews in Microbiology*, vol. 24, pp. 195–218.
4   National Research Council 2004, *Biotechnology Research in An Age of Terrorism*, The National Academies Press, Washington, DC.

- host range mutants are developed to understand the mechanism of host range restriction and to develop methods of blocking host–pathogen interaction, the first step of infection
- transmissibility mutants can be created to understand the mechanisms that control transmissibility, to develop methods of reducing transmissibility
- altering virulence can be used to identify the genes encoding virulence factors, and interfering with these factors that may limit infection and pathogenesis of disease.

Therefore, any comprehensive genetic analysis of virulence will eventually be confronted with the possibility of an 'experiment of concern'. The question for researchers is whether to continue to pursue the line of inquiry or to find an alternative approach. I now want to consider these issues in relation to my own work.

The experiment: Nutrient transport mutants of *Mycobacterium tuberculosis*

Microbes use small molecules such as sugars and amino acids to survive, and pathogens that live in their host's cells have a multitude of ways to acquire these nutrients from their environment. In the case of *M. tuberculosis*, their environment is the inside of mammalian white blood cells, where they replicate quite happily; indeed, the bacterium's method of pathogenesis is to reside within the very cells designed to protect the host against them![5]

Our experimental approach was to create mutants of *M. tuberculosis* that were defective in uptake of a certain class of small nutrients called peptides. These are short (2–6) chains of amino acids that are transported by efficient transport systems called ABC transporters.[6] We were able to construct mutants of *M. tuberculosis* lacking the genes that encode one of these transporters.

Simultaneously, we also discovered that a small tri-peptide called glutathione had the unusual and unexpected property of being toxic to *M. tuberculosis*.[7] This meant that in the presence of this small peptide, *M. tuberculosis* could not grow. The concentration of glutathione in mammalian cells varies considerably but is usually quite high; under some conditions, the concentration is much higher than the toxic levels that kill *M. tuberculosis*. We wondered how this

---

5   Pieters, J. 2008, 'Mycobacterium tuberculosis and the macrophage: maintaining a balance', *Cell Host and Microbe*, vol. 3, pp. 399–407.

6   Detmers, F. J. M., Lanfermeijer, F. C. and Poolman, B. 2001, 'Peptides and ATP binding cassette peptide transporters', *Research in Microbiology*, vol. 152, pp. 245–58.

7   Green, R. M., Seth, A. and Connell, N. D. 2000, 'A peptide permease mutant of *Mycobacterium bovis* BCG resistant to the toxic peptides glutathione and S-nitrosoglutathione', *Infection and Immunity*, vol. 68, pp. 429–36.

situation could have developed: that a bacterium that spends its entire life cycle in mammalian cells does so in the presence of a small molecule, glutathione, that is toxic! We characterised this toxicity, and showed that in mouse-derived cells, when the glutathione levels were high, *M. tuberculosis* did not grow well, but when the glutathione levels were low, the bacteria grew very efficiently.[8]

We were, however, now at a crossroads. First, we showed that glutathione is toxic to *M. tuberculosis*, and we found glutathione at toxic levels in all mammalian cells. Second, we had created a mutant of *M. tuberculosis* in the laboratory that was lacking a peptide transporter and incapable of accumulating small peptides, including glutathione. Therefore, this mutant should have been resistant to the toxic effects of glutathione. We showed that this was indeed the case.

But if glutathione is toxic to *M. tuberculosis* and present in all mammalian cells then it might play a natural role in limiting *M. tuberculosis* growth. This was our hypothesis. The logical next step was to see whether our mutant strain survived better than its normal parent strain during intracellular growth in white blood cells in tissue culture. It did. In fact, the mutant strain grew to three to five times higher levels than normal parent strains when inoculated and grown in cultured macrophage cell lines. Our experimental approach following a logical series of steps had led us to create a hyper-virulent mutant of *M. tuberculosis*. We had generated a strain that survived in host cells better than its normal parent strain, as a result of the deletion of a single gene encoding one component of a basic nutrient transport system. Interestingly, the dual-use implications of our work were brought to my attention only during a discussion of that issue with members of the Controlling Dangerous Pathogens Project at the Center for International and Security Studies at Maryland (Elisa Harris and colleagues). Apparently, I had partitioned my thinking about these matters, and separated the experiment itself from the possible ethical concerns associated with it, despite my own decades-long involvement with the issue of biological weapons and the responsibility of scientists in preventing a biological arms race.

Virulence factors

Defining virulence is not a simple matter. By a general and superficial definition, virulence is the degree of 'pathogenicity' of a biological agent, and pathogenicity

---

8   Venketaraman, V., Dayaram, Y. K., Amin, A. G., Ngo, R., Green, R. M., Talaue, M. T., Mann, J. and Connell, N. D. 2003, 'Role of glutathione in macrophage control of mycobacteria', *Infection and Immunity*, vol. 71, pp. 1864–71; Venketaraman, V., Dayaram,Y. K., Talaue, M. T. and Connell, N. D. 2005, 'Glutathione and nitrosoglutathione in macrophage defense against *Mycobacterium tuberculosis*', *Infection and Immunity*, vol. 73, pp. 1886–9.

is the ability of an organism to cause disease. Both terms are relative and highly dependent on context. Furthermore, virulence and pathogenicity are frequently used interchangeably.

Myriad factors contribute to the virulence of an infectious agent. Virulence factors that cause severe damage to the host are called toxins. Some toxins act directly on the host cell such as listeriolysin, which punches holes in the membranes of host cells and breaks them open.[9] Many viruses reproduce to such high numbers in cells—grabbing bits of the host cell membranes as they go—that the cells just break apart.[10] Other toxins act directly on host cells to alter their function severely. For example, one component of anthrax toxin destroys the proteins that orchestrate the immune response and cause swelling in the tissues surrounding the site of infection, while another component inactivates a small protein, called G-protein, which regulates the movement of ions across membranes; its inactivation leads to severe oedema.[11] Other toxins block protein synthesis, so the cell can no longer carry out normal functions (diphtheria toxin),[12] or interfere with neurotransmitter activities, as with the clostridial toxins, botulinum toxin and tetanus.[13] In some, but not all, cases elimination of the gene encoding the toxin will render the bacterium completely harmless ('avirulent'); this is true with *Clostridium botulinum*, the causative agent of botulism. Conversely, expression of the gene encoding the botulinum toxin in an alternative bacterial host would lead to the creation of a very dangerous agent. Ken Alibek, a former Soviet bioweapons researcher, claimed that the Soviet offensive program devoted considerable effort to splicing the botulinum toxin genes into other bacteria for delivery.[14]

But in the case of *Yersinia pestis* (plague),[15] for example, and *Burkholderia mallei* (glanders)[16]—both select agents—there are many small, secreted virulence factors that work together to damage the host; elimination of one or two of the genes encoding the complex array of 'effectors' will not lead to inhibition of virulence, rather, just defects in the ability to cause disease ('hypo-virulent',

9   Schnupf, P. and Portnoy, D. A. 2007, 'Listeriolysin O: a phagosome-specific lysin', *Microbes and Infection*, vol. 9, pp. 1176–87.

10   Kaminskyy, V. and Zhivotovsky, B. 2010, 'To kill or be killed: how viruses interact with the cell death machinery', *Journal of Internal Medicine*, vol. 267, pp. 473–82.

11   Moayeri, M. and Leppla S. H. 2009, 'Cellular and systemic effects of anthrax lethal toxin and edema toxin', *Molecular Aspects of Medicine*, vol. 30, pp. 439–55.

12   Iglewski, W. J. 1994, 'Cellular ADP-ribosylation of elongation factor 2', *Molecular and Cellular Biochemistry*, vol. 138, pp. 131–3.

13   Popoff, M. R. and Bouvet, P. 2009, 'Clostridial toxins', *Future Microbiology*, vol. 4, pp. 1021–64.

14   Alibek, K. and Handleman, S. 1999, *Biohazard*, Random House, New York.

15   Viboud, G. I. and Bliska, J. B. 2005, '*Yersinia* outer proteins: role in modulation of host cell signaling responses and pathogenesis', *Annual Review of Microbiology*, vol. 59, p. 689.

16   Sun, G. W. and Gan, Y. H. 2010, 'Unraveling type III secretion systems in the highly versatile *Burkholderia pseudomallei*', *Trends in Microbiology*, vol. 18, pp. 561–8.

or 'attenuated'). By the same token, expression of these individual effector molecules in other bacteria would not likely lead to creation of a more virulent organism, without the context of the entire genome for full virulence.

The many faces of virulence?

The experimental models used to identify and characterise these virulence factors include bacterial culture conditions, infection of mammalian cells in culture and small animal/non-human primates.[17] Each of these approaches has its limitations and can affect the way in which overall virulence is expressed. Indeed, the terms hypo-virulence, hyper-virulence, avirulence and antivirulence[18] are all common in the medical literature. Yet the use of each of these terms must be considered in its context.

Returning to the experiments described above, we had identified a mutant strain of *M. tuberculosis* that exhibited hyper-virulent characteristics only in the context of the cell culture model—that is, mammalian cells grown in petri dishes in incubators. We were at another crossroads: should we go ahead and perform an experiment in an animal infection model of TB? Our laboratory was under some pressure from funding agencies to demonstrate that glutathione resistance was relevant in the larger scale model of mouse infection, which is more closely relevant to human disease than the tissue culture model we used for our more preliminary experiments.

We decided that the experiment was worth performing. We felt that if we could show the relevance of our observation to TB disease in the whole animal, we could convince the reviewers of our grant proposals that manipulation of the glutathione system might be exploited to limit TB disease. Fully mindful of the safety risks, we consulted our Institutional Biosafety and Institutional Animal Use and Care committees, and we performed a risk analysis that weighed the risks and benefits of the experiment. We concluded that knowing the answer to this question was important for the project's direction and that our biosafety plans were sufficient to carry out the experiment safely. Among our discussion points was the observation that there is a high degree of variability among clinical strains of *M. tuberculosis*, and our new strain was well within the range of virulence observed in cell culture models. We were not dealing with a strain of unusually extreme virulence. So we designed and executed a mouse infection experiment under biosafety level three (BSL3) containment conditions with

---

17   O'Toole, R. 2010, 'Experimental models to study human tuberculosis', *Advances in Applied Microbiology*, vol. 71, pp. 75–89.

18   Casadevall, A. and Pirofski, L. 2002, '"Anti-virulence" genes—further muddling the lexicon?' *Trends in Microbiology*, vol. 111, pp. 413–14.

appropriate safety and security measures. We found that there was no difference in the survival or replication rate of the mutant compared with the normal *M. tuberculosis* strain (unpublished observations). Thus, the hyper-virulent behaviour we observed in the tissue culture model was not carried through in an experimental system closer to the human disease state.

Our increased awareness of the dual-use implications of altering virulence significantly impacted on our deliberations concerning additional testing of our 'hyper-virulent' mutant of *M. tuberculosis*. On the one hand, we were under considerable pressure from our grant reviewers to demonstrate hyper-virulence in the mouse model. If we did *not* demonstrate hyper-virulence then they might not consider funding out project. On the other hand, our deliberations in the laboratory about the nature of virulence and pathogenicity increased our appreciation of the complexity of infectious potential and the inherent dual-use nature of our research. These are the kinds of careful discussions that we would urge scientists to have at many stages of their work.

Part II: Ethical Frameworks and Principles

7. Moral Development and Ethical Decision-Making

Judi Sture

A variety of approaches aimed at mitigating the intermittent friction between science and society and the risks of malign use of modern scientific advances has been defined: ethics,[1] the responsible conduct of science,[2] self-governance by scientists,[3] and top-down initiatives from policymakers and other authorities.[4] These approaches have allowed for a number of perspectives on the challenges that biosecurity poses to society and biotechnology today, but none has considered the pre-existing values of stakeholders as they interact with these challenges.

I would suggest that scrutinising the cultural pressures, forces and processes that influence the development of the values that individuals bring to their scientific work and policy views will shed further light on possibilities for engagement with dual use. The lack of attention to this so far is possibly due to the apparent lack of recognition of the role played by what may be called 'moral development' in ethical decision-making processes. We need to ask whether the private and the public value sets held by individuals clash; is it possible to hold multiple value sets in our private and professional lives and still manage to operate effectively in one or both spheres? These are some of the questions that need attention in approaching the potential challenges involved in deciding what is good, bad, right or wrong in the context of scientific work and its effects both inside and beyond the laboratory door.

The processes by which we learn and adopt 'moral' stances have been recognised and studied for more than 100 years. An influential work on what he termed 'folkways'[5] by W. G. Sumner, professor of political and social science at Yale University in the late nineteenth and early twentieth centuries, described and interpreted the development of both 'individual habits' and 'group customs' as they develop into morality-mediated usage in specific groups. He believed that moral values and their associated behaviours achieve the status of 'unwritten

1   Selgelid, M. J. 2010, 'Ethics engagement of the dual use dilemma: progress and potential', in B. Rappert (ed.), *Education and Ethics in the Life Sciences: Strengthening the Prohibition of Biological Weapons*, ANU E Press, Canberra, pp. 23–34, <http://epress.anu.edu.au/education_ethics/pdf/ch01.pdf>.

2   National Research Council/Institute of Medicine 2006, *Globalization, Biosecurity and the Future of the Life Science*, The National Academies Press, Washington, DC.

3   Royal Netherlands Academy of Arts and Sciences 2008, *A Code of Conduct for Biosecurity: Report by the Biosecurity Working Group*, KNAW, Amsterdam.

4   For example, Friedman, D. 2010, 'Israel', in Rappert, op. cit., pp. 23–34.

5   Sumner, W. G. 1906, *Folkways: A Study of the Sociological Importance of Usages, Manners, Customs, Mores, and Morals*, Ginn & Co., Boston, <www.gutenberg.org>.

laws' for the individual and the group because 'popular usages and traditions, when they include a judgment that they are conducive to societal welfare ... exert a coercion on the individual to conform to them, although they are not coordinated by any authority'.[6] I would argue that this is a relatively accurate description of the position of most scientists and science stakeholders today in the early years of the twenty-first century.

It is important that recognition of the role of pre-existing cultural folkways (national/ethnic, economic, religious, social and technological values and so on) and within these, domestic (familial) folkways or value sets, is factored in to any consideration of how individuals reach a state of moral and ethical maturity in their private and professional lives. This is particularly important in science, where the rapid speed of advance typically outstrips the development of appropriate ethical responses needed to deal with it and in which it is typically supposed that all practitioners share a common set of professional values and perspectives by virtue of their identity as scientists.

This chapter looks at the issues of moral development and ethical decision-making in terms of the publicly stated group value sets and individual, privately held value sets. Later in the chapter, I describe the case of a 'conflicted scientist' as he tries to make sense of a number of professional/personal value conflicts that face him in his daily work life. This unfortunate individual faces a set of daily challenges that create an uncomfortable dissonance in him; he needs to find a way in which to move forward without disadvantaging himself. Such ways forward cannot be identified unless we have a better understanding of the origins of the dissonance. By focusing on biological, psychological and anthropological research into the development of value systems, within the context of cultural identity, it is possible to highlight possible mechanisms that may usefully be explored in further research.

Terminology

Many authors use the term 'morals' to refer to ethical practice, standards or beliefs; indeed it is common to find that the words 'morals' and 'ethics' are used interchangeably. The very fact that there is this lack of clarity and consensus in this vocabulary indicates a lack of understanding of the potential differences between private and public values held by both individuals and groups. Because this chapter focuses on the potential variance between private and public value sets, I have chosen to be specific in the way I use these terms. For the purposes of this chapter, the term *morals* will be used to refer to the attitudes and values

---

6   Ibid., p. 4.

learned in childhood, pre-professional and private life, and the term *ethics* to refer to the attitudes and standards learned and required in the context of professional life.

The word 'moral(s)' tends to be heavily laden with culture-specific values and especially religious meaning, both of which are highly influential in forming our early values and views. These values may belong to a value set separate to that held professionally (although religious values may also direct ethical values). Furthermore, many people think of 'morals' as the attitudes they learned from their personal and family lives, with 'ethics' being something outside the domestic sphere: 'a morality for the private sphere that is personal and subjective, and a morality for the public sphere that is impersonal and objective.'[7]

The English word 'mores'—meaning values, norms and, occasionally, virtues—is derived from the Latin plural noun *mores*. 'Mores' refers to the traditions and customs practised by specific groups, which relate directly to the subject of this chapter. These traditions and customs are derived from accepted behaviours and practices in a group rather than from actual laws, and consist of a shared understanding of the manners and conduct that are acceptable to and expected by the group. In short, either one conforms to the traditions and customs of the group or one cannot function effectively as a full member. Indeed, an individual whose values clash with the predominant values of the group may find that s/he is not welcome in it at all.

'Folkways', according to Sumner, 'are habits of the individual and customs of the society which arise from efforts to satisfy needs'. These folkways can be understood as informal mores and, while not generally enforced in any way, they tend to be perpetuated by members of the group emulating the behaviours and attitudes of older or more established members. Once folkways are established, they become, essentially, a value set that is shared by members of the group and is maintained by group consent rather than by formal governance. Clearly this process has implications in the biosecurity context when we are seeking to develop, in effect, a new norm of biosecurity to which all scientists and science stakeholders can subscribe.

In this chapter, the term 'culture' refers to the 'culture of origin' of the individual (as in the ethnic sense), as well as to nationality and to religious, economic, technological and linguistic groupings to which the individual may belong. All of these dimensions impact on one another, having differing degrees of influence depending on the context in which the individual is acting at any one time. One of the outputs of such cultural influence is a value set, or group of value sets.

7   Catchpoole, V. M. 2001, 'A sociobiological, psychosocial and sociocultural approach to ethics education', PhD, Centre for the Study of Ethics, School of Humanities, Queensland University of Technology, Brisbane, p. 274.

These may also be defined as beliefs, attitudes and perspectives on the world. Clearly, these in turn impact on actions and behaviour patterns. Clashes occur when an individual is faced with some conflict between the influences of one or more cultural value sets in a given situation. Which value set should have precedence?

'Ethics' may be defined as (potentially) universal standards and commitments that support justice, care, welfare and rights;[8] among other definitions, they may also be defined as a 'principled sensitivity to the rights of others'.[9] The irony here is, of course, that most people tend to believe that their own 'moral' (private and/or family derived) standards are also defined in the same way. Variations in these have a role in defining scientists' and policymakers' ethical standards and norms and must be recognised as key variables in any debate about what needs to be done to prevent the hostile use of the life sciences. Nevertheless, bearing in mind that cultural variations exist, it is probably reasonable to agree that any desired socially responsible 'ethical' approach involves 'the development of adequate … commitments that inform actions'[10] and that this includes 'an understanding of what is beyond self-interest, and how we may justify that to reasonable people and put it into practice'.[11] Even so, these aspirations, while striving to be comprehensive and culture neutral, still incorporate approaches that may not be prioritised by all cultures and therein lies part of the difficulty in seeking a common set of values or norms. Everyone thinks that their existing moral outlook is 'commonsense', and that their professional ethical values are also 'commonsense'. It is in this balancing act between the variation in moral and ethical values as defined by diverse cultural backgrounds that we need to find a way forward that allows individuals and groups to engage equally and openly in the context of biosecurity as it challenges the needs of science in the way it meets the needs of society.

Private morals and public ethics: Cultural aspects of moral development

Given that cultural origins vary significantly, while providing a fundamental platform of values from which each individual views the world, it follows that everyone has at least one value set that influences the ways in which they see, understand (or believe they do) and treat other people. Most people have multiple value sets, and apply them variously in the different contexts of their lives—for example, we may adhere to one value set in the family or domestic

8   Laupa, M. and Turiel, E. 1995, 'Social domain theory', in W. M. Kurtines and J. L. Gewirtz (eds), *Moral Development: An Introduction*, Allyn & Bacon, Boston, pp. 455–73.
9   Gilbert, N. 2001, *Researching Social Life*, Sage, London, p. 45.
10   Catchpoole, op. cit., p. 7.
11   Singer, P. 1993, *How Are We to Live? Ethics in An Age of Self-Interest*, Text, Melbourne.

sphere, but another in the work context; we may employ yet more value sets in our relations with people we do not like, or with people we need, such as our bank manager or our doctor.

How we learn about morality: Mechanisms of development

This section introduces two processes of moral learning, which may be applied in understanding the development of ethical thinking in the context of professional science.

The development of a moral or ethical perspective presupposes the *biological capacity* to do so. Growing up, we process moral learning as part of our social heritage through our ability to symbolically represent the moral to ourselves, which involves a relational aspect between ourselves and others:

> [C]ulture serves to complement the process [of learning morality] in important ways. Rather, we are led to moral experience and insight. Real morality can't be forced on people, nor can they be fooled into having it, nor do they just act on their 'moral instincts'. Real morality does not simply bubble up from beneath, nor is it imposed from the outside. In each one of us, it must be discovered anew. The discovery process may require great mental and emotional effort and may bloom only in the right climate, but human beings see morality, recognize it, regardless of what it is that they want or need or love or hate or feel compelled to do.[12]

We learn what is good and what is bad primarily in relation to ourselves as individuals, but also, through social development, what is good and bad for 'the others'. These 'others' will initially be the close 'others' of family and friends. Later, this can develop also into a sense of shared experience, and care or empathy with the distant other: the outsider or even the enemy, the one who is not a close other. Perhaps this conscious decision to have empathy for the enemy or the outsider, the different one, is a measure of our ethical maturity.

Edward Wilson[13] suggested that genetics and evolutionary processes shape the human mind at least as much as culture, that a predisposition to religious belief is genetically driven in humans and that acting altruistically through religious belief can confer survival advantages to those willing to change their views. Wilson also described the dark side of our predisposition to tribal affiliations—xenophobia—and suggested there is a need to 'globalise the tribe'.[14] In terms of the debate on dual use and biosecurity, it may be useful to consider the relative

---

12   Goodenough, U. and Deacon, T. W. 2004, 'From biology to consciousness to morality', *Tradition & Discovery: The Polanyi Society Periodical*, vol. 30, part 3, pp. 6–20.
13   Wilson, E. O. 1978, *On Human Nature*, Harvard University Press, Cambridge, Mass.
14   Wilson, E. O. 1998, *Consilience: The Unity of Knowledge*, Alfred A. Knopf, New York, p. 269.

positions of those defending the practice of research from security-inspired interventions and those urging a change in attitude to the nature of scientific freedom. These may be considered as belonging, at least in the eyes of some, to different 'tribes' and thus be subject to 'intertribal' conflict.

In order to consider any change of allegiance (in our context, in terms of changing what s/he thinks is 'ethical'), an individual or group must be able to reflect on and evaluate a range of cultural factors that pertain to the situation, and then make an informed decision as to whether change of allegiance is the right choice. The change may incur a short or a long-term cost, but the ability to consider the change and the costs is part of a moral capacity, leading to the application of moral agency. Essentially, the individual is faced with the choice of allegiance or reprioritisation of values.

Griffiths[15] considered that emotions play a key part in our moral response system, and therefore in our capacity for moral agency, while Izard and colleagues suggested that emotions have an adaptive role in equipping individuals to act appropriately in order to survive.[16] We can already see this in action in our prioritisation and negotiation of values between the private morals and public ethics in any given situation.

Wilson's views have been criticised by the evolutionary biologists Stephen Jay Gould and Richard Lewontin, who oppose ideas of genetic determinism, and the anthropologist Marshall Sahlins, who holds to the view that cultural factors are a powerful driver of people's behaviours, distinct from genetic influences. Other researchers, however, have also supported a biological mechanism in the development of values. The evolutionary biologist Robert Trivers[17] suggested that reciprocal altruism (all parties helping each other in order to survive) is an evolutionary predisposition that offers the best chance of survival of the individual and the species. Moore,[18] however, has pointed out that it is not always easy to distinguish between altruism and selfish acts disguised as altruism. Richard Dawkins[19] suggests that some biological determinants are non-altruistic and reflect a predisposition to look after oneself, rather than considering possible self-sacrifice for the good of others. Williamson[20] argues that the human concepts of shame and guilt function in evolutionary terms as restraints on social behaviour, and points out that human self-interest is an

15   Griffiths, P. E. 1998, *What the Emotions Really Are: The Problem of Psychological Categories*, University of Chicago Press, Chicago.
16   Izard, C. E. 1984, 'Emotion–cognition relationships and human development', in C. Izard, J. Kagan and R. B. Zajonc (eds), *Emotion, Cognition and Behavior*, Cambridge University Press, Cambridge, pp. 17–37.
17   Trivers, R. 1971, 'The evolution of reciprocal altruism', *Quarterly Review of Biology*, vol. 46, pp. 35–57.
18   Moore, J. 1984, 'The evolution of reciprocal sharing', *Ethology and Sociobiology*, vol. 5, pp. 5–14.
19   Dawkins, R. 1976, *The Selfish Gene*, Oxford University Press, Oxford.
20   Williamson, D. 1998, 'Mixed feelings', *The Australian Review of Books*, June, pp. 14–15, cited in Catchpoole, op. cit., p. 60.

obvious factor in survival: those of our ancestors who looked after their own interests survived to reproduce themselves, and those who did not, failed to do so.

Work by Krebs and Janicki[21] indicated that the evolutionary development of norms, or memes as Dawkins may refer to them (which may be defined as culturally transmitted units of data such as ideas, habits and so on), has resulted in four identifiable ways in which people operate ethically

1. they try to get others to invoke the norms that they themselves hold

2. they try to tailor their norms to others so as to enhance their persuasive impact

3. recipients tend to adapt to the norms that are most advantageous to them

4. people in different sorts of relationships will preach different sorts of norms.

These 'operating mechanisms' are, according to Krebs and Janicki, biologically determined in that they are evolutionary adaptations to ensure survival. Whether these mechanisms are truly biological or are solely culturally determined artefacts is open to debate.

The development of reciprocity, empathy and other relationship-related characteristics may be considered as part of a *psycho-cultural mechanism* of learning morality and ethics. Our capacity to consider the needs of others is enhanced by our cultural interactions with other people at a psychological level. Most people learn how to make and sustain friendships, work relationships and marriage relationships based on cultural concepts such as respect, interdependence, empathy, cooperation and negotiation, amongst others. All of these are partial foundations of a moral and ethical perspective, even though they vary in how they are prioritised and expressed culturally.

Research suggests that humans can exhibit empathy (the capacity to intellectually understand how another person feels) from an early age, which is potentially open to development by other cultural factors. Work by a number of psychologists has shown that even very young children have the capacity for empathy.[22] This runs contrary to earlier thought by Piaget,[23] who believed that very young children were essentially egocentric and unable to perceive the world from the perspective of others. Hoffman devised a set of levels in the development of empathy encompassing the first 12 years of life.[24]

---

21 Krebs, D. L. and Janicki, M. 2004, 'Biological foundations of moral norms', in M. Schaller and C. Crandall (eds), *Psychological Foundations of Culture*, Lawrence Erlbaum Associates, Mahwah, NJ, pp. 125–48.

22 For example, Hoffman, M. L. 1988, 'Moral development', in M. H. Bornstein and M. E. Lamb (eds), *Social, Emotional and Personality Development. Part III of Developmental Psychology: An Advanced Textbook*, Lawrence Erlbaum Associates, Hove, UK, pp. 497–548; Zahn-Waxler, C., Radke-Yarrow, M. and Wagner, E. 1992, 'Development of concern for others', *Developmental Psychology*, vol. 28, part 1, pp. 126–36.

23 Piaget, J. 1932, *The Moral Judgment of the Child*, Kegan Paul, London.

24 Hoffman, M. L. 1988, 'Moral development', in Bornstein and Lamb, op. cit.

A major player in the development of moral theory learning was Kohlberg.[25] He developed a six-stage model of moral development, with successive stages being more able to affect moral reasoning. His model focuses particularly on the concept of justice and he believed that this development continues throughout adult life. He further developed his model to accommodate 'soft' developmental-level changes in adults over the age of about thirty years, and acknowledged that his stages do not provide a complete description of adult development from stage four onwards[26] (see Table 7.1). It must be noted, however, that Kohlberg's stages refer not to specific beliefs but to *underlying modes of reasoning*,[27] and as such, there is no guarantee that an individual will reach the highest level of reasoning. Also, it should be noted that Kohlberg's work focuses on moral thought, not moral action, and we can probably all accept that the two do not always go together.

Table 7.1 Kohlberg's Six-Stage Model of Moral Reasoning

Level 1: Pre-conventional morality	Stage 1: Punishment–obedience orientation (avoiding punishment for doing 'wrong', but only because it is good for you)
	Stage 2: Instrumental relativist orientation (how will this decision/action be good for me and my needs?)
Level 2: Conventional morality	Stage 3: Interpersonal concordance (being 'good' to please other people because that is the 'right' thing to do)
	Stage 4: Law and order orientation (abiding by laws for the good of society, not just for the resulting benefits to yourself)
Level 3: Post-conventional morality	Stage 5: Social contract orientation (varying personal values are recognised but are mediated by social agreement and a democratic approach in society as a whole; the law is respected but is subject to change if it is not just)
	Stage 6: Universal ethical principle orientation (right decisions/actions are defined by ethical principles that are universal and consistent, which emphasise justice, reciprocity, rights and respect for human dignity in society)

Source: Kohlberg, L., Levine, C. and Hewer, A. 1983, *Moral Stages: A Current Formulation and A Response to Critics*, Karger, Basel.

Kohlberg's model also appears to provide evidence of the ways in which people from different cultures make decisions,[28] with different cultures passing through the stages of the model differently, usually in the same order, but not

25   Kohlberg, L. 1971, 'Stages of moral development as a basis for moral education', in C. M. Beck, B. S. Crittenden and E. V. Sullivan (eds), *Moral Education: Interdisciplinary Approaches*, University of Toronto Press, Toronto, pp. 23–92; and Kohlberg, L. 1981, *The Philosophy of Moral Development: Moral Stages and the Idea of Justice*, Harper & Row, San Francisco; and Kohlberg, L., Levine, C. and Hewer, A. 1983, *Moral Stages: A Current Formulation and A Response to Critics*, Karger, Basel.
26   Kohlberg et al., op. cit., p. 6.
27   Kohlberg, L. and Gilligan, C. 1971, 'The adolescent as a philosopher: the discovery of the self in a postconventional world', *Daedalus*, p. 1051 ff.
28   Harkness, S., Pope Edwards, C. and Super, C. M. 1981, 'Social roles and moral reasoning: a case study in a rural African community', *Developmental Psychology*, vol. 17, part 5, pp. 595–603.

always to the same end. This means that people in any given culture only need to achieve the level of reasoning that is necessary for their culture to operate effectively.[29] Nevertheless, even when different cultures have different attitudes to an issue, they will express the same reasoning methods when coming to their differing conclusions. I would suggest that individuals and groups may operate at several levels simultaneously, depending on the multiple contexts they are operating in at any one time (home and work, friends and work, friends and home, and so on), and this may be a mechanism by which scientists manage any clash between their private morals and public ethics. In addition, I would suggest that although we may like to think we are all operating professionally in Kohlberg's 'Stage 6', we are probably not, or at least, not all the time and in all circumstances.

Kohlberg's work has been subject to criticism, including that his work is culturally biased. Simpson[30] said that Kohlberg's stage model is essentially based on the Western philosophical tradition and that it is inappropriate to apply it to non-Western cultures without adequate reflection on their different moral outlooks; however, the model appears to remain a useful indicator of the development of moral reasoning per se, and I would suggest that as long as cultural variations are effectively and appropriately considered as variables when applying the model, it retains useful credibility.

We must also consider the psychological *effects* or outcomes of our personal cultural upbringing. A range of cultural factors may be explained by psychology as well as by anthropology and the social sciences. Characteristics of cognition, motivation and social interaction may answer questions about the origins and development of any culture, and subsequently those subject to the influences of a culture. For example, a society that values the rights of the individual will produce people who think differently about any given scenario to people who originate from a society that does not place the rights of the individual high on the social agenda. Likewise, cultural views on gender, equality, age, family structure, education, work, marriage, religion and science all influence us heavily as we grow up, embedding within us a standard of 'normality' by which we measure the rest of the world. This is the basis of some of the criticism of Kohlberg's stage model. It is probably only when we leave the confines of our own cultural context that we are challenged to focus on our own reactions to 'the others', and have to negotiate ways forward that enable us to coexist in a mutually acceptable way. This can, of course, include the transition from the private, domestic sphere to the professional, public sphere.

---

29 Ibid.
30 Simpson, E. L. 1974, 'Moral development research', *Human Development*, vol. 17, pp. 81–106.

Private morals, public ethics and dual-use/biosecurity education

When considering the biosecurity and dual-use dilemma therefore, we need to recognise that life scientists bring to their existing professional ethical perspectives a range of prior moral perspectives that influence the ways in which their professional ethical views and behaviour develop. There may even be multiple levels of prior professional ethics and norms involved in this equation. A practising scientist may have been required to adhere to different norms in different laboratories and workplaces;[31] s/he may have been required in one place to prioritise safety over economics, but then be pressured to reverse this elsewhere; leaders in one workplace may resist post-qualification research training to the detriment of the team, whereas leaders in another may promote such training—the scientist finds himself in the new workplace with his 'old' norm of resistance to training in place, and as a result, he finds he is viewed by the new team as backward and unwilling to change, unless of course he manages to revise his view on training. A number of cultural expressions may affect this process, and although as scientists, individuals may all tend to publicly express similar value sets, as scientists, they will be modified by the underlying culturally mediated values of their earlier life experiences. This will occur at both the individual and the group levels.

In order to understand this more fully, it is useful to consider why certain behaviours and attitudes develop. Private morality, according to work by Kohlberg, Wilson, Piaget, Catchpoole and others (above), involves characteristics of altruism, care, empathy and a sense of justice. Krebs and Janicki[32] suggested that the way in which these are prioritised and acted on by a given group may be an adaptive mechanism, with people choosing the adaptation that best helps them to tackle given cultural problems and manage their own responses in such a way as to enable them to work. For example, scientists' views may conflict with the institutional views of the workplace in which they operate, but in order to keep their position and retain a reasonable lifestyle, they have the choice to either conform or be released from their position, perhaps with other penalties as well as the commensurate economic ones. There is arguably a tendency to overlook such cultural pressures and to insufficiently recognise the forces acting on scientists. An example of this may be seen in a situation that has been recognised under the Biological Weapons Convention (BWC): sometimes even well-intentioned international prohibition regimes can lead to confusing situations. The BWC prohibition norm's allowance for biotechnology research carried out for peaceful purposes has actually enabled a number of countries

---

31 See, for example, Bezuidenhout, Chapter 17 in this volume.
32 Krebs and Janicki, op. cit.

to 'cloak their biological weapons programmes within seemingly legitimate facilities'.[33] Scientists working in such conditions are faced with the need to juggle their private values and professional ethics in order to reach a state of equilibrium in which they can continue to work with the least possible stress and dissonance to them personally.[34] This example offers a partial illustration and explanation of the different 'ethical' values and behaviours that we see apparently exhibited around the world today—in both developing and developed countries. 'Cultures' (national, political, ethnic and religious, in some cases) prioritise the values that enable them to best tackle the challenges they face, and at the micro-level, the individuals involved in these circumstances have to prioritise and compromise their own values within these contexts in order to survive. Martin wrote an excellent commentary on this, focusing on those workers who chose not to compromise their own moral and ethical values.[35] I will return to this later.

So how do we draw together common ethical values from such disparate moral backgrounds within the context of the risks of the dual use of scientific activity? It may be argued that scientists already share a common culture, but Crandall and Schaller[36] have highlighted professional/personal value clashes even within (and between) scientific cultures. They suggested that most scientists publicly state that they hold to an explicit set of values including a commitment to sharing universal truths, to sharing data with colleagues, to being disinterested in the sense of being non-judgmental and open to whatever the evidence suggests, and to a sceptical outlook that distrusts anything not supported by empirical data and observation. They also found, however, that a hidden agenda is apparent in the work of most scientists, which clashes with the explicit values of the scientific method. They suggested that in order to survive in the competitive world of science, scientists must employ a set of hidden values, including

---

33 Atlas, R. and Somerville, M. 2007, 'Life sciences or death sciences: tipping the balance towards life with ethics, codes and laws', in B. Rappert and C. McLeish (eds), *A Web of Prevention: Biological Weapons, Life Sciences and the Governance of Research*, Earthscan, London, pp. 15–33.

34 See the examples of 'Experiments of concern' given in the Fink Report, which states that review of research activity by the scientific community would be necessary if it involves (amongst other processes) demonstrating: how to render a vaccine ineffective; how to enhance resistance to antibiotics; or how to alter the host range of a pathogen. All of these activities could be legitimate elements of peaceful research, but who on a team engaged in such work is likely to raise any alarm if funding for the research and the team could be put at risk? See National Research Council 2004, *Biotechnology in An Age of Terrorism*, [Fink Report], The National Academies Press, Washington, DC, <http://www.nap.edu/catalog/10827.html>. See also the discussion of the ethical implications of publishing the reconstruction of the 1918 flu virus when it was published in 2005. See Tumpey, T. L., Basler, C., Aguilar, P., Zeng, H., Solorzano, A., Swayne, D., Cox, N., Katz, J., Taubenberger, J., Palese, P. and Garcia-Sastre, A. 2005, 'Characterization of the reconstructed 1918 Spanish influenza pandemic virus', *Science*, 7 October, pp. 77–80. This paper was accompanied by an editorial in *Science* discussing the decision to publish the research, bearing in mind concerns about the risks of misuse of it.

35 Martin, B. 2007, 'Whistleblowers: risks and skills', in Rappert and McLeish, op. cit., pp. 35–47.

36 Crandall, C. S. and Schaller, M. 2004, 'Scientists and science: how individual goals shape collective norms', in Schaller and Crandall, op. cit., pp. 201–24.

resistance to new ideas that challenge the status quo, the selective forgetting of data that contradict personal views, favouring of work supporting their own views and a generally conservative approach to their practice. Conflicts between the personal and professional value sets will obviously arise here.

This conflict may be explained psychologically by cognitive dissonance theory,[37] which describes how, in situations in which an individual's behaviour or beliefs must conflict with beliefs that are integral to his or her self-identity (for example, when being asked to professionally prioritise a value that is not equally prioritised in his private value set), a conflict will arise. This conflict is referred to as cognitive dissonance. In order to resolve this dissonance, the individual can either leave the situation that presents the source of the conflict (in this case, the professional role) or reduce his attachment to the private value that is being challenged or compromised. If he does the latter then he may ease the conflict further by emphasising the positive aspects of the professional value that is challenging him. Cognitive dissonance is clearly an important factor in the decision-making process.

Three key psychological strategies are commonly used to alleviate cognitive dissonance. The individual may decide to seek more supportive beliefs or ideals that could outweigh the dissonance; in essence, this may mean 'masking' or suppressing the conflicting value and his response to it. This would, in turn, potentially cause longer-term psychological stresses. The individual may convince himself that the challenging belief is not really important, and can therefore be ignored or at least given less priority. He may also begin to manipulate or moderate the belief that is challenging him and seek to bring it more into line with his other beliefs and action patterns. These strategies involve rationalisation: the tendency to develop their own explanations as to why they have made their choices in relation to the challenging belief or action.

Festinger's cognitive dissonance theory has not been without its detractors. Bem[38] suggested that individuals' attitudes may be changed by self-observation of their own behaviours (rather than by any response to feelings of discomfort in a challenging situation), followed by consideration of what attitudes caused those behaviours, thus producing emotional responses. For some time it was considered that cognitive dissonance and self-perception theories were in competition; however, work by Fazio et al.[39] identified attitude-congruent

37   See Festinger, L. 1957, *A Theory of Cognitive Dissonance*, Row & Peterson, Evanston, Ill.; and Festinger, L. and Carlsmith, J. M. 1959, 'Cognitive consequences of forced compliance', *Journal of Abnormal and Social Psychology*, vol. 58, pp. 203–10.

38   Bem, D. J. 1967, 'Self-perception: an alternative interpretation of cognitive dissonance phenomena', *Psychological Review*, vol. 74, part 3, pp. 183–200; and Bem, D. J. 1972, 'Self-perception theory', in L. Berkowitz (ed.), *Advances in Experimental Social Psychology*, Academic Press, New York, pp. 1–62.

39   Fazio, R. H, Zanna, M. P. and Cooper, J. 1977, 'Dissonance and self-perception: an integrative view of each theory's proper domain of application', *Journal of Experimental Social Psychology*, vol. 13, part 5, pp. 464–79.

behaviour and attitude-discrepant behaviour, with self-perception theory explaining the former and cognitive dissonance theory explaining the latter. Attitude-congruent behaviour was defined as any position within an individual's scope of acceptable behaviour and attitude-discrepant as any position within an individual's scope of rejected behaviour.

The conflicted scientist: An example

Now that we have considered biological and psychological explanations of moral development, let's look at the scientist we mentioned at the start of the chapter. This hypothetical, recently graduated biochemist was appointed to his junior post as a research assistant on a team in a commercial organisation developing vaccines to counter an emerging tropical disease.

Let's imagine that his team leader and his employers prioritise economic and scientific advantages over what they *perceive to be* the requirements of biosecurity; indeed, the organisation believes it is already accommodating biosecurity concerns through its biosafety policy. In as much as it has been addressed at all, biosecurity as a concept is generally considered to be irrelevant in the organisation's laboratories because the management does not believe that their work presents any biosecurity risk. Moreover, it is vital that this organisation maintains scientific and economic leads over its competitors.

Our scientist has been brought up at home to believe that it is socially necessary to abide by 'the rules' in life, so that everyone can live together in the community in some sort of respectful harmony. At school and at university, he modified this 'necessity' into a more flexible 'suggestion', meaning that he continued to abide by the rules (which he is quite good at doing, at least in as much as it benefits him), but felt able to ignore some as it suited him. In his childhood, he learned that respect for the quality of life of others, as well as his own, is a key value, and that he should always do his best to ensure the safety and wellbeing of others. He learned respect for his elders, politeness and the need to recognise his own position in the 'food chain' both socially and at home. At work as an adult, he believes that his leaders adhere completely to what is often known as 'the scientific method' (not an official term, but one that implies a commitment to empiricism, objectivity, truth, openness and so on; whether one accepts this or not is an interesting discussion in itself), as he believes he has seen them do so regularly. He generally accepts the values that his scientific leaders pass on to him both consciously and unconsciously. He has somehow forgotten the one or two instances in which he noticed his leaders taking shortcuts in an experiment and ever so slightly exaggerating a finding in a paper. Seeing himself as a qualified and employed scientist, his priorities are twofold: accepting and promoting the 'scientific method' as an independent, value-free and objective concept, and making sure he gets on well with his workmates and bosses.

He can see that his organisation's new vaccine has considerable economic potential as well as humanitarian value; however, he is unexpectedly challenged by the need to manage his response to some new values—in this case, a new version of the biosecurity norm—which have been presented to him through an international report[40] he has just read. He wishes to persuade his colleagues of the new perspective that he has recognised because he is so persuaded by the argument that he has prioritised this 'new' ethical approach at a high position in his own professional value set. He is concerned that one or more of the processes involved in the vaccine research and development could be used against human populations if they were to be used maliciously. He sees this very clearly once he has thought about it, and is concerned that no-one at work has considered this before. How can he proceed?

He tries to tell his colleagues about his concerns. One or two of them agree that there may be some theoretical risk, but generally take the view that 'it will never happen' and tell him that he should forget about it. Some point out to him that this new process will generate a lot of money for the organisation (and long-term employment) and he should not raise problems with it. He tries to forget his concerns, but as he continues to work on the new process, he sees more and more chances for it to be misused. Other researchers from different branches of the organisation have had access to the laboratories during the work, and no confidentiality mechanisms were in place that would have prevented them from seeing the work, so the processes could already be repeated at other sites.

Our scientist's research team leader is planning to publish the work, albeit not in full, so as to protect the organisation's economic advantage. Even so, key processes and outcomes can be relatively easily worked out by anyone with a basic knowledge of genetics and vaccine science from what is to be published. In order to advance his position further in the discipline, his team leader is also going to give some more detail in an international conference podium presentation, where he will take questions and presumably give some answers. Our scientist goes to see the team leader to ask him if it would be possible to restrict the amount of information given out in the publication and at the conference, as he has concerns about security and what could happen if the information fell into the 'wrong hands' (delegations from certain countries with security problems will be present). His team leader is shocked and annoyed at the suggestion that he could possibly be involved in work that could be used to make, in effect, a biological weapon. He challenges our scientist to show him

---

40   For example, National Research Council, op. cit.

what is wrong with the work, and goes on to say that he cannot be responsible for what someone else may do with 'his' work. Our scientist is not sufficiently up to speed with *all* the possible ways in which the work may be misused, but he suggests that certain potential outcomes of the research could be used against the interests of society. He finds it difficult to understand that his colleagues and team leader cannot see the potential problems that may arise from the team's work on interfering with the susceptibility of organisms to vaccines; they do not seem to be concerned that it could be used against human communities. The team leader then falls back on the argument of scientific freedom and the duty of the scientist to share and to replicate work with other scientists. Moreover, he also points out that the research is going to generate a major income stream for the organisation. He suggests to our scientist that he might wish to consider his *junior* position on the team, and points out that contracts will soon be up for renewal. Over the next few weeks, the team leader begins to give more work to other colleagues on the team and less to our scientist, who does not receive an invitation to the conference at which the leader is to speak.

Our scientist manages to find a friend who is quite senior at another workplace, with whom he shares his concerns in a general way. This new friend suggests that he could think about what it is about his own values that is causing him so much stress in this situation. He has another friend, who is not a scientist, who says that he should ignore the stress and what others say, and simply focus on himself and what he can do to change himself in order to manage better at work. Our scientist is now in an ethically challenging situation and does not know which way to turn. What should he do?

We can review our scientist's position using the concepts that we have been looking at in this chapter. His upbringing has given him a family-based, culturally derived value set (a bio-cultural mechanism) that encourages him to abide by the rules in order to maintain his social position and to gain advancement socially as he 'fits in' with his 'tribes' both at home and at work (Wilson's theory). He has grown up with this approach, which has worked so far, but if he has to challenge it beyond certain narrow limits, he feels emotionally stressed (Griffiths' and Izard's theories). He is aware that it is necessary to balance competing values and tries his best through life to manage the periodic clashes between the need to fit in and the need to maintain his own values (the theories of Sahlins, Gould and Lewontin on cultural drivers overcoming genetic drivers). He feels that it is possible to manage this sort of balance by everyone 'giving and taking' in order for everyone's needs and values to be accommodated (Trivers' reciprocal altruism). He is aware that some people are manipulators and will take when appearing to give, but as long as it all 'balances out' he is happy and assumes that everyone else is as well (Moore's criticism of Trivers, and the work of Mauss on gift-giving as a means to self-promotion). He is even of the mind, at times, that Dawkins' ideas on genes as drivers can be related to humans in the sense of 'every man for himself', although he does not like this view of life, as it seems uncaring and inhumane.

He has a number of options for dealing with these questions (psycho-cultural mechanisms). He is vaguely aware that he manages the periodic clashes between his personal and professional value sets by balancing his unconscious feelings of guilt and shame for 'doing wrong' if and when he steps out of line in the family or at work (Williamson's theory). His mechanisms for managing these emotional reactions are those described by Krebs and Janicki—he always manages, usually unconsciously, to use one of their four ways to make sure he can operate in 'the right way' in any given situation. He is, unconsciously, at different stages of Kohlberg's model depending on which context he is operating in at any given time—for example, at home, at work or at play.

Now that he has been confronted with the challenge at work that I have described above, he is faced with a situation of cognitive dissonance. In order to remain at work and to flourish in the way he wishes to professionally, he must now either conform to the prevailing view at work by giving up his concerns (or at least stop talking about them) or choose to leave and seek work elsewhere. As a junior member of the team, he does not think he is in a position to influence the team himself. If he were higher up the 'food chain', he may feel that this is a third option.

Our scientist now has four options to mitigate his cognitive dissonance at work, thereby relieving himself of any sense of ethical responsibility about the possible outcomes of the scientific activity of his organisation. First, he may choose to look for some other beliefs that will help him to mask the dissonance so he can pretend it does not exist; he could do this by saying to himself that his actions alone will have no effect in the 'big picture' as he is only a junior researcher: 'what do I know?' Second, he may decide to go along with his colleagues and persuade himself on some level that 'it will never happen'; he could do this by actively deciding that he is worrying over nothing—he may be the only person who has seen the potential problems, so maybe he is blowing it all out of proportion and he is 'worrying about nothing'. Third, he may try to manipulate, in his head, the challenges that the dissonance is causing, so that the views of his leader and colleagues do not cause him so much difficulty; he may do this by reassessing the value that he places on the views of his colleagues and leaders, deciding that their views are not as important to him as he previously thought. Fourth, he may choose to be a 'whistleblower' and speak publicly in some way about his concerns.

The first three approaches involve our scientist in a process of rationalisation whereby he will be able to invent and justify his own explanation as to how he has reached his eventual position of relative ethical comfort (or discomfort). Unfortunately, all these actions are likely to lead to further stress, which may debilitate him in the future. In the first instance, he is effectively degrading his own value in the team as a scientist and as an individual. In the second, he is subduing his own intelligence and ignoring what his own values and intellect are telling him. In the third, he is downgrading the value of those to whom he naturally looks for learning. All of these rationalisations ultimately mean compromise of his personal and professional values and are highly likely to result in further dissonance.

In terms of self-perception theory, all three strategies will simply move him to various stages along the continuum of attitude-congruent (acceptable) and attitude-discrepant (unacceptable) behaviour. If he makes himself comfortable at one point, this means that he will thereby throw up a further dissonance that requires another relocation on the continuum. This may progress to such an extent, raising more and more stress, that our scientist may feel that he is in an impossible situation if he is unable to convince his colleagues and leaders to take his concerns about the research seriously and to act on them.

The fourth approach—to become a whistleblower—is, unfortunately, not a stress-free option, even though it may assuage his moral and ethical concerns. Our scientist may well 'blow the whistle' on the research, and be the cause of it being abandoned or amended by his organisation. So far so good—he may have prevented a possible biosecurity threat from translating into an unwelcome action; however, in the process of doing this, if we read Brian Martin's work on whistleblowing,[41] we find that he is highly likely to be ostracised by his workmates, sidelined (at best) in the organisation and, more likely, fired from his post as a result. If not fired, he may find that his contract is not renewed as the organisation is 'restructured'. Either way, he now has no job, no references and little hope of continuing to work in the field he had chosen to spend his life in. This prevents or seriously delays him buying his first house, starting a family and advancing up the career ladder. Maybe he should have kept his mouth shut after all?

A way forward: Memes and norms

Given that modern-day whistleblowing is such a dangerous option for the individual to take, and given also that we accept the concept of biosecurity risks being inherent in biotechnology research, it seems that we need to find a way forward that reduces the necessity for whistleblowing to take place and

---

41 Martin, op. cit.

reduces the risk of biosecurity lapses occurring. One such way is to develop and foster a new norm of biosecurity; this would become, in time, a part of everyday scientific practice, just as the biosafety norm is hopefully embedded in daily activities.

This is best considered in the context of the ways in which scientists carry out their daily activities and approach potential challenges in their work. We have seen earlier in this chapter that there are two possible psychological mechanisms—bio-cultural and psycho-cultural—that may be usefully pursued further when attempting to identify, understand and predict the changing perspectives of scientists when faced with the challenges of biosecurity in their professional and personal lives. We should therefore consider these potential explanations in the derivation and development of value sets or norms when developing a new ethic of biosecurity and dual use. What we are faced with is, essentially, a relativist scenario not only between the differing values held by different people in regard to specific concepts, but also in the differing values held by individuals themselves in terms of specific concepts through conflict between their private moral values and their public ethical values.

Now that we have seen the challenges faced by our hypothetical challenged scientist (albeit supported by real-life examples in Brian Martin's work), let's look at how cultural ideals and values can be transmitted effectively, thus, hopefully, avoiding or reducing the chances of his situation being repeated in real life.

Norms may be defined as formal or informal rules that govern what is acceptable in terms of behaviours, attitudes, beliefs and values within a group or culture. They may be explicitly stated or implicitly approved by the majority; they may lead to warnings or reproach for those transgressing minor normative rules, or to more severe punishment for those guilty of the transgression of norms considered to be of great consequence. Norms develop over time, and are subject to a range of cultural influences. They are often closely tied to religious and social forces, and are commonly associated with group or cultural identity. Once this cultural identification element takes hold, it may be particularly difficult to dislodge or change a norm. Considering that such norms are usually part of our childhood development and therefore our moral learning process, we can see how they may challenge later professional ethical values that may clash with them. So how do we develop and implement a new norm? Maybe it is better to start with a smaller task and look at what Dawkins refers to as memes, which may be referred to as 'infant norms', for want of a better term.

Memes[42] are units of cultural transmission that may be skills, knowledge, beliefs, attitudes, behaviours or values. They are spread through observation and social learning[43] and have been called 'cultural software'[44] that may be passed on as they are or in modified forms. Dawkins has, in the past, cited the fashion for wearing a baseball cap back-to-front as an example of a meme that has spread through certain sections of society. One could argue that a well-entrenched meme becomes, in effect, a new norm, if it lasts long enough. Crucially, unlike genes and norms, which can take long periods to change, memes can, apparently, change very quickly. It may follow then that ethical attitudes, knowledge and characteristics could also be transmitted as memes, which, in passing from one person to another, will change slightly with each transmission, allowing development and swift cultural change to occur:

> The power of human reason made possible in part by the memes we possess, is also the power to mutate those memes and create something new from something old. We are not simply inheritors of a zealously guarded patrimony but entrepreneurial products of a new cultural software which will help constitute future generations of human beings.[45]

In order for old memes to be transmuted into new memes, however, there must be a dominant host culture that allows change to take place.

In order to enfranchise and empower all people there needs to be a mutual exchange between equals. If some individuals or groups are marginalised or excluded for any reason, be it at the political, legal, social or economic levels, there is no basis for trust, mutual respect and cooperation.[46]

This is clearly the level at which we need to operate when developing a new norm of biosecurity as it is embedded in the responsible conduct of science. In any moves to take on board shared common ethical values, it is vital that we do as much as possible to support all actors in the biotechnology field so that no-one is left behind and marginalised. Such help does not need to be financial, although it could be, at least in part. Help could be provided in the form of intellectual, educational, training or other relevant forms of support as well. Once a norm has become established, recognised and shared among growing numbers within a community, it is important that we define appropriate penalties to be faced by those breaking the norm. Next, we need to agree on the implementation of those penalties; without imposed penalties, a norm, by its very nature, is effectively

---

42 Dawkins, op. cit.
43 Catchpoole, op. cit., p. 77.
44 Balkin, J. M. 1998, *Cultural Software: A Theory of Ideology*, Yale University Press, New Haven, Conn., and London, p. 43, cited in ibid., p. 265.
45 Ibid.
46 Catchpoole, op. cit., p. 267.

toothless. In today's world, it may be argued that the economic penalty is the one of greatest effect; however, penalties need not be just economic—but that is another argument for another day.

Let's now consider the education route as a norm developer. Education is itself a memes transmitter. We absorb new memes through the values set by those who pass the new values on to us. Over time, new memes can become new, deep-rooted norms. Social norms (such as the baseball cap) may start as memes, developing into expectations of behaviour, then into controls that regulate behaviour and identity (the baseball cap again). Philosophical norms imply some obligation or duty, carrying threatened or real sanctions if they are broken. There is cultural pressure to abide within the norm, so culture provides a setting in which the norm may survive and develop.

Ethics and ethical attitudes may be passed on as memes (quick-changing, initially superficial cultural artefacts) and then develop over time into norms (more deep-rooted values and attitudes within a culture). Ethics, as norms, can be implicit or explicit. They can vary from group to group and place to place and can change over time. It is this capacity to change that we must address in order to develop a new ethics of biosecurity so that it becomes a dominant scientific norm. The biosafety norm is based on the principle of containment and safe management of harmful biological agents, but in such a way as to allow ongoing scientific practice; the biosecurity norm needs to be built on the principle of the containment and safe management of biological knowledge and processes, also allowing ongoing scientific practice. Just as the biosafety norm (when implemented properly) draws a line in the sand over which practice cannot step, so must the biosecurity norm.

So how may we develop research ethics as memes-into-norms and persuade others to sign up to them in the biosecurity context? This would mean taking a norm of, say, 'no harm', and defining it in practical ways in order to teach the principle. Catchpoole believes that new norms can be taught and developed as memes. She suggests that education can choose which memes to transmit and which it will not.[47] Educational transmission of a commonly held value, such as care for others, could be 'inserted' as a norm as part of the building of a multicultural ethic of biosecurity, with a common value set, around it:

> Having care for one another and our world involves finding a balance between the needs of self, close others and global others in ways that respect basic human rights and yet which seek to encourage more than a minimalist commitment to having care for one another … By delineating the nature of care itself and recognising the ubiquitous nature of power

---

47   Ibid., p. 126.

within relationships, an extended ethic of care provides a set of trans-cultural memes for both guiding and evaluating the development of the ethical form of life in the private and public spheres.[48]

The memes of ethics could therefore be taught and learned as a form of cultural transmission in a multicultural context if they are presented from the perspective of a commonly held value such as care for others, which few cultures would dispute.[49] At the same time, we need, however, to recognise that a range of cultural pressures may be brought to bear on the implementation of a 'no harm' or 'care for others' value. It is one thing to state a belief in 'do no harm to others' publicly, but at the same time, be driven by a more pressing value to actually prepare to do the opposite—we only need to look at those countries in violation of the BWC with their 'peaceful purposes' cloak hiding offensive weapons programs to see real-life cases in action. This prioritisation of pressures is to be the subject of further research.[50] One could argue that the norm of 'no harm' is so routinely violated in science (for example, in the use of animals in product testing) that it is diminished as an argument and is no longer worth pursuing or developing in practice; however, this in itself is an ethical debate: we either accept the norm of 'no harm' as being something to aspire to or give in to an anarchic world in which 'care for the other' no longer has a place worth arguing for. I would suggest that it is 'inside' the norm of 'no harm' that much future research is required. We can readily see that many individuals and groups apply this norm differently, simply by the way in which they define who is to be actually protected from harm, in what circumstances and why. This is the area in which we seem to make our moral and ethical judgments based on culturally appropriate norms with which we face everyday life.

Conclusions

This chapter has shown that we need to appreciate the variety of private moral backgrounds of individual scientists as we seek to identify common values with which to form a new norm of biosecurity within the public culture of ethical science.

It is probable that there is a biological foundation underpinning our moral development from infancy into adulthood. On top of this genetic and physiological framework, we can identify a range of psychological processes that further

---

48 Ibid., p. 47.
49 Ibid., p. 186.
50 Further culture-based research is being planned by Judi Sture and Masamichi Minehata of Bradford Disarmament Research Centre, looking at how scientists and educators prioritise and manage pressures in their everyday work.

develop within us an increased capacity for empathy and consideration of the rights of others. Some of these processes have been identified and categorised by Kohlberg, Hoffman and others. An increasing body of work has shown that psychology is a powerful factor in the development of culture, and that it is possible to understand the nature and growth of cultures and cultural changes through the lens of psychology. This is an area ripe for further investigation in the biosecurity context.

Each individual scientist engaged in work or research that has or may have potential for misuse is faced with a moral-ethical dilemma of his own. His private moral value set(s) may come into conflict with his publicly required professional value set(s). Because they are culturally derived and embedded, private moral values can be difficult to change, as they are usually driven or controlled by one of five expressions of culture—social organisation, economics, technology, religion or language—all of which are interlocking and may be hard to shift (these expressions of culture are commonly used by archaeologists in understanding culture).[51] We need to identify the key drivers from among these five expressions that influence the private morality that scientists bring to their professional ethics, in order to challenge them and move forward to a common value set. In addition, we need to recognise how these drivers also inform public ethics in the workplace. By considering a number of existing theories such as cognitive dissonance and self-perception, against a biological background of genetic and evolutionary tendencies to altruism, emotional reaction or self-interest, it may be possible to delve into the processes that give rise to the professional ethical value set(s) of scientists in any given cultural setting and that inform their prioritisation of each of the five expressions of culture.

Against the background of a call for increasing awareness of dual-use issues,[52] we may consider the following fields of activity and inquiry in order to move forward. First, we need to help scientists recognise their own value sets and identify how they may clash with both the values of the science they practice and the need to protect against the risk of dual use. Second, we need to review our institutional organisations and look for hidden curricula (confounding values that compete implicitly with required values) in science at governance and teaching levels that may be impeding the search for and implementation of a common value set in biosecurity and dual use. Third, in our aim to embed a new ethics of biosecurity and dual use within the culture of life sciences, we need to consider a range of psychological drivers that influence others and

---

51   See Renfrew, C. and Bahn, P. 1991, *Archaeology: Theories, Methods and Practice*, Thames & Hudson, London.
52   United Nations 2008, BWC/MSP/2008/5, p. 6.

ourselves. Fourth, we need to look out for threats or undue influence from the five archaeological expressions of culture, as these may exert overt or covert pressure on scientists.

Finally, by identifying even a single commonly held value, such as care for others, it may be possible to start laying the foundations of a new value set that will underpin our efforts in building a sustainable capacity in biosecurity in the life sciences.

8. Responsible Dual Use

John Forge

Introduction

This chapter addresses the *moral* dimension of dual-use research. To set the scene, I will begin by explaining what I take dual use to be. I understand a dual-use item to be something that has both a good or neutral (neither good nor bad) use or application and a bad use. Three different categories of dual-use items can be distinguished: research, technologies and artefacts.[1] These are clearly different sorts of things. Research is an activity, while technology is a form of knowledge—knowledge of the techniques for the production of artefacts—whereas artefacts are objects. But it is also clear that these items are related: research aims to give us technology, which in turn produces artefacts.[2] It is the last that normally has the immediate impact on the individual, be this good or bad. For instance, research has led to methods for mass-producing cheap ammonium nitrate, which has a 'primary' use as a fertiliser and a 'secondary' use a bomb-making material. It is the substance—the artefact—ammonium nitrate that is the fertiliser/bomb component, not the technology or the research. The import of this is that when we talk of threats or risks associated with dual use, we need to distinguish between different levels of risk or threat, depending on the nature of the dual-use item, and between different ways of dealing with these. For instance, substances like ammonium nitrate can be physically contained, but the knowledge of how to manufacture it cannot.

My suggested definition of a dual-use item is therefore as follows:

> An item (knowledge, technology, artefact) is dual use if there is a (sufficiently high) risk that it can be used to design or produce a weapon, or if there is a (sufficiently great) threat that it can be used in an improvised weapon, where in neither case is weapons development the intended or primary purpose.[3]

---

1   This is the classification that I gave in Forge, J. 2010, 'A note on the definition of dual-use', *Science and Engineering Ethics*, pp. 111–18. This paper was written as a response to a challenge by David Resnik to come up with a definition of dual use that is neither too narrow nor too broad. I do not claim that my contribution is the last word, but I do think it is along the right lines. I also think that it is enough by way of a starting point for the present discussion. I would, however, note that the sense in which artefacts are said to be objects is a special one—see ibid., p. 115.
2   There are of course other kinds of research and other kinds of technologies besides those that aim to produce artefacts.
3   Forge, op. cit., pp. 111–18.

There are several comments in order. First, I equate the bad—secondary, unintended, and so on—use with weapons development. It is certainly possible to broaden the discussion of dual use to include other harmful activities, but I think these would be more controversial.[4] I should stress that by 'weapons development' I include all aspects of the design, production, testing and commissioning of both weapons themselves and all the ancillary structures, platforms and processes that are necessary for their use. Second, the reference to *risk* in the definition points to the possibility of the production of a new or improved weapon. The (by now) familiar examples of techniques of synthesing pathogens are thus risky in this sense. Threats differ from risks here in that they signify the intention to do something harmful, rather than the possibility that something harmful might happen. I have coupled risk with the design of (new) weapons but threats with improvised weapons. States threaten one another, as North and South Korea are doing at present, though the threat here is usually conditional.[5] In regard to dual use, however, I think we are more concerned at the moment with sub-state actors, such as terrorist organisations, using or adapting existing items for improvised weapons, as has been done with ammonium nitrate, than with states arming themselves with improvised weapons, as states do not usually need to improvise weaponry.[6] Finally, research that is intended to produce a weapon and which, fortuitously, has a friendly civilian application, is not dual use according to my definition: the 'bad use' is always the unintended one.

My focus here is with dual-use research, the dual-use item that appears furthest removed from any immediate impact on the individual, and with the responsibilities of those who undertake research that is, or could be, dual use. So if a particular research project has some prospect or likelihood of leading to a bad outcome then what do we expect of those who would conduct the project? Perhaps if the researcher does not intend for the bad outcome to come about, or if she does not foresee that it will, she has no responsibility for it? There are well-known discussions of the doctrine of the double effect that could have some application here.[7]

Another issue is if the likely bad effects seem to be outweighed by the prospect of good outcomes. The consequentialist who sees moral action as reducible to the production of what, on the whole, is the best outcome is likely to be sympathetic to this possibility. I shall touch on these issues in what follows, but

4   And this is very much in the spirit of the early definition of 'dual use' given by the Office of Technology Assessment in the United States and subsequent definitions given by bodies such as the National Research Council. See ibid., p. 112.
5   *If* North Korea shells South Korean islands again *then* the South will retaliate.
6   The extent to which we should therefore be cautious about undertaking dual-use research becomes important. The relevance of the precautionary principle, which would seem to have application to dual use, is the subject of other chapters in this volume.
7   See Suzanne Uniacke's Chapter 10, in the present volume, for a discussion of this.

my main aim is to apply a particular way of understanding the responsibilities of researchers that I proposed in a recent work, *The Responsible Scientist*,[8] to the question of dual use. These responsibilities are both 'backward looking' and 'forward looking', having to do with what researchers are responsible for in the past and what we want to hold them to in the future. Both sorts of responsibility are relevant to dual use. The upshot of the account that emerges is a cautious and restrictive response to dual-use research; other accounts will have other implications. I will not assume much familiarity with the topic of responsibility or with moral philosophy, so much of what follows in section one will be exposition. I hope the implications of this will become clear in sections two and three. As a final comment by way of introduction, what follows is put forward as a contribution to the *ongoing* discussion of dual use. The topic is still, I think, at a stage where a variety of different approaches are worth considering, in regard to both definitions and the moral and policy dimensions. I hope the account of responsibility that I have developed can stand as one such approach.

Responsibility and research

There are two kinds of responsibility: 1) responsibility for past actions, actions that an agent has already completed, or omitted to do—namely, backward-looking responsibility; and 2) responsibility for future actions or omissions—namely, forward-looking responsibility. As we shall see, these concepts are related—and both are relevant to scientific research. It is easiest to begin with a simple example that has nothing to do with science: suppose it is the job of a railway guard to make sure everyone is on the train and the doors are closed before signalling to the train driver to leave the station. If the guard fails to realise that a disabled person will take extra time to get on the train or if he omits to make sure that a carriage door is closed then he will be held responsible for any harm that subsequently occurs. This would not be true of a member of the public. We would hope that a member of the general public would try to assist the disabled person or close the carriage door, but, unlike the guard, she has no 'special responsibility' to do so. In this example, the special responsibility in question can be termed *role responsibility*. It is (a kind of) forward-looking responsibility—that is, one that obliges the person in the role to conform to certain expectations about his or her future actions in a certain sphere: train guards, for example, are supposed to make sure trains are safe. There are other jobs and professions that define roles that are such that the performers are supposed to safeguard others from harm. These are not all of a piece. For instance, those of the medical and legal professions are more complex than those of train guards, lifeguards, and so on, which is one reason these roles have an

---

8   Forge, J. 2008, *The Responsible Scientist*, Pittsburgh University Press, Pittsburgh.

associated professional ethic. An important point here is that certain failures to foresee or omissions to act that result in harm to others can only be attributed to the agent if there is a corresponding forward-looking responsibility. I am not obliged to keep an eye on toddlers at my local pool, but the lifeguard is, and if she omits to do so then she is open to blame.

In my book, I argued that scientists have a kind of role responsibility, even, and perhaps especially, scientists who claim to be doing pure research.[9] I used this to ground a wide account of backward-looking responsibility for science, one that held that scientists could be held responsible for failures to anticipate where their work might lead, and for omissions to act in certain ways, in addition to being responsible for what they intended to do and for what they foresaw would be outcomes of their work. These are, of course, complicated issues.[10] For instance, it might be asked how could a scientist know where his research might go, what kind of applications it could have in the future? How could Newton know of all the applications of universal gravitation or Einstein of relativity? I did not try to argue that all scientists can foresee all applications of their work, for that would have been a hopeless task, but I did show that there are circumstances where this is possible. For example, had all the information about the properties of the uranium nucleus available in 1939 and 1940 been published, there was a significant probability that this would have led to a more vigorous effort on the part of the Nazis and others to develop an atomic bomb. And as the gap between pure research and applied research narrows, as it is surely doing more and more in an age when funding for science is increasingly tied to applications, it is easier to look ahead. Indeed, to continue the theme of funding, research proposals are very often obliged to list possible applications as outcomes. I will come back to this point.

The basic reason all scientists have a kind of role responsibility is simply because science affects people (SAP) and does so in profound ways. I contend that these effects are not always good. I believe that all weapons of mass destruction, especially nuclear weapons, have done us no good, are terrifyingly dangerous, and that it would have been better had the Manhattan Project never been set up and no-one else had attempted to make atomic bombs.[11] Like all these issues, this is somewhat (but I think in this case not very) controversial: some think that the dropping of the atomic bombs on Japan was a good thing because it ended

---

9   I understand 'scientist' in a broad sense, not restricted to the natural sciences. Most of the examples in the book are from physics, which is my area of expertise.

10   So complicated that I needed to devote two chapters to deal with them. One of the important steps in the argument here was to show that we can be responsible for what we don't know—we can be responsible for being ignorant, which is at odds with the classical account of the matter by Aristotle. See Forge, 2008, pp. 107–9.

11   I agree with the assessment that it was only in wartime that the vast expenditure would have been justified for the first nuclear weapons project, so without the Manhattan Project there may well have been no nuclear weapons.

World War II, and that nuclear weapons prevented the Cold War from becoming a hot one. As we learn more about the intentions of the former Soviet Union, the latter proposition is becoming increasingly less plausible. (I am also willing to argue, though not here, that *all* scientific research for the ends of weapons innovation is wrong.[12]) Whatever one makes of these particular examples, even someone with blind faith in science would be hard put to maintain that *everything* scientists do has turned out for good. It should be stressed that it is *not* part of the present position that when something has gone wrong then this has been done intentionally or that scientists are always to blame. And it is far from the present position that science either has no effect on us or is bad overall. Clearly, science has done a great deal of good, as is clear when we review the history of medical science.[13]

Once we accept SAP, it seems clear that we would like science to somehow maximise the good outcomes and minimise the bad ones. The first step, or maybe the preliminary step, is to get scientists to accept SAP; by this I mean all scientists and not just those working in applied areas. Surely this should not be too difficult and is probably already true of the majority of scientists. That all scientists should be aware that their work can affect others can be argued with reference to what I call the changed context of science. Indeed, since World War II, all working scientists really should know that no scientific research is pure in the sense that it cannot in principle affect people. Once this is established, we get on to the hard choices: just what kind of research should scientists be encouraged to do, and what to avoid? Philosophers can have some influence on the policy process—at least they can in theory—by suggesting some general guidelines. What I have tried to do is to provide a set of moral principles that could be used to inform the choices made by scientists themselves, though policymakers could also use them. There are different ways in which this can be done, given what I have said so far about responsibility, and in the rest of this section, I will outline the way I prefer.

The reader may be aware that there are two traditions in moral philosophy—traditions that seem sharply opposed. According to the consequentialist tradition, the moral import of an action or choice resides solely in its consequences. The nature of the act itself, or of how we describe or characterise it, counts for nothing. Evidently, there will be issues here for the consequentialist, resembling those discussed above, in regard to how consequences stretching into the future can be determined and weighted with respect to their moral significance. Leaving these aside, a consequentialist normative ethic will specify some

---

12 See my *Designed to Kill: The Case against Weapons Research*. Dordrecht: Springer, 2012.
13 But even here there is no universal agreement: some strange folk even think that the triple antigen vaccination is wrong. My general point here is that science is a 'mixed blessing' and that is why it is important to determine just what the responsibilities of the scientist are, Forge, 2008, op. cit., pp. 28–31.

property, to be taken as 'the good', and then require moral agents to maximise this property by their actions. Famous nineteenth-century consequentialists like Bentham and Mill took happiness to constitute the good, and thus moral action was such as to maximise happiness, the agent's included. The calculus has to include consequences where some were made unhappy, and this means that all sorts of scenarios can be constructed in which happiness is maximised but some small number of unfortunates is made utterly and forever miserable. Some of these scenarios are portrayed as counterexamples to the theory. My own account is non-consequentialist in that it does not hold that consequences are all that matter.[14] I won't make any further criticism of the consequentialist tradition here.

Non-consequentialist moral philosophy is usually expressed as a set of norms or *rules*. What these are, how these are understood and how they are justified distinguish different non-consequentialist accounts. Modern accounts do not usually interpret the rules as absolutely binding, and here there is a break with the Kantian tradition. Kant famously—or notoriously—said that one should not lie, not even to mislead a maniac looking for one's best friend with malicious intent. The rules contained in modern accounts are thus often said to be prima facie, or to lay down prima facie duties: they are to be obeyed 'in the first instance', unless there are very good reasons not to obey them. As for the content of these rules, I follow those like Bernard Gert who think that they should forbid harming but not require any positive or benefiting action. Here there is evidently a considerable difference with the traditional consequentialist. One good reason to accept our view is that it is possible to impartially refrain from harming everyone, but one cannot impartially benefit everyone. If I have a limited amount of help to give, I must favour some, but moral action is supposed to be impartial. Gert's moral philosophy, which I follow closely here, does not exclude trying to do good; however, good action, which is understood to be equivalent to the prevention of harm, is taken to be the content of moral *ideals*, and hence agents are not bound by rule to prevent harm. It has to be said that these are quite subtle matters and it is hard to do justice to them here. Another way to put the difference between the rules and the ideals is as follows: agents need to justify breaking the rules—recall that these are not absolutely binding—but only encouraged to act in accord with the ideals. Gert refers to his account as *common morality*.

To conclude this discussion of the moral basis of responsibilities of the scientist, we should first note that this way of doing moral philosophy makes a lot of sense. No person in her right mind wants to be harmed: *everyone*, wherever

---

14   Note that non-consequentialism is not the contrary of consequentialism, which would be that consequences do not matter at all.

they come from, can agree with that.[15] Thus everyone can accept a moral code that proscribes harming, and agree that everyone would be better off with less harming. Doing good in some positive sense is much more difficult to universalise. Moreover, striving not to harm others is surely within the reach of everyone, while looking to maximise the good consequences of one's actions looks just too hard. Now scientists are in a special position because of the far-reaching implications of their work, primarily through technology.[16] Therefore it seems fair to attribute to them special responsibilities, and in line with the proposed moral position, these should enjoin scientists in the first place not to do research that has harmful outcomes. I have taken this to mean above all (and at the very least) that scientists should not engage in weapons research;[17] however, I acknowledge that scientists should be encouraged to do research that will have beneficial outcomes. In *The Responsible Scientist*, I therefore maintained that the responsibilities of the scientist are two tiered. The first 'tier' comprises this demand to avoid research that can have harmful outcomes, while the second encourages scientists to do research that prevents harm. Research in the biomedical sciences could well fall into the latter category, as could research that breaks down racial and sexual stereotypes and biases. These two tiers correspond to the moral rules and the moral ideals of common morality. Most of this section has been devoted to the forward-looking responsibility of the scientist, and I will come back in the final section to see how this applies to dual-use research. The next section, on the other hand, will be concerned mainly with backward-looking responsibility.

Dual use and responsible research

Our interest here is with dual-use research, and not with other dual-use items, although of course dual-use research is worrisome precisely because it can lead to the latter. To begin with, then, suppose that a given line of research, R_1, is already established as dual use. By this I mean that it is *known* that its implications are such that there are bad uses as well as good or neutral ones. It may seem strange that anyone would work on such a project, and indeed it may be hard to find many real-life examples; however, the enrichment or reprocessing of spent nuclear fuel rods to make weapons-grade uranium and plutonium, and also to make new, and high-power, fuel rods counts as an example, and thus

---

15 Gert holds that common morality gets purchase on all those who are rational in the sense of accepting certain basic beliefs, such as that I am mortal, I can be harmed, I have interests, and so on. See Gert, B. 1998, *Morality: Its Nature and Justification*, Oxford University Press, Oxford.

16 Science as a body of ideas has had, and still has, a huge impact on us; witness the work of the likes of Galileo and Darwin.

17 See Forge, J. 2007, 'What are the moral limits of weapons research?' *Philosophy in the Contemporary World*, vol. 14 (Spring), pp. 79–88; and Forge, 2008, op. cit. In *Designed to Kill* I argue that no one should engage in weapons research.

experiments or even theoretical work on this topic count as dual use. There are two reasons a scientist might think it is morally acceptable for her to work on this kind of project in applied nuclear physics. She might think that in fact there is no bad use, or that if she only *intends* to work on enrichment or reprocessing insofar as it has a benign use in civilian power reactors and that it deals with the disposal of hazardous waste, that is okay. The first of these reasons raises an important general issue about dual use—namely, the basis on which the uses in question are classified.[18] The second raises issues that concern attributions of backward-looking responsibility. I will address this second matter first.

The second issue can be dealt with by what I call the wide view of backward-looking moral responsibility, given in *The Responsible Scientist*. The wide view holds that scientists are responsible for what they foresee as well as for what they intend (and they can also be responsible for what they should have foreseen). The view expressed by our scientist above—that one is only responsible for what one intends—is what John Mackie called the straight rule of moral responsibility and which I call the standard view. The latter term is appropriate, as the view seems to have slipped in as a kind of orthodoxy, being endorsed by Peter Strawson and John Austin, both, like Mackie, renowned Oxford philosophers, but without convincing argument. At first sight it seems plausible, because intentions are given as reasons for action: 'Why do you work on this project?' 'To make better nuclear fuel rods and remove hazardous nuclear waste.' The obvious question, then, is why we should only be responsible for actions that we have a reason for doing.[19] For those who do think that responsibility must be tied to reasons for action, and I am not one of this number, it is easy to see how a foreseen but not intended outcome or side-effect of an action can be included in a reason for action as an explicit qualification. In this example, it can be done by adding 'although I am aware that reprocessed fuel can be used in weapons and this did not make any difference to my choice'. By incorporating foreseen outcomes in this way, we get what I have called the modified standard view.[20] How we might manipulate or change the standard view is really beside the point, for it is simply a commonplace that we *do* hold people responsible for what they foresee as well as what they intend, in spite of what Mackie, Strawson and Austin might

---

18   I raised the issue of the role of values in the definition of dual-use items and how different sets of values could lead to different judgments about what counts as a dual-use item in Forge, 2010, op. cit., p. 117. I have more to say about this question below.

19   In some ways it would be clearer if this were expressed in terms of interests. Our researcher has an interest in disposing of hazardous waste, but no interest in making material for nuclear weapons; her reason for undertaking the project cites the former as her reason for action. Nevertheless, she sees that her work contributes to the latter end.

20   The modified standard view, and why it is not enough, is discussed at length in Chapter 5 of Forge, 2008, op. cit. Also, it is one thing to recognise that something is a commonplace and quite another to give it a convincing philosophical rationale—hence the length of Chapter 5 and the chapters before and after it.

have thought. And hence the response of our scientist that any other outcomes of this work are beside the point because they are not what she intends will not wash.[21]

On what basis do we decide whether something has both good and bad outcomes or uses, and hence whether it is a dual-use item? And if there are alternatives that lead to different judgments, which do we chose? These are important questions, ones that I cannot address fully here. But for a start we can all agree that bad outcomes are those that are, or have the means to be, harmful. Common morality of course would have this implication. If we think of our moral philosophy as expressing values then common morality elevates the value of not harming above all others, and would not, for example, allow the harming of a few to benefit many. Again, this is surely something that most of us would accept: if we could benefit at the expense of others being harmed then most of us would not accept such a deal if it were offered to us (at least I hope we would not). So I assume again that something like the values expressed by common morality provide a plausible and realistic value system. We will see in a moment how this can affect classifications of items as dual use. First, however, note that there is a difference between harmful acts, acts that directly harm moral subjects, and outcomes of research—the topic of the present discussion—which are such that they could be the means to harm. Provision of the means to harm is not the same as harming, and if we are to try to attach the kind of strong moral prohibitions about harming suggested here to activities that do not directly harm, it seems that we need some further argument.

In fact I don't think that this is too hard to supply in cases of research whose *explicit* aim is to design new and better weaponry.[22] Scientists who engage in this kind of work need *justification* for what they do: they need to give reasons why it is acceptable for them to engage in an activity whose objective is to provide tools whose purpose is to harm others. The typical justification is that we need weapons to defend ourselves from others, from the bad guys. This is similar in form to the typical justification that we give when we feel we need to harm others—namely, that we need to do so to *prevent* harm to others and ourselves. The acceptability of that justification will depend on the circumstances: clearly, inflicting a great deal of harm to prevent a small amount of harm will not do. The harm inflicted should in some way be commensurate with that prevented. But this judgment is more difficult to sustain when it comes to deciding whether making new and improved weapons—namely, making the means to harm rather than harming directly—is justified. Are they really necessary? Will others

---

21 The challenge, of course, is, as in the previous footnote, to provide the philosophical argument.
22 Not too hard, but quite lengthy: in Forge, 2008, op. cit., pp. 155–8, I argue for a means principle that is such as to transfer responsibility for bad outcomes using artefacts—weapons, for instance—to the designer of the artefact.

acquire them and use them for evil purposes? What alternatives are there to weapons research? I think these questions are extremely hard if not impossible to answer in ways that support weapons research (namely, yes, no, none), and so I think that weapons research is an activity that should be undertaken only in exceptional circumstances.[23]

In this way we can see how to go about addressing questions about matters that are morally suspect, such as weapons research. When the topic is dual use, one of which involves weapons, for instance, the questions become more complicated, but I think we can now see just where the complications arise. The assumption is that R1 is such that we know it can have bad uses. Any research that can be harmful needs to be justified, and so far we have focused on justifications that do not take into account any good use (that is not a product or function of the bad use). A good use, on the present system of values, is one that prevents harm; in the case of R1 this is the disposal of hazardous waste. The role this will play is thus as part of the justification of the project. What the present system of values and justifications requires for dual-use research is that the bad use is 'offset' by considerations that appeal to the good use, and any other relevant considerations, and that these good uses must involve the prevention of harm. Where the 'good' use does not prevent harm, where it is 'neutral' as evaluated on the prevent system, it cannot figure in the justification of the project. In regard to a project like R1, it may therefore even be useful to proceed as follows: justification needs to take the form as if it were the bad use that is the main object of the investigation—that this is foreseen but not intended is irrelevant here—and the good use is to be cited as a reason *not* to abandon the project.

Dual use, responsible research and uncertain outcomes

Now we need to introduce a further complication. Suppose project R2 is such that it might have a bad use, if it were to fall into the wrong hands, if it turned out in a certain way, if certain people used it as a point of departure for other research, and so on—in other words, R2 is risky. Consider a response here resembling that made above about intention—that it is not the researcher's job to look ahead and consider other possible uses of her research. None of these potential outcomes needs to be taken into account; only the project itself and its 'scientific outcomes' need concern the researcher. As we have seen already, this attitude is at odds with the responsibilities of the scientist in this day and age. Science affects us and our environment, whether it is set up as pure

---

23 So exceptional as to never obtain in fact, though this is not the occasion to try to make that case.

or applied, and all scientists are obliged to try to look ahead and see where their research is going. Everyone has responsibilities when they take on the role of the scientist, and in this way scientists resemble our train guard.[24] But again there are difficulties: how are scientists able to look into the future and see where their research will lead, for are not outcomes unpredictable? This matter was raised above, where I conceded that some research is unpredictable but also maintained that some is not. Moreover, in the present context where science is heavily sponsored and funded, it can only operate where it promises to give useful outcomes. Were this not true then institutions such as defence departments would not spend more money than anyone on scientific research.[25] Science is now far too big and expensive to be left undirected. Of course, when scientists propose research projects and say what they hope to achieve, they maintain that these are desirable outcomes, that their methods are proper and so on. They do not, as a rule, list undesirable outcomes; these will not normally be the focus of attention of researchers, but I do not see any reason they could not be. I conclude that the present account of the responsibilities of scientists has the implication that they are obliged to do their best to look ahead and try to see if their work will lead to any bad outcomes.

These kinds of assessments involve judgments about risk and threat, and are therefore more or less uncertain, as indeed are good outcomes, because all research has some uncertainty associated with it. The present account implies a cautious and conservative approach—and this may not be welcomed by all researchers. Consider the following familiar scenario: a research project is aimed at uncovering the structure of a pathogen with the aim of finding a preventative therapy or cure for the condition that it causes; however, this knowledge could also be used for making the pathogen more virulent and resistant to the very measures that the project was designed to put in place. The present account, as we saw in section one, sees the prevention of harm as something to be encouraged. This is the second tier of the two-tiered account of the forward-looking responsibility of the scientist, and we can all agree that the prevention of harm is a good thing. But the first tier forbids scientists from doing harmful research, and that strongly proscribes any bioweapons research, including bioweapons research that is unintended. Where does this leave the research project? It does not follow that it should not be undertaken, but it does follow that justification must be given. And here an assessment of the costs versus the benefits is needed, as well as control of the results.

---

24 I discuss the case of Frederic Joliot-Curie, who published nuclear data in 1938, against the urging of Szilard. Fortunately, the data were incorrect and it did not seem to interest the Nazi scientists; in Forge, 2008, op. cit. Joliot-Curie's actions were irresponsible even back then.
25 For many years the US Department of Defense has spent more money on scientific research than anyone else.

If there is some risk that the project will be used for bad ends then it seems clear that not only should the results not be published, but also they should be tightly controlled. Michael Selgelid[26] has raised the question of the censorship of research in regard to dual-use questions, and has advocated a moderate and balanced position. The present account will certainly suggest stricter controls (perhaps another unwelcome implication). From the perspective of common morality, agents are free to do whatever they wish, as long as this does not harm others. Thus, scientists are free to do whatever research they like, but the freedom to research, or research itself, is not given special value or status. When there is a risk that research will cause harm then it is firmly proscribed. And when research also has benefits then it should be conducted with effort to minimise harms, and if this entails censorship then that is what is required for the project to go ahead. Indeed, it seems that this is precisely the view that makes the most sense. The response that the whole purpose of publication is to make ideas and results open to all members of the community who can then build on them will not do here. Networks of respected colleagues can be informed, but the risks of open publication may be too great with dual-use research.

Conclusion

I conclude that the account of science and responsibility put forward in *The Responsible Scientist* has relevance for the ethics of dual-use research, and in two sorts of ways. In the first place, the theory of responsibility ties the scientist to the outcomes of his or her work, and does so more tightly than other viewpoints. For instance, an account that incorporated the straight rule and did not see any special sort of forward-looking responsibility for scientists would have quite different implications; however, I think that the topic of dual use shows that we want a wide-ranging account of responsibility, and we want researchers to think carefully about what they do. The second way in which the account is relevant is in the way it sees responsibility as being discharged or cashed out. This is with respect to the no-harming ethic of common morality. I suspect that this will be the more controversial aspect of the present proposals.

---

26 Selgelid, M. 2007, 'A tale of two studies; ethics, bioterrorism, and the censorship of science', *Hastings Center Report*, vol. 37, no. 3, pp. 35–43.

9. An Expected-Value Approach to the Dual-Use Problem

Thomas Douglas

In this chapter I examine how expected-value theory might inform responses to what I call the dual-use problem. I begin by defining that problem. I then outline a procedure, which invokes expected-value theory, for tackling it. I first illustrate the procedure with the aid of a simplified schematic example of a dual-use problem, and then describe how it might also guide responses to more complex real-world cases. I outline some attractive features of the procedure. Finally, I consider whether and how the procedure might be amended to accommodate various criticisms of it. The aim is not to defend the procedure in its original form, but to consider how far we must deviate from it to evade the criticisms that I consider. I seek to show that, though it is necessary to make substantial ammendments to the procedure, we need not eschew a role for expected-value theory altogether. Even if one accepts the criticisms I discuss, it is possible to defend a role for expected-value theory in responding to dual-use problems.

The dual-use problem

Scientific work can typically be used in both good and bad ways, and sometimes it is difficult or impossible for a scientist to prevent the bad applications without also forestalling the good ones. For example, if a scientist publishes her findings, thus enabling good applications of those findings, she may then have little power to prevent bad applications of the same results. Scientists may thus face a trade-off between enabling certain good applications of their work and preventing certain bad ones. In a few cases, the risk of bad applications may be so high that there is a genuine dilemma about how to resolve this trade-off. Consider the following hypothetical scenario:

> It's 2030. Designer viruses are now used to treat some cancers and infectious diseases. But they're expensive and difficult to manufacture. You discover a new, cheap way to produce synthetic viruses using out-of-date benchtop DNA synthesisers that are now ubiquitous, even in developing countries. You're excited about the discovery and hoping to publish it in *Nature*. You think it could bring a wide range of medical treatments, not to mention research tools, within the grasp of the

developing world. However, there's a catch: every major military and terrorist group in the world has access to these obsolete synthesisers. It would take only one malevolent agent and one such machine to produce enough vaccine-resistant smallpox virions to devastate humanity.[1]

Arguably, it is genuinely unclear in this case whether you ought to publish your discovery. If this is so, you may be faced with what has been termed the *dual-use dilemma*.

As a tentative starting point, we might define the dual-use dilemma as the quandary that arises whenever: 1) a scientist faces a choice about whether or not to produce or disseminate some scientific output, such as a piece of scientific knowledge, or a physical technology; 2) that output could be used in both good and bad ways, and the agent is not in a position to altogether eliminate the risk of bad use without also reducing the likelihood of good use; 3) there is therefore a trade-off between enabling the good application(s) and preventing the bad one(s); and 4) it is consequently unclear whether producing the scientific output is the right thing to do, morally speaking.

This definition captures some commonplace ideas about dual-use dilemmas; however, there are, I think, good reasons to broaden the definition somewhat before embarking on any ethical analysis.

First, though the dual-use dilemma is typically presented as a problem that arises *for scientists* and *in relation to particular scientific projects*, similar problems may arise for others in a position to support or impede particular projects, and for those faced with decisions about what general stance or policy to take on scientific work, or some particular class of scientific work. For example, a policymaker considering whether to institute censorship of scientific journals may also face a trade-off between enabling good applications of scientific findings and preventing bad ones, though in this case the choice does not relate to any particular scientific project. Given the similar structure of this kind of ethical problem to paradigmatic examples of the dual-use dilemma faced by individual scientists, it seems sensible to consider them together.

Second, as defined above, the dual-use dilemma arises because of a tension between *two* ethical considerations: reasons to enable good applications of some scientific output, and reasons to prevent bad ones. But in real-world cases, there will be a number of other ethical considerations bearing on decisions relating to the production or dissemination of scientific outputs.[2] For example, a scientist

---

1 Adapted from Douglas, T. and Savulescu, J. 2010, 'Synthetic biology and the ethics of knowledge', *Journal of Medical Ethics*, vol. 36, no. 11, pp. 687–93.
2 Buchanan, A. and Kelley, M. 2013, 'Biodefense and the production of knowledge: rethinking the dual-use problem', *Journal of Medical Ethics*, vol. 39, pp. 195–204.

could have a reason to pursue scientific knowledge about some phenomenon regardless of how that knowledge is likely to be used; she could have a reason to pursue the knowledge because it would be intrinsically valuable, or simply because she previously promised to do so. Similarly, a policymaker could have reasons (not) to censor scientific publications that are independent of how the censored information would have been used—for example, she might have reasons not to censor the information in order to respect freedom of speech. An adequate formulation of the dual-use dilemma should accommodate the possible existence of reasons besides those normally taken to generate the dilemma.

Third, according to the above definition, the dual-use dilemma involves making a choice between exactly two alternatives. This feature is suggested by the term *dilemma*. But it might be thought that such binary choices are in fact rather rare. Often when we think there are only two alternatives on the table there are actually more. For example, a scientist may think she is faced with a choice between either publishing her research in full or not, when in fact there may be several other options—for example, publishing elements of the research, or publishing the research conditional upon its distribution being restricted. Similarly, a policymaker may believe that there is a simple choice between censoring scientific journals and not, when in fact there are many different types of censorship (for example, self-censorship by journal editors, censorship by peer reviewers, censorship by external agencies). It is tempting to say that, when presented with a dual-use dilemma, an agent should in the first instance try to find a way to *escape* the dilemma by finding some third, superior course of action. Only if this is impossible should he grasp one horn of the dilemma.

Recognising the existence of further alternatives, however, does not necessarily resolve the problem. Strictly speaking, when there are more than two alternatives available, there can be no *dilemma*. However, there may still be a quandary, since alternatives more likely to enable good applications may also be more conducive to bad applications or be more problematic in other ways (for example, they may be more resource intensive). We can imagine a case in which the more likely an option is to allow good applications relative to alternative options, the more likely it is to also enable bad applications. It may thus remain unclear which option to select.

I wish to include within the scope of the discussion that follows ethical problems that are similar to paradigmatic dual-use dilemmas but are not faced by individual scientists, do not relate to specific scientific projects, involve choices between more than two alternatives, and involve ethical considerations besides reasons to enable good and prevent bad applications of scientific outputs. I will do this by introducing the broader idea of a *dual-use problem*. I define this as the quandary that arises whenever: 1) an agent faces a choice between two or more alternatives that will influence the production and/or dissemination of some scientific output, such as a piece of knowledge or a physical technology; 2) that scientific output could be used in both good and bad ways; 3) there is a trade-off between enabling the good application(s) and preventing the bad one(s), since any alternative that increases the likelihood of the good use, relative to some alternative, also increases the likelihood of the bad use; and 4) given this trade-off and other normative considerations bearing on the choice, it is unclear which alternative(s) it would be morally right for the agent to take.

Introducing an expected-value approach

Having just suggested that the dual-use problem should be understood broadly, I am now, temporarily, going to narrow down on a simplified example of a dual-use problem, so that we have something more tractable to work with. The example is a schematic version of a putative dual-use problem faced recently by scientists working on H1N1 avian influenza.[3]

Suppose that a scientist is deciding whether to embark, with her assistants, on a research project that will investigate whether it is possible to mutate the H1N1 virus so as to make it transmissible by air between humans. (Assume that existing variants of the virus cannot be transmitted in this way.) A likely outcome of the project is that the scientists will indeed discover a method for rendering the virus air-transmissible—that is, that will acquire knowledge that would enable others to create such a virus. This knowledge could, let us suppose, be used to develop a vaccine against the air-transmissible virus that will save 5000 lives, or to facilitate a highly lethal act of biological terrorism in which an air-transmissible variant of H1N1 is released and kills 100 000 people. Suppose (implausibly) that these are the only two possible applications of the knowledge, and that there are no other moral considerations bearing on the scientist's decision whether to embark on the project. To further simplify the case, suppose that the development of the vaccine and the lethal biological attack are events that can occur at most once each. Finally, suppose also that knowledge about how to render H1N1 air-transmissible will certainly not be produced by anyone other than the scientist and her team, so that if the scientist decides not to pursue the project, the lethal H1N1 attack will certainly not occur, but nor will a vaccine against the air-transmissible variant of H1N1 be developed.

The scientist is faced with the choice of taking a gamble. She can gamble on bringing about the good outcome—the development of the vaccine—but it is possible that the bad outcome, the lethal attack, will occur instead or as well. Taking the gamble generates four possible future states of the world

---

3   See, for a brief description of the case, Evans, N. G. 2013, 'Great expectations—ethics, avian flu and the value of progress', *Journal of Medical Ethics*, vol. 39, no. 4, pp. 209–13.

- state I: vaccine developed; no bioterrorist attack
- state II: bioterrorist attack; no vaccine developed
- state III: vaccine developed and bioterrorist attack
- state IV: no vaccine developed; no bioterrorist attack.

If the scientist does not take the gamble, state IV is guaranteed.

Assuming (again, implausibly) that the values of these states of the world can be assessed simply by tallying up the number of lives saved by the vaccine and subtracting the number of lives lost in a bioterrorist attack, these states can be assigned the following values

- state I: 5000 lives saved − 0 lives lost = 5000
- state II: 0 lives saved − 100 000 lives lost = −100 000
- state III: 5000 lives saved − 100 000 lives lost = −95 000
- state IV: 0 lives saved − 0 lives lost = 0.

Suppose that the probability of developing a vaccine is rather high if the project goes ahead—say, 0.5 (50 per cent). On the other hand, suppose that the probability of the devastating bioterrorist attack, if the project goes ahead, is low—say, 0.05 (5 per cent). Then the probabilities of each of these states coming about will be

- state I: 0.5 x 0.95 = 0.475 (47.5 per cent)
- state II: 0.5 x 0.05 = 0.025 (2.5 per cent)
- state III: 0.5 x 0.05 = 0.025 (2.5 per cent)
- state IV: 0.5 x 0.05 = 0.475 (47.5 per cent).

States II and III are both very bad. In terms of lives saved and lost, they are more bad than states I and II are good. This counts in favour of abstaining from the scientific project; however, given the probabilities specified above, states I and IV are much more likely to come about than states II and III. This militates in the opposite direction. How should *A* decide whether to take the gamble? One approach would be to calculate the *expected value* of the gamble, and proceed only if it exceeds the expected value of the alternative—namely, obtaining state IV with certainty. (Readers familiar with expected value theory may which to skip to the next section, 'Complicating the Picture'.)

The expected value of an individual state is given by multiplying the value of that state by its probability of occurring. The expected value of a gamble over multiple possible states is given by summing the expected values of the individual states. We can think of the expected value of the gamble as a weighted average of the values of the individual states that may result, with the weights given by the probabilities of those states coming about.[4]

To illustrate, suppose that you are considering whether to engage in a gamble that involves tossing a coin. If the coin turns up heads, you win $10, if it turns up tails, you lose $5. The probability of a heads is 0.5, and likewise for tails. If you don't take the gamble, you win or lose nothing. Thus, the expected value of the gamble will be given by

probability(heads).value(heads) + probability(tails).value(tails)

$$= 0.5 \text{ x } \$10 + 0.5 \text{ x } -\$5$$

$$= \$2.50.$$

The expected value of this gamble is $2.50. On the other hand, the expected value associated with declining the gamble is $0, as, with certainty, one wins nothing and loses nothing. Since the expected value of the gamble exceeds that of the alternative, the expected-value approach under consideration will advise you to take the gamble.

Now let's return to our schematic dual-use problem. The gamble faced by the scientist has four possible outcomes: I, II, III and IV. So the expected value of the gamble, V_{gamble}, will be given by

$$V_{gamble} = p(I).v(I) + p.(II).v(II) + p(III).v(III) + p(IV).v(IV)$$

where $p(X)$ is the probability of state X and $v(X)$ is the value of state X. (A difference with the coin toss example, however, is that in this case, the values of the states are supposed to reflect their overall value, not simply their *value to the agent*, in this case, the scientist. Our question is not whether the scientist ought, from a self-interested point of view, to undertake the research, but whether it would be morally right for her to do so. Insofar as the value of an outcome bears on the moral rightness of the action that brings it about, it is usually thought to be the overall value that matters.)

Plugging in the values and probabilities that we assigned to these gambles, we get the following result:

$$V_{gamble} = 0.475 \times 5000 + 0.025 \times -100\,000 + 0.025 \times -95\,000 + 0.475 \times 0$$

$$= -2500.$$

---

4 Some use the term 'expected value' in a way that presupposes that the values of individual states are measured in monetary terms. Expected value can then be contrasted with 'expected utility', which is instead determined by the amount of utility contained in each state of the world. I use 'expected value' in a way that is neutral between different metrics of value such as monetary value, utility and health. I thus regard 'expected utility' as a species of expected value.

That is, an expected 2500 lives will be lost, on net. On the other hand, the expected value of the alternative—obtaining state IV with certainty—is given by:

$$V_{alt} = v(IV)$$

$$= 0.$$

The approach under consideration would advise taking the gamble if and only if V_{gamble} exceeds V_{alt}, and against taking the gamble if and only if V_{alt} exceeds the gamble. Given the probabilites and values that we have used, Valt exceeds V_{gamble} so the approach will recommend against taking the gamble—that is, against pursuing the scientific project.

Complicating the picture

The schematic dual-use problem set out above was highly simplified, but the same basic approach could be taken to more complicated, actual dual-use problems. Some revisions and qualifications will, however, be necessary.

For example, in actual cases, there will typically be some chance that the scientific output in question will come about, and be used in good and/or bad ways, even if the agent in question does not produce it: someone else might do so. In producing the output, then, the agent is not replacing a certain outcome, in which there is no chance of the good and bad applications in question, with a gamble in which these outcomes might occur. Instead, she is replacing one gamble with another. To accommodate this, we will need to amend the expected-value approach so that it compares the expected values of two different gambles. In fact, in most cases, since there will be many different actions open to the agent, we will have to compare the expected values of multiple gambles: there will be a different gamble posed by each alternative.

In addition, each gamble will typically involve many more possible outcomes than in the simplified example set out above. In an actual case where a scientist is considering whether to try to develop an air-transmissible variant of the H1N1 virus, there will be a variety of different ways in which the resulting knowledge, if the project is successful, might be used. In addition to being used to develop a vaccine against the new variant or to facilitate a bioterrorist attack, it might also be used, for example, to publicly demonstrate the ease with which terrorists might create a biological weapon, thus encouraging politicians to take further steps to protect against such an attack, or to advance the understanding of what makes viruses air-tranmissible in a way that allows the development of vaccines for existing air-transmissible diseases. Moreover, there will typically be many different forms that each of these kinds of outcome could take. For example, there are many different levels of severity (in terms of both lives lost and other negative consequences) that a bioterrorist attack could have. In addition, though we assumed above that each outcome would occur either once or not at all, in actual cases many of the important good and bad applications of scientific output could occur multiple times. Finally, the development of a technology or piece of scientific knowledge might have good and bad effects that are unrelated to how it will subsequently be used (for example, one good 'effect' might be that there is now more intrinsically valuable knowledge; one bad 'effect' might be that valuable research funding is consumed).[5] These complications all serve to greatly multiply the number of permutations of different states of the world that could result from the scientist's decision, compared with the simplified dual-use problem outlined above.

Moving from schematic examples to the real world would thus greatly complicate any attempt to apply expected-value theory in order to resolve dual-use problems. However, it does not obviously raise any 'in principle' difficulties. One can still envisage an approach to such problems that would: 1) identify the possible outcomes of each alternative course of action; 2) identify the possible 'states of the world', each consisting of combinations of these outcomes;[6] 3) explicitly assess the values and probabilities of these states of the world; 4) use these probabilities and values to calculate the expected value associated with each alternative; and 5) select the alternative associated with the highest expected value. I will call this approach the expected-value procedure or EVP.

Strengths of the expected-value procedure

The EVP has several attractive features.

First, it captures some commonsense intuitions—for example, that decisions should be made by weighing the upsides and downsides of the various choices, that upsides and downsides should be given equal weight, and that, other things being equal, one should prefer an alternative with more valuable upsides, more likely upsides, less disvaluable downsides, or less likely downsides.

Second, the expected-value approach is consistent with a major school of ethical thought: consequentialism. Consequentialism is often taken to hold that an

---

5   The outcomes that I describe here would perhaps not normally be described as 'effects' of the action since their relationship to the action is arguably constitutive rather than causal. There is, however, arguably nothing to prevent us from thinking of them as effects in order to include them within the scope of the expected-value approach.

6   I leave it open whether, in identifying possible outcomes and states of the world, an agent applying this procedure should seek to list *all possible outcomes/states, the most likely outcomes/states, all reasonably foreseeable outcomes/states* or some other subset of possible outcomes/states.

act is morally right if and only if its consequences will be at least as good as those of any alternative action. One variant of consequentialism holds that an action's *actual* consequences are what determine its rightness. Others hold that the action's *foreseen, foreseeable* or *likely* consequences are what are important. Proponents of these latter variants typically hold that the goodness of a set of foreseen/foreseeable/likely consequences is determined by its *expected value*. Their variants of consequentialism thus imply that an action is right if and only if it is associated with at least as much expected value as any alternative. Thus, if one applies the expected-value procedure outlined above when faced with a dual-use problem, one can think of oneself as explicitly trying to identify and adopt the course of action that is morally right according to some variants of consequentialism.[7]

Third, the expected-value approach has a track record. The methodology used in the EVP is a methodology that has been widely employed, often under the label 'risk analysis', 'risk management' or 'cost–benefit analysis', to inform major corporate and government decisions—for example, decisions between alternative power generation projects or mining operations.[8]

Criticisms of the expected-value procedure

Many criticisms of expected-value theory, of its application to major public and private decisions, and of consequentialism might also be adduced against a proposal to use the EVP to confront dual-use problems. In this section I discuss six such criticisms, in each case considering whether and how it might be possible to ammend the EVP to accommodate the criticism.

My aim is not to defend the EVP tooth and nail, by rebutting all objections that might be raised against it. Indeed, I do not comment on whether the criticisms I discuss are persuasive. Rather, the aim is to explore *how far* we must deviate from the procedure in order to evade these criticisms assuming they are persuasive. My conclusion will be that we have to deviate from it to a rather great degree; the approach I end up defending bears little resemblance to the EVP. However, it does, I will suggest, retain at its heart an important role for expected-value theory.

---

7 Note that consequentialists differ on what makes a given consequence good or valuable. The most well-known variety of consequentialism—utilitarianism—holds that the goodness of a set of consequences is determined by the total amount of welfare or wellbeing that it contains, but other variants of consequentialism allow that other things besides wellbeing may be of value. For example, a fairer distribution of wellbeing may be better than a less fair one.

8 For an early, but somewhat controversial, application of expected-value theory to the assessment of risks from nuclear power generation, see Rasmussen, N. C. (chair) 1975, *Reactor Safety Study: An Assessment of Accident Risks in U.S. Commercial Nuclear Power Plants*, WASH-1400, US Nuclear Regulatory Commission.

This strategy is motivated by the thought that some authors have, in other areas, been too quick to move from the thought that a straightforward expected-value approach faces fatal problems to the claim, often implicit, that expected-value theory is not relevant at all to the problem at hand. Insofar as my argument succeeds in showing that, at least in relation to the dual-use problem, one can accommodate the most important objections to a straightforward expected-value approach without giving up *entirely* on expected-value thinking, I hope I will have established that such a swift rejection of expected-value theory is not justified in relation to the dual-use problem.

1. No adequate metric

An initial criticism of the EVP focuses on the need, in that procedure, to evaluate various possible states of the world. It denies that there is any adequate metric of value for making such evaluations.

In order to apply the expected-value procedure, we will need a cardinal metric of value that can be used to evaluate the various states of the world that might follow from different responses to a dual-use problem.[9] Moreover, we will want this metric to assign a value to every state of the world that is taken as an input into the EVP, and to capture all of the morally significant features of each state. Arguably, there is no such metric. One possibility would be to find some cardinal metric of individual wellbeing. We could then sum the wellbeing of all individuals in each state of the world and use this measure of aggregate wellbeing to compare different states of the world. A problem, however, is that it is controversial whether there is a common cardinal scale on which we can measure and sum the wellbeing of different people. Some deny that there is any cardinal metric that allows the wellbeing of different people to be measured in a reliable way,[10] others question whether interpersonal comparisons of wellbeing have any meaning at all, implying that, as a matter of principle, there could be no such metric.[11] In addition, an aggregate wellbeing metric might fail to capture some morally significant features of each state of the world. Some would argue that, holding aggregate wellbeing constant, differences in the distribution of wellbeing across individuals are morally significant. For example, some argue that it is worse if a fixed amount of wellbeing is distributed less equally, justly

---

9   A cardinal metric is one that preserves orderings uniquely up to linear transformations. In a cardinal metric, the interval between two points on the scale has a consistent meaning: it means the same regardless of where those points fall on the scale.

10   See, for example, Jevons, W. S. 1970 [1871], *The Theory of Political Economy*, Penguin Books, Harmondsworth, UK, 'Introduction'.

11   See, for example, Robbins, L. 1935, *An Essay on the Nature and Significance of Economic Science*, 2nd edn, Macmillan, London; Arrow, K. J. 1963 [1951], *Social Choice and Individual Values*, 2nd edn, Yale University Press, New Haven, Conn., p. 9. Robbins allows that interpersonal comparisons of wellbeing have meaning as disguised normative judgments, but denies that they have any descriptive meaning.

or fairly.[12] Similarly, others would argue that, in assessing the value of a given state of the world, it matters not just what distribution of wellbeing it contains, but also how that distribution came about.

2. Empirical and evaluative uncertainty

A second criticism targets both the evaluation of different states of the world and the assignment of probabilities to them.

In actual settings where dual-use problems arise, there will often be significant uncertainty about the probabilities and values of possible outcomes, and indeed about what outcomes are even possible. For example, we may simply lack any good evidence of how likely a project investigating whether to render the H1N1 virus air-transmissible between humans is to succeed in realising that goal. Similarly, we may lack any good evidence of how likely it is that knowledge about such a virus would be used to produce biological weapons, of how likely it is that such weapons would be deployed, and of what the likely effects of their deployment would be. Finally, as an example of *evaluative* uncertainty, we may lack evidence on how much disvalue the effects of a terrorist attack, even if they could be fully specified, would have. Since, when faced with such uncertainty, we will simply have to make imperfect estimates of probabilities and values, our expected-value calculations will often give incorrect results.

Accommodating objections 1 and 2

Though both of these problems are serious and may give us some reason to deviate from the EVP, it can be argued that neither of these concerns gives us a clearly decisive reason to reject the use of expected-value theory altogether.

In response to the first of these concerns, it can be argued that, even if there is no *perfect* value metric for employing in the expected-value procedure, there are nevertheless *adequate* ones. Perhaps there is no cardinal metric of value that captures *all morally significant features* of all of the different states of the world that we wish to compare. But there are cardinal metrics that we could use to evaluate *some* of the morally important features.

One approach, then, would be to simply retain the EVP and employ the best cardinal metric that we have. We might thus fail to capture some morally significant considerations, but we would quantitatively balance as many considerations as it is possible to quantitatively balance.

---

12   See, for example, Temkin, L. 1993, *Inequality*, Oxford University Press, New York; Persson, I. 2008, 'The badness of unjust inequality', *Theoria*, vol. 69, nos 1–2, pp. 109–24.

It might be objected that this strategy would give disproportionate weight to outcomes that can be rated on a cardinal scale at the expense of morally important considerations that cannot be rated in this way. It would allow cardinally measurable considerations to fully determine what course of action to take, with other considerations given no influence at all; however, this problem could be avoided by modifying the EVP so that the expected values that one calculates on the basis of one's cardinal metric are treated as only an *imperfect indicator* of what course of action to pursue. We could proceed by calculating expected values using the best cardinal metric that we have, and then use the values produced as an imperfect indicator of which course of action to take, though we might sometimes wish to choose a course of action associated with lower expected value than another because we believe it has virtues that are not captured by the metric we have used.

Thus, suppose we wish to compare the value of a world in which a robust vaccine for air-transmissible H1N1 is discovered, but there is also a major bioterrorist attack, with a world in which neither of these things happens. I suggested above that the scientist could compare the value of these with states of the world by tallying up the number of lives lost and saved. In fact, there are likely to be better cardinal measures available. For example, she could instead look at the total number of quality-adjusted life-years (QALYs) that would be enjoyed by the whole population in each possible state of the world resulting from each of the available courses of action. This would allow her to capture at least some of the possible effects of her actions on morbidity, as well as mortality. Clearly, the QALYs measure would still be imperfect and would not capture many evaluatively important effects of the two alternatives. For example, it would not capture non-health-related outcomes such as the inconvenience and loss of privacy caused by security measures that would be introduced as a result of a major terrorist attack. But a determination of the expected level of QALYs associated with each course of action nevertheless plausibly provides *some indication* of which of these courses of action the scientist should pursue. The scientist could use expected QALYs as an indicator of whether to pursue the H1N1 research, and then consider whether factors not captured by this measure would justify deviating from the course of action that it suggests. Exactly how these other factors would need to be taken into account would of course need to be determined. It is not clear how we should weigh considerations that cannot be quantified using a shared cardinal measure. But this problem is not specific to attempts to use expected-value theory to respond to dual-use problems. If there are factors that cannot be rated on a shared cardinal measure, *any* proposal for how we ought to respond to dual-use problems will need to determine how these should be taken into account.

The concern about uncertainty can, I think, be accommodated in a similar way. As with the concern about metrics, this worry does not take issue with the basic idea underpinning EVP—the idea that our choices when confronted with dual-use problems should reflect the expected value of the available alternatives. Rather, it points out a practical difficulty with trying to implement this idea. But it remains plausible that *trying* to calculate the relevant expected values would be desirable. We could calculate expected values for the available courses of action using our best estimates of the values and probabilities of the relevant outcomes, insofar as it is possible to assign values or probabilities at all. We could then decide what course of action to pursue on the basis of these estimates of expected value.

It might be argued that this approach would yield implausible results in some cases. It is possible that two courses of action could have the same expected value, though the probabilities and values of the outcomes associated with one course of action are highly uncertain, and the probabilities and values associated with the other course of action are certain. Arguably, in this case one should prefer the latter course of action.

But again, this problem could be accommodated by treating the expected values as only an imperfect indicator of what to do. This would retain some role for expected values, but would allow us to favour a course of action with a lower expected value than another because the other course of action could cause outcomes whose probability or value is less certain.

The concerns about the lack of a cardinal metric and uncertainty suggest that the EVP may not be able to capture all of the considerations that are relevant to how one should act when faced with a dual-use problem: it does not capture outcomes that cannot be evaluated on a common cardinal scale, nor does it capture differences in the certainty of the values and probabilities of the possible outcomes. It is, however, still plausible that we should adopt a procedure that incorporates the most important elements of the EVP. More specifically, it remains plausible that we should conscientiously calculate expected values using the best metric available and take this as at least an imperfect indicator of what to do. Call this the *restricted* expected value procedure or rEVP.

There are, however, two further criticisms that suggest it might be unwise even to adopt the rEVP.

3. Bias

In attempting to identify and assign probabilities and values to different outcomes, agents might have systematic tendencies to neglect or exaggerate some considerations. For example, people might overstate the disvalue of negative outcomes relative to the value of positive outcomes, or they might exaggerate the probability of the most extremely valuable or disvaluable events occurring. In some cases, the direction of a bias may depend on the particular circumstances and psychology of the person(s) applying the rEVP. For example, people may be inclined to relatively exaggerate the (dis)value or probability of outcomes that they have experienced in the recent past, and so are associated with readily available mental images.[13] Or they might be prone to exaggerate the (dis)value or probability of outcomes that will primarily affect them rather than others (self-serving bias).[14] They may also be prone to underestimate the disvalue of harms that affect large numbers of people (scope insensitivity)[15] and susceptible to framing effects in which the way an outcome is presented affects the probability or value assigned to it.[16] Given these biases, the application of the rEVP could be expected to give misleading results in many cases—that is, the expected values that it takes as indicators may often be incorrect.

4. Demandingness

Applying the rEVP comprehensively and conscientiously might require substantial time, effort, expertise and financial resources—all resources that could otherwise be spent on other worthwhile activities. It might also be psychologically burdensome in the sense of requiring those who apply it to override their natural inclinations (suppose that a journal editor strongly committed to academic freedom decides, on the basis of the rEVP, that she must heavily censor a submitted article). Applying the rEVP might, due to the presence of such costs, have negative overall consequences even if it leads decision-makers to make the best choices. The benefits of making the best decisions might be outweighed by the costs associated with the means via which those decisions were reached.

---

13 Lichtenstein, S., Slovic, P., Fischhoff, B., Layman, M. and Combs, B. 1978, 'Judged frequency of lethal events', *Journal of Experimental Psychology: Human Learning and Memory*, vol. 4, no. 6, pp. 551–78; Sunstein, C. 2007, *Worst-Case Scenarios*, Harvard University Press, Cambridge, Mass., p. 6.

14 Babcock, L., Loewenstein, G., Issacharoff, S. and Camerer, C. 1995, 'Biased judgments of fairness in bargaining', *American Economic Review*, vol. 85, no. 5, pp. 1337–43.

15 Desvousges, W. H., Johnson, F. R., Dunford, R. W., Boyle, K. J., Hudson, S. P. and Wilson, N. 1993, 'Measuring natural resource damages with contingent valuation: tests of validity and reliability', in J. A. Hausman (ed.), *Contingent Valuation: A Critical Assessment*, North-Holland, Amsterdam, pp. 91–159; Fetherstonhaugh, D., Slovic, P., Johnson, S. and Friedrich, J. 1997, 'Insensitivity to the value of human life: a study of psychophysical numbing', *Journal of Risk and Uncertainty*, vol. 14, pp. 238–300.

16 Tversky, A. and Kahneman, D. 1981, 'The framing of decisions and the psychology of choice', *Science*, vol. 211, no. 4481, pp. 453–8.

Accommodating objections 3 and 4

The problem of bias is likely to be a problem for many approaches besides the rEVP. Any plausible approach to dual-use problems will require us to identify, evaluate and assess the probability of at least some possible outcomes. Some may permit us to do so in an implicit, subconscious way rather than an explicit, conscious way—for example, rather than adopting the rEVP we could simply have experts make decisions based on their intuitions about the case at hand. This does not, however, necessarily render such approaches less susceptible to bias. Indeed, it may render them more susceptible. The process of assigning values and probabilities may make our biases more obvious and allow us to correct for them. Nevertheless, it may be possible to reduce the risk of bias by deviating from the rEVP. Suppose it were known that people quite generally tend to understate the probability of extremely bad states of the world, even when they are aware of this and attempt to correct for it. We could to some extent compensate for this bias by making some changes to the rEVP. We could require, for example, that the most disvaluable states be given greater weight than other states even if we judge their probability of occurring to be the same.[17]

Similar thoughts apply to the problem of demandingness. Again, this problem is likely to apply to many approaches besides the rEVP; however, it may be possible to reduce the burdens of deciding between alternative strategies by deviating from the rEVP. One approach would be to apply simple heuristics (for example, 'unless there is a clear and present risk of misuse, proceed as if there were no risk'), and resort to the rEVP only in hard cases—for example, cases that are not covered by the simple heuristics, or cases in which heuristics conflict. Alternatively, the problem of demandingness could be mitigated by ensuring that the rEVP is applied by those able to apply it at least cost. For example, it may be less burdensome for a committee of scientific, health and security experts to assess the values and outcomes of synthetic biology research than it is for individual scientists to do the same. Regulatory bodies could be charged with applying rEVP, whilst individual scientists follow a simple heuristic such as 'aggressively pursue your scientific goals unless this violates a regulation'.

Concerns about bias and demandingness may justify deviation from the rEVP. But note that neither justifies a wholesale rejection of the basic ideas underpinning that procedure. In particular, both are consistent with retaining the idea that *the right course of action to take, when faced with a dual-use problem, is (the) one associated with at least as much expected value as any available alternative* (call this the expected-value *criterion* or the EVC). The concerns simply show that the *procedure* by which we make decisions when faced with dual-use problems should not necessarily be to explicitly calculate expected values. Explicitly

---

17   This might, for example, involve applying a multiplier (> 1) to the most negatively valued outcomes.

attempting to satisfy the expected-value criterion might not be the best way of in fact satisfying that criterion, or might involve costs that outweigh the benefits of satisfying it.

This distinction between *criteria of rightness* and *decision procedures* has long been emphasised by consequentialists when presented with concerns about demandingness and bias and other related objections. Consequentialists have argued that their theory provides an abstract description of which acts are right and which are not. It does not provide a procedure via which to decide how to act.[18] In fact, consequentialism is consistent with adopting some other ethical theory as a procedure for guiding decisions. In an ideal world where we responded to our evidence in a perfectly rational and costless way, we *could* simply apply consequentialism as a decision procedure. But the actual world is not like that. In the actual world it may, according to consequentialists, be best *not* to make decisions by explicitly calculating which act will have the best consequences.

Drawing a distinction between criteria of rightness and decision procedures allows us to retain the basic idea underpinning EVP (and thus rEVP) while deviating from these procedures. More importantly, it provides us with a higher standard—the EVC—against which we can measure alternative decision procedures. Other things being equal, we should prefer a decision procedure that comes closer than some alternative procedure to recommending the courses of action that are right according to the EVC—the courses of action associated with the highest expected value. In an ideal world, the best decision procedure would be the EVP or rEVP. In the actual world, it may be some amendment of these procedures, or even a wholly different approach. Either way, it is plausible that we should keep the EVC in mind as a standard by which to judge competing decision procedures.

I now turn to consider two criticisms that suggest that even the EVC will need to be rejected.

5. Rational risk aversion

The expected-value procedure, and the criterion that underpins it, is blind to the distribution of value across possible states of the world. Suppose that a

---

18   See, for example, Austin, J. 1954 [1832], *The Province of Jurisprudence Determined*, H. L. A. Hart (ed.), Weidenfeld, London, p. 108; Mill, J. S. 1985 [1861], 'Utilitarianism', in *Collected Works. Volume X*, John M. Robson (ed.), University of Toronto Press, Toronto, ch. 2, pp. 203–60; Sidgwick, H. 1907, *The Methods of Ethics*, 7th edn, Macmillan, London, p. 413; Bales, R. E. 1971, 'Act-utilitarianism: account of right-making characteristics or decision making procedure?' *American Philosophical Quarterly*, vol. 8, pp. 257–65; Parfit, D. 1984, *Reasons and Persons*, Clarendon Press, Oxford, pp. 24–9, 31–43; Railton, P. 1984, 'Alienation, consequentialism, and the demands of morality', *Philosophy and Public Affairs*, vol. 13, pp. 134–71, at pp. 165–8. The terminology 'decision procedure' and 'criterion of rightness' is due to Bales, op. cit.

scientist is considering whether to publish a scientific finding that has a 50 per cent chance of being used in a good way and a 50 per cent chance of being used in a bad way. Suppose further that the value of the good outcome is 100, and of the bad outcome is −100. Then there will a 25 per cent chance of realising a state of the world with a value of 100 (good outcome, no bad outcome), a 25 per cent chance of realising a state with a value of −100 (bad outcome, no good outcome), and a 50 per cent chance of realising a world with a value of zero (because both the good outcome and the bad outcome occur or neither occurs). Suppose that the scientist does not publish the result: there is a 100 per cent probability that neither the good nor the bad outcomes will occur. In this case, the expected value associated with publishing the result is zero, as is the expected value associated with not publishing it. Thus, the EVC will hold that both courses of action are right. But, arguably, the right thing to do in this case is to abstain from pursuing the project since this alternative is associated with lower (indeed zero) risk: if the scientist abstains from the project, a world with a value of 0 will obtain with 100 per cent probability, whereas if she pursues the project there is an expected value of zero, but a significant risk of winding up in a world whose value is −100. If we ought to be risk averse, as some have argued,[19] we should reject EVC in favour of an alternative criterion that penalises riskier alternatives.

6. Agent-relativity

The EVC is also blind to the way in which an agent posed with a dual-use problem contributes to a good or a bad outcome; it focuses only on the probability and value of the outcome. But, according to many nonconsequentialist ethical theories, an agent's relation to an outcome is also important. Consider a case in which a scientist performs and publishes some piece of research in synthetic biology that is intended to develop a new cure for cancer but could also be misused in unjustified biowarfare. It might be argued that this possible negative outcome should be discounted relative to at least some of the other outcomes because

1. It is not caused by the scientist, but is merely allowed to occur. In contrast, at least some positive outcomes of the research—for example, the production of intrinsically valuable knowledge—might be said to be *caused* by the scientist.[20]

2. It is not intended, but is merely foreseen, by the scientist. By contrast, the possible positive outcome of developing a cure for cancer is intended.[21]

---

19 See, for example, Hansson, S. O. 2003, 'Ethical criteria of risk acceptance', *Erkenntnis*, vol. 59, no. 3, pp. 291–309.

20 This claim could be grounded on the doctrine of doing and allowing.

21 This claim could be grounded on the doctrine of double effect. See Suzanne Uniacke's Chapter 10, in this volume.

3. The occurrence of the bad outcome depends on the subsequent immoral actions of another agent: the person who actually engages in unjustified biowarfare. Thus, even if the scientist's actions resulted in unjustified biowarfare, the scientist would not be the primary wrongdoer, but would at most be an *accomplice* to the wrongdoing. By contrast, some other outcomes of the work, such as the production of intrinsically valuable knowledge, may occur without requiring the intervention of other moral agents. The scientist might be said to be the principal agent implicated in bringing about these outcomes.

Accommodating objections 5 and 6

If these criticisms are well founded, the EVC should be rejected since the right strategy to adopt when faced with a dual-use problem will *not* necessarily be the one that offers the greatest expected value. It may, for example, be the one that minimises *risk*, or that minimises the likelihood of *intentionally causing* harm.

I am not in a position to assess the arguments for and against agent-relativity or the rationality of risk aversion here. But I will briefly consider how a proponent of the EVC might seek to accommodate these criticisms. She might argue that, even if we accept agent-relativity and that it is rational to be risk averse, it may be that the EVC still serves as a useful starting point—as a default position from which alternative criteria and procedures for assessing responses to dual-use problems might be derived. Thus, we would assume that all equally likely and equally valuable outcomes should be given the same weight *unless* some sound nonconsequentialist argument could be supplied for prioritising or discounting outcomes to which we bear a certain relationship. Similarly, we would take a risk-neutral approach unless a sound argument for risk aversion could be found.

One reason to take an expected-value-based approach as a starting point is that it captures certain core considerations while taking all other possible factors to be irrelevant until proven otherwise: everyone should agree that in confronting dual-use problems, it matters what good and bad outcomes might result, how likely they are, and how good or bad they are. An expected-value approach captures the relevance of these considerations. Some might argue that other factors, such as risk and the agent's relationship to the good and bad outcomes, are also important; but this would be controversial, so it might seem appropriate to adopt, as a default position, a decision procedure that does not take them into account.

A second reason to adopt an expected-value approach as a point of departure is that the formal framework of expected-value theory may provide a helpful tool for thinking about other approaches as well. As we saw above, the expected value of a gamble over a number of different states of the world can be thought of as a weighted average of the values of those states, with the weights given by their probabilities. But we could also weight the values using further factors to accommodate agent-relativity or rational risk aversion. Consider a nonconsequentialist approach that takes into account the *intentions* of an agent posed with a dual-use problem as well as the consequences of her decision. It may be helpful to think of such an approach as a variant of the expected-value approach that weights the values of different possible states of the world according to both their probability *and* whether they were intended.[22] Adopting such a formalised approach may help to ensure that we are clear and explicit about precisely *how much* the intentions of the agent matter.

Conclusions

I have outlined an approach to dual-use problems that involves: 1) identifying possible outcomes of each alternative course of action; 2) identifying possible states of the world that might result, each consisting of a combination of these outcomes; 3) explicitly assessing the values and probabilities of these states of the world; 4) using these probabilities and values to calculate the expected value associated with each alternative course of action; and 5) selecting the alternative with the highest expected value.

This expected-value procedure is attractive because it captures some commonsense intuitions about how to respond to risks and benefits, is consistent with a major school of ethical thought (consequentialism), and has a track record in government and corporate decision-making. It is, however, also susceptible to several criticisms, which may justify deviation from it. Criticisms adverting to the lack of an adequate metric for ranking states of the world and to empirical and moral uncertainties suggest that we may need to regard expected values as providing only an imperfect indication of which course of action to pursue. Criticisms appealing to bias and demandingness suggest that, even if the right course of action to take when faced with dual-use problems is always the one that maximises expected value, it may be best to adopt a decision procedure that does not involve calculating expected values at all. Finally, criticisms based on agent-relativity and the rationality of risk aversion go even further: they suggest that in some cases the right course of action will *not* be the one that maximises expected value.

I have not assessed whether these criticisms are persuasive. But I have tried to show that, even if they are, there may be some value in retaining the expected-

---

22 The expected value assigned to an alternative would then be relative to the agent making the decision between alternatives; it would no longer be a value that could be ascribed from any standpoint.

value criterion as a default position from which deviations must be justified. Thus, there may remain an important role for expected values in thinking about dual-use problems.

Further questions

Suppose that we do keep the EVC in mind, at least as a starting point for further discussion. Two important further questions arise.

First, what outcomes should be included among the positive and negative outcomes of a course of action? For example, should the production of knowledge be included as a positive outcome independent of any positive applications of that knowledge? This will depend on the controversial question of whether knowledge has any intrinsic value.

A second question is, to the extent that *anyone* should ever explicitly assess the expected value associated with a particular strategy for preventing misuse, *who* should do it? For example, should the decision be made by some form of expert committee, or should it be made by a political or representative body?

I leave these questions as potential topics for future discussion.

10. The Doctrine of Double Effect and the Ethics of Dual Use

Suzanne Uniacke

Is the doctrine of double effect (DDE) relevant to dual-use dilemmas? Can consideration of the DDE make a significant contribution to the ethics of dual use? Several writers assume that dual-use dilemmas are instances of double effect.[1] Certainly this particular connection is strongly suggested by the terms 'dual use' and 'double effect' and also by the structure of dual-use dilemmas. A dual-use dilemma is said to arise when an action or activity such as research in the life sciences or publication of that research can have both good and bad effects: alongside its intended good outcome, for example, the enhancement of knowledge or human improvement, there can also be a foreseeable bad effect— for example, assisting in the production of bioweapons. Since the doctrine, or principle of double effect as it is sometimes called, is specifically intended as a guide to decision-making in ethically difficult cases where an action or course of action with an intended good effect can also produce a foreseen bad effect, it is natural to consider whether the moral guidance that the doctrine offers in cases of double effect is also applicable to dual-use dilemmas.

In order to answer the questions posed above we need an understanding of the DDE and its intended role in practical ethics. I provide these in the next section of the chapter. According to the account of the DDE that I shall outline, dual-use dilemmas are not paradigm instances of double effect. Nonetheless, as I go on to explain, because dual-use dilemmas have significant features in common with typical instances of double effect, some important considerations that can arise in relation to the DDE and its practical application are important to providing a satisfactory account of the ethics of dual use.

The doctrine of double effect

The DDE is concerned with the application of normative moral theory to practical moral problems. It takes its rationale from the view that certain actions are morally objectionable because of the types of actions they are. Those who invoke the DDE in particularly difficult cases hold that what an actor intends— that is, what she aims to achieve by her action—can itself be highly significant

---

1 See, for example, Briggle, A. 2005, 'Double effect and dual use', in C. Mitcham (ed.), *Encyclopaedia of Science, Technology and Ethics. Volume 2*, rev. edn, Macmillan, London, pp. 543–6.

to the morality of what she does. This view can take absolutist and non-absolutist forms. In its absolutist form, it holds that certain types of actions, such as intentionally killing an innocent person, are intrinsically wrong and always morally impermissible (absolutely prohibited). A non-absolutist version maintains a very strong moral constraint against certain types of actions such as those that involve the intended killing of an innocent person; on this view, such actions are always intrinsically morally objectionable even if in extreme circumstances they might be necessary in the absence of any morally acceptable alternative.[2]

Not all prominent contemporary normative moral theories regard an actor's intention as directly morally significant in this way. An entirely outcome-oriented (consequentialist) theory such as utilitarianism, for instance, does not share the assumptions on which the DDE is founded and thus regards the DDE as irrelevant to the moral evaluation of actions that have both good and bad effects (including instances of dual use). Furthermore, some of those who accept or are sympathetic towards the moral assumptions behind the DDE nevertheless regard the DDE itself as problematic in some respects. There is an extensive critical philosophical literature on issues surrounding the DDE, which obviously cannot be explored here.[3] Since the principal purpose of this chapter is to address the relevance of the DDE to the ethics of dual use, I shall assume as a modus operandi that there is a morally significant difference between a bad effect of an action that the actor intends and one that she foresees but does not intend and that this distinction can be directly relevant to the moral evaluation of some actions that have both good and bad effects. This assumption can be disputed but it represents a widely accepted view.

A key to understanding the nature and purpose of the DDE can be found in its origins. St Thomas Aquinas appealed to a distinction between an intended, as opposed to a foreseen, effect of an action in his explanation of why it can be permissible to kill another person in self-defence.[4] Homicide in self-defence posed a problem for Aquinas because he held the more general position that it is never permissible for a private person to engage in intentional killing. Does this view imply that I cannot legitimately use lethal force on an unjust attacker if it is necessary to save my own life? In responding to this question, Aquinas claimed that the act of fending off an attacker can have two effects: a good effect (saving one's own life), which the self-defending actor intends, and also

---

2   This latter position is sometimes referred to as threshold deontology. See Alexander, L. and Moore, M. 2007, 'Deontological ethics', in *Stanford Encyclopedia of Philosophy*, Stanford University Press, Stanford, Calif., <http://plato.stanford.edu/entries/ethics-deontological/> (viewed 6 February 2012).

3   See, for example, McIntyre, A. 2009, 'Doctrine of double effect', in *Stanford Encyclopedia of Philosophy*, op. cit., <http://plato.stanford.edu/entries/double-effect/> (viewed 6 February 2012).

4   Aquinas, St Thomas 1966, *Summa Theologiae*, vol. 38, Blackfriars edn, Eyre & Spottiswood, London, 2a, 2ae, 64, article 7.

a bad effect (the attacker's death), which the actor can foresee but need not intend. Thus, Aquinas maintained, homicide in genuine self-defence does not contravene an absolute prohibition against intended killing.

The distinction between an action's intended effects and those effects that the actor foresees but does not intend is central to what later came to be known as the doctrine of double effect. Subsequent to Aquinas's discussion of homicide in self-defence, the DDE was developed as part of natural-law reasoning about the morality of a range of actions that can have both good and bad effects, where the bad effect is something that it would be wrongful to intend. (Those who invoke the DDE in cases of double effect need not share Aquinas's view that homicide in self-defence against an unjust attacker is always unintended killing or that it must be justified as such.[5] In contemporary applications of the DDE the foreseen bad effect in question is usually the death of an innocent person.) In its traditional form, the DDE is now a general principle of practical ethics that holds that in some circumstances and under specific conditions it is morally permissible to cause a foreseen bad effect of a type that is always morally wrong to intend. These conditions are

1. the act itself must be morally good or at least indifferent

2. the agent must not positively will (intend) the bad effect

3. the good effect must be produced directly by the action, not by the bad effect

4. the good effect must be sufficiently desirable to compensate for the bad effect.[6]

Although the DDE has its origins in Thomistic and natural-law ethics it has been taken up more widely as part of secular moral thinking by those who maintain that whether or not an actor intends to bring about a particular bad outcome, such as an innocent person's death, is itself a morally significant feature of her action that can make a difference to the morality of what she does. For instance, the DDE is frequently invoked in a number of prominent contemporary contexts, most notably in relation to the ethics of war and to issues of medical ethics. These applications include the foreseen killing of noncombatants as an incidental effect of aiming at a military target, the use of triage in hospital emergency rooms and on battlefields, some cases of risky surgery, and the administration of increased doses of pain relief that suppress respiration.[7] It is perhaps worth pointing out here that within its natural-law context, the DDE was developed

5   For a critical discussion, see Uniacke, S. 1994, *Permissible Killing: The Self-Defence Justification of Homicide*, Cambridge University Press, Cambridge, ch. 4.
6   A fuller statement can be found in the *New Catholic Encyclopedia*, 1967, vol. 4, McGraw-Hill, New York, pp. 1020–2.
7   See Uniacke, S. 2007, 'The doctrine of double effect', in R. Ashcroft et al. (eds), *Principles of Heath Care Ethics*, 2nd edn, Wiley & Sons, Chichester, UK.

and is characterised as a guide to decision-making in morally difficult cases where an action will have both good and bad effects; however, many recent secular philosophical discussions seem to assume that the DDE's plausibility depends on its ability to give a definitive answer about morally permissible action in every conceivable problem to which it might be applied.

As noted earlier, the DDE derives from the view that irrespective of their overall effects on the world, some actions are morally objectionable as the types of actions they are—for instance, as instances of intentional killing. There are obvious practical difficulties for this view, especially (but not exclusively) in circumstances of emergency where the alternative actions available to an agent are severely restricted. Consider the following example. My car's brakes fail on a hill and I need to swerve to the left in order to avoid oncoming traffic and also some children who are on a pedestrian crossing in front of me. If I swerve to the left I will hit and could seriously injure or kill a pedestrian who has just stepped out from the curb. Is it permissible to swerve the car in these circumstances? (Since I will hit oncoming traffic and the children if I do not swerve, if it is not permissible to swerve, will I act impermissibly whatever I do?) In this particular example the DDE says it is permissible to swerve. My swerving the car is not in itself a morally impermissible type of act. (The morality of swerving a car depends on the circumstances and on why I swerve.) If I swerve the car to the left in these particular circumstances, this would have two distinguishable, independent effects: an intended good effect (avoiding oncoming traffic and the children) and also a foreseen bad effect (hitting a pedestrian) that is strictly incidental to what I intend to do in swerving the car.

According to the DDE, a foreseen effect is strictly incidental to the actor's intention if it is not part of what the actor aims to achieve either as an end or as a means of achieving an end in the circumstances.[8] If at the time of swerving the car I were to ask myself whether my intention would be fulfilled if, against expectation, I happen not to hit or kill the pedestrian, I can honestly answer yes. The presence and position of the pedestrian and any harm he might suffer are irrelevant both to my swerving to the left in order to avoid oncoming traffic and the children and also to how I achieve that aim in the circumstances. (It would be misleading not to point out here that there are vexed philosophical issues about intention and its moral significance in relation to the DDE.[9] To address those deeper issues would take us too far into a discussion of the DDE itself. Suffice to say that despite these critical issues there are cases of double

---

8 Incidental effects of an action are not necessarily bad or unwelcome. Consider, for example, taking on hard manual labour solely in order to earn money and becoming physically very fit as a result.
9 See, for example, Kamm, F. M. 2000, 'The doctrine of triple effect and why a rational agent need not intend the means to his end', *Proceedings of the Aristotelian Society, Supplementary Volume*, vol. 74, no. 1, pp. 21–39; McIntyre, A. 2001, 'Doing away with double effect', *Ethics*, vol. 111, no. 2, pp. 219–55; and McIntyre, 2009, op. cit.

effect, including the example just described, where a foreseen bad effect of an action is clearly incidental to what the actor aims to achieve in acting as she does.) While the actor's intention is central to the guidance offered by the DDE it is not the sole consideration. For an act of double effect to be permissible the intended good effect must also be sufficiently morally weighty in comparison with the foreseen bad effect to warrant causing the bad effect. The DDE's fourth condition requires a judgment of proportionality. This particular condition could permit my swerving the car in order to avoid hitting a greater number of innocent people in the circumstances described above, but it would not permit, for example, my swerving a car in the direction of a pedestrian in order to avoid hitting a dog or damaging property.

An obvious question is why we should bother with a distinction between intention and foresight in circumstances such as these. Why not simply consider something like the DDE's fourth condition and compare the probable outcomes of the available alternatives (for example, to swerve or not to swerve) and say that someone who is faced with making such a decision should act so as to save as many innocent people as possible? Those who think that the DDE is morally significant in such cases would reply that if we simply appeal to a principle that tells us to save as many innocent people as possible and do not *also* invoke a distinction between an effect that an actor intends as opposed to an effect that she (merely) foresees, we commit ourselves to killing some innocent people *as a means* of saving others, and this is morally unacceptable. It is one thing that I put the life of a pedestrian in grave danger in swerving the car in order to avoid hitting a greater number of people; it would be another thing to cause the death of innocent person as a means of preventing harm to a greater number by, for example, using an innocent person as a human shield.

A related point is that according to the DDE, actions of double effect that meet its four conditions are morally permissible. (They are not necessarily the right thing to do all things considered or what the actor is morally required to do.) If under the fourth condition the foreseen bad effect is disproportionate in relation to the intended good effect then the action under consideration is held to be impermissible. The fourth condition does not represent an all-things-considered judgment based on the action's predicted or actual *overall* outcome. This point of clarification is important in the context of a discussion of the application of the DDE to the ethics of dual use because dual-use dilemmas are often explicitly discussed solely in cost/benefit terms, as being entirely a matter of whether the overall likely benefits of an activity's good use can be 'traded off' against and would outweigh the risks of its malevolent or negligent use. As with actions of double effect, however, the basic question posed by a 'dual-use dilemma' is, I take it, whether it would be morally permissible to engage in the activity in question given the risks of its misuse, and not whether it would be morally right or morally obligatory to do so.

Double effect and dual use

It is easy to see why one might regard dual-use dilemmas as instances of double effect. A dual-use dilemma arises when an action or activity has both an intended good use and a possible malicious or negligent use. Structurally this looks very much like a case of double effect; however, dual-use dilemmas are also dissimilar to paradigm instances of double effect in a number of respects. For instance, in the above example my decision whether to swerve the car is a forced choice in circumstances in which I have little time for deliberation and extremely limited options. This is the situation in many instances of double effect whereas dual-use dilemmas are usually not like this. But we can set this particular dissimilarity aside; although the DDE is often applied to situations of forced choice where the actor has little time for reflection or practical manoeuvre, this is not necessary to double effect, and very familiar applications of the DDE in medical contexts and in war do not display these features.

Other dissimilarities between double effect and dual use are, however, more significant. In typical instances of double effect the actor foresees the bad effect as morally certain or highly probable; the bad effect is usually also an unavoidable and a direct effect of the action that produces the good effect. (This is true of my swerving the car in the above example, for instance.) These particular features have implications for what is required under the DDE's four conditions, fuller statements of which say that in coming to a judgment under condition (4) a foreseen bad effect that is morally certain to occur should weigh more heavily (ceteris paribus) than one that is merely probable; they also specify that if the actor could obtain the intended good effect without also producing the foreseen bad effect he should do so. (A reasonable interpretation of this latter directive will inevitably involve considerations of comparative cost and risk, but it clearly implies that, for example, if I swerve the car I must try to warn the pedestrian of the danger by, say, tooting the car's horn.) By contrast, an actor facing a dual-use dilemma who foresees a bad use as a possibility is often unable to ascertain the probability of its occurring; this is partly because the bad use, if it occurs, is indirect and dependent upon the further agency of another person.

The resolution of dual-use dilemmas must obviously address the extent to which an actor can reasonably foresee the harmful use of something that she intends for a good purpose. Such judgments are notoriously difficult, particularly because they involve the actions of others, and this epistemological difficulty is prominent in the literature on the ethics of dual use in relation to the so-called precautionary principle.[10] What is less often highlighted is that dual-use dilemmas also involve the complex question of the *moral* significance of another person's further agency in relation both to the degree to which I ought to take this into account in deciding what it is permissible for *me* to do and to the degree to which by acting in a certain way I am morally obliged to try to prevent the further agency of another person or its bad effects. Above I have brought out this moral complexity by identifying a number of dissimilarities between dual-use dilemmas and typical cases of double effect. Might considerations that inform the DDE and the debate surrounding its application nonetheless assist us in reasoning about this moral issue in relation to dual use? I will suggest in the next section that they can.

Prior to that discussion, as a crucial first step, we must ask whether the DDE is applicable to cases in which the foreseen bad effect is due to the agency of another person. (If it is not then dual-use dilemmas are not instances of double effect.) I see no reason in principle why it cannot be. Consider a modified version of the earlier example in which my car's brakes fail, but where I will hit the pedestrian only because when I swerve to the left he will be jostled into the path of my car by someone else. We can also think about the DDE in relation to a case of duress. Say the Gestapo threatens to torture or kill an innocent hostage unless I tell them the whereabouts of a Jewish family in hiding. In resisting this threat I foresee but I do not intend the injury that the Gestapo will inflict on an innocent hostage, even if I could prevent it by giving in to the threat. In a third type of case, we might apply the DDE to my decision to work as a bartender, where in taking up such a position I foresee that despite taking due care, at some future point I will almost certainly serve alcohol to someone who will then drive while intoxicated and possibly injure or kill someone.

If the DDE is indeed applicable to examples such as the three just described, in which a foreseen bad effect is due to the agency of another person, we cannot conclude simply on this basis that the DDE is therefore applicable to dual-use dilemmas. This is because there are also various respects in which each of these three examples differs structurally from a dual-use dilemma. For instance, in the first, modified car-swerving example, although the bad effect is contingent upon the intervening agency of another person (who jostles the pedestrian into the path of my car), nonetheless I will *directly* hit the pedestrian by swerving

---

10   This is a general principle that states that if an action or activity has a suspected risk of causing serious harm to the public or to the environment, if experts disagree about whether the action or activity will or is likely to cause this harm, it is up to those proposing to undertake the action or activity to prove that it will not cause this harm. The implication of the principle is that without such proof they should not proceed.

my car.[11] In the second, duress example, the act in question (my refusal to reveal the family's whereabouts to the Gestapo) is a *negative action*, as opposed to something that I do; furthermore, even if I could protect the innocent hostage by telling the Gestapo what they want to know, the injury inflicted on the hostage will be *wholly* due to the actions of another agent (the Gestapo). In the third, bartender example, the action in question is my assumption of a longer-term *activity*, as opposed to an individual action of double effect. Nonetheless, the third example is probably structurally the closest to an instance of dual use of the three examples, in that it involves positive action on my part with a foreseen bad effect that is both indirect and due to the agency of another person. More importantly, the third example also shares a crucial feature with dual-use dilemmas that is missing in the second, duress example, but is present in the first, modified car-swerving example. And this is that in the circumstances as they actually occur, my positive action plays a specific causal role in bringing about the bad effect—namely, it *enables* or *aids* another person to do something wrongful (to jostle the pedestrian into the path of my car/to drive while intoxicated/to create a bioweapon). (In the second example, on the other hand, my refusal to give in to the Gestapo's threat does not enable them to torture or kill the innocent hostage: it does not provide them with the means.) For this reason, in the first and the third examples the description of the foreseen bad effect of my positive action includes *enabling* or *aiding* the wrongful action of a third party, and this is also true of dual-use dilemmas where what I do (for example, publish research) enables or aids its malicious or negligent use.

The fact that in each of these cases the further actions of another agent are properly regarded as a foreseen bad *effect* of my action presses the following question: under what conditions am I responsible for the actions of another person in the sense that I must regard what he or she does as an effect of what I do for which I can be accountable? This is a broad and complex question that clearly cannot be addressed in detail in this chapter. Nevertheless, in the next section I shall identify considerations relevant to the DDE that can also shed important light on how we should answer this question in the case of dual-use dilemmas.

Responsibility for a foreseen bad effect

The DDE does not say that a person is not responsible for the foreseen bad effect of his action provided this effect is unintended and proportionate to an intended

---

11   Whether we would regard the jostling as (merely) part of circumstances in which I swerve the car, as opposed to a primary cause of the pedestrian's being hit, could depend on whether it was accidental or inadvertent (simply a result of a crowded footpath, for instance), as opposed to an intended, negligent or malicious action.

good effect. On the contrary, according to the DDE the actor is responsible for the bad effect in an important sense—namely, that she must take it very seriously into account in deciding how to act and she must morally account for having brought it about or not prevented it when she might have done otherwise.[12] As an upshot of this responsibility for the bad effect, the actor must try to achieve the good effect without the bad if she can possibly do so. It might uncritically be assumed that this type and degree of responsibility for the bad effect are due to the fact that in typical cases of double effect although the bad effect is unintended it is both a foreseen and a direct effect of what the actor chooses to do. Certainly these features strengthen the case for attribution of responsibility for the bad effect. But they are not necessary to such attribution. In the first (modified car-swerving) and the third (bartender) examples discussed earlier, the foreseen bad effect of my action incorporates the agency of another person, and this is something that I must take into account in deciding how to act and for which I can be called to account. To be sure, in the first example, the bad effect is directly and strongly connected to my swerving the car, but the connection is less direct in the third example in which my serving someone alcohol might be a sufficient but not a necessary condition of his being intoxicated and also an indirect cause of his driving in this condition. In both these examples, attribution of responsibility for the bad effect to my own agency is due in large part to the enabling role that my positive action plays in the further agency of another person. And this will also be true of cases of dual use.

At this point I would like to suggest that I am responsible for the bad effect of another person's further agency, in the sense that I must take it into account in deciding how to act and I can be accountable for it (and will bear a degree of culpability for it in the absence of justification or excuse) if at the time of my own prior action: 1) I can reasonably foresee that another agent will bring about this effect; and 2) in acting as I do I provide someone else with the means or the opportunity to bring about this effect.[13] Discussions of dual-use dilemmas tend to focus on the first of these two conditions—that is, on the extent to which another person's further agency is reasonably foreseeable and on the probability of the suspected bad effect actually occurring. In drawing attention to the significance of the second condition, I hope to emphasise that where these two conditions are met, although the bad effect is indirect and due to the agency of

---

12   Adherents to the DDE regard our obligation to prevent harm as generally less stringent than our obligation not to do harm. Nonetheless, they also hold that it can be wrong not to prevent harm to an innocent person, particularly where this is an intended effect of one's inaction. It can be important to the permissibility of some instances of non-prevention of harm (for example, some decisions not to rescue) that the harm although foreseen is not intended.

13   Note these two conditions say 'if' and not 'only if'. Some people would also regard me as responsible in this sense for what the Gestapo does to the innocent hostage in the second example. But a person's responsibility for a foreseen outcome of her negative action is a much more contentious issue, especially where it also involves the further agency of someone else, and needn't be taken up here.

another person, my contribution to bringing about the bad effect is *strong* in virtue of the fact that I both *foresee* and *enable* it. Under these two conditions, I am responsible for the bad effect in the robust sense outlined above. Moreover, responsibility in this sense also generates an *agent-related* obligation on my part to try to prevent the bad effect if possible. Although the bad effect, if it occurs, will be an indirect effect of my action that is due to the further agency of someone else, because it is nonetheless something that I have enabled or aided, my obligations in this regard are stronger than any general obligations one might have to prevent harm or wrongdoing by others.

Concluding remarks

I have maintained in this chapter that dual-use dilemmas are not paradigm instances of double effect and I have identified significant similarities and dissimilarities between the two. All the same, our critical thinking about whether the DDE and its conditions are indeed relevant to dual-use dilemmas has served to highlight some morally important aspects of the ethics of dual use that might otherwise not receive sufficient critical attention. These include the relevance of a distinction between foresight and intention to the moral permissibility of an action with both good and bad effects; the conditions under which we are accountable for the further agency of other people who bring about a bad effect; and the relatively stringent agent-related obligation to prevent harm or wrongdoing by others where one's action would provide them with the means or the opportunity. A suitably morally complex ethical evaluation of dual use needs to address these various issues. But in a more sceptical vein we can ask whether we need to refer to the DDE in order to do this.

The DDE is centrally concerned with the moral relevance of a distinction between intention and foresight, since this distinction is held to be crucial to permissible action in cases of double effect. It is clear in the case of a dual-use dilemma that the foreseen bad use is *not* intended by the agent who faces the dilemma (although it may well be intended by the further agent who engages in the bad use). According to the DDE, the fact that an actor does not intend a foreseen bad effect of her action can be highly significant to the moral permissibility of what she does. This then will also be a consideration in deliberation about the permissibility of activities that are potentially dual use. The DDE also emphasises that in cases of double effect the fact that the bad effect is foreseen is itself morally significant to the permissibility of the action. The actor must take the bad effect seriously into account in her deliberations and her action must be justified in terms of a good effect that is sufficiently morally important to warrant causing the bad effect. And as I have argued above, this can be so in cases of double effect where the bad effect will be indirect and due to the further agency of another person. Furthermore, we can be responsible in the sense of being accountable for the harmful or wrongful actions of other people if by our own actions we have enabled or aided them. This means that our obligation as agents not to do harm extends to the obligation not to provide others with the means of doing harm; while not an absolute obligation, it is nonetheless a strong one. It also means that our obligation to prevent harm being done by others becomes a more stringent, agent-related obligation when the harm in question is something that we ourselves have provided others with the means of doing. These obligations are particularly salient to the ethics of dual use.

11. Unknowns in Dual-Use Dilemmas

Michael Smithson

Dual-use dilemmas are defined as a consequence of the potential for the same piece of research to be used for harm and for good. Miller and Selgelid[1] advise that 'fine-grained ethical analyses of dual-use research in the biological sciences would seek to *quantify* actual and potential benefits and burdens, and actual and potential recipients/bearers of these benefits and burdens. These analyses would also identify a range of salient policy options.' Desirable as such quantification may be, the path to it is obstructed by several yawning abysses in the form of unknowns. If unresolved or ignored, these unknowns can render fine-grained analysis and quantification impossible or arbitrary.

This chapter investigates these unknowns and presents some approaches for dealing with them or, at least, taking them into account. These approaches are grounded in subjective expected utility (SEU) theory, whose primary tenet is that 'rational' agents weigh up the potential consequences of acts by summing the products of the probability of every possible outcome and its utility. At least some of the probabilities and utilities might be based on subjective assessments, whence the 'S' in SEU. SEU is employed here as a prescriptive or benchmark framework. My primary intent is to ask what an SEU-rational agent would conclude or choose, so that human decision-makers can knowledgeably decide whether to take the agent's advice on board or reject it.

Our survey of unknowns comprises three sections

1. dilemma structures

2. state space indeterminacy

3. imprecision and biases in judgments.

The first section examines dual-use dilemmas from the viewpoint of the standard social dilemmas framework. The primary purpose is to ascertain when a dual-use dilemma is a mixed-motive game and therefore a genuine dilemma, and when it is a trade-off. Dilemmas pose difficulties for rational self-interest that trade-offs do not. The second section begins with the observation that dual-use dilemmas often are not limited to considering just two possible uses and may instead involve an indeterminate number of uses. Likewise, the number of response options also may be a matter of choice. In other words, the use and response of

---

1   Miller, S. and Selgelid, M. 2007, 'Ethical and philosophical consideration of the dual-use dilemma in the biological sciences', *Science and Engineering Ethics*, vol. 13, p. 542, emphasis in the original.

state spaces are indeterminate. Both state spaces have consequences for decision-making and these are elaborated in this section. The third section begins by pointing out the dangers in restricting judged probabilities and utilities to falsely precise representations and describing the decisional consequences of imprecision. It then brings in psychological considerations such as tendencies towards overconfidence in predictions and confirmation bias.

The discussion of these issues involves a bit of mathematics and some technical definitions. Both the mathematics and technicalities are necessary, and these are far from merely 'academic' matters. Dilemmas were not well understood until a mathematical framework was developed for describing them and distinguishing them from non-dilemmatic trade-offs, and that development transformed the economics of public goods and common-pool resources as well as our ideas about individual versus collective rationality. It also provided heretofore unachievable insights into the evolution of cooperation, and has spawned a vast literature in economics, political science and psychology. Likewise, not knowing what all the possible outcomes are is commonplace in real-world decision-making in the face of a largely unknowable future—and is routinely ignored by decision-makers and standard decision frameworks. A systematic consideration of the consequences thereof includes asking which options (and how many) should be 'on the table' for responding to dual-use dilemmas. The framework presented in this chapter has been applied to debates about this issue in law, medicine and related policy matters. Finally, a careful assessment of the consequence of uncertainty requires understanding that there are different kinds of uncertainty, with distinct consequences for reasonable decision-makers. A growing literature on this topic includes demonstrations of its practical impact in domains such as insurance.

One reviewer of an earlier draft of this chapter declared that I indulged in 'make-believe' that mathematics can contribute to our understanding of dual-use dilemmas and what to do about them. I have three rebuttals to this dismissive remark. First, as I hope the previous paragraph has made abundantly clear, the mathematics in this chapter are about as far from 'make-believe' as it is possible to be; they concern some of the most important uncertainties in real-world decision-making. Second, without even the simple mathematics that informs this chapter, it is nigh well impossible to properly understand and deal with those uncertainties; words simply will not suffice. Finally, failing to distinguish a trade-off from a dilemma, failing to take into account the fact that we do not know what new technologies will be invented and yet often underestimate the likelihood of their emergence, not bothering to think carefully about whether we need two, three, 10 or 1000 alternative ways of responding to dual-use dilemmas, and ignoring the ways in which we are systematically biased in our judgments about unknowns—now *that* would be genuine 'make-believe' indeed.

Dilemmas or trade-offs?

When are the dilemmas actually social dilemmas, as opposed to trade-offs? Genuine social dilemmas are harder to resolve than trade-offs. They also present a fundamental difficulty for rational, self-interested agents because the pursuit of self-interest in a social dilemma leads to the destruction of the common good. Moreover, the structure of a social dilemma partly determines the approaches needed to resolve it.

First, social dilemmas are social. They involve a game structure comprising at least two decision-makers. Some 'dual-use dilemmas' do not readily yield such a game structure because they are cast as single-agent decisions. An example is the concern that research conducted for beneficial purposes might be used by secondary researchers or other users to construct bioweapons. If these users would not be able to exploit the research if it were not conducted then the situation reduces to a single-agent decision:

Research → potential benefits and risk of exploitation

versus

No research → potential costs and no risk of exploitation.

While this decision may be difficult, it is not a social dilemma or even a dilemma in the sense of 'damned if you do and damned if you don't'. Instead, this is arguably a trade-off wherein each option combines potentially strong positive and strong negative consequences.

Dual-use dilemmas can become social dilemmas involving multiple agents if the decisions made by each agent alter the consequences for all of them. Biological research as an arms race is perhaps the most obvious example. For instance, if researchers in country A revive an extinct pathogen and researchers in country B do not, country A temporarily enjoys a tactical advantage over country B while also risking theft or accidental release of the pathogen. If country B responds by duplicating this feat then B regains equal footing with A, but has increased the overall risk of accidental release or theft. For country A, the situation has worsened not just because it has lost its tactical advantage but also because the risk of release has increased. Conversely, if A restrains from reviving the pathogen then B may play A for a sucker by reviving it. It is in each country's self-interest to revive the pathogen in order to avoid being trumped, but the collective interest resides in minimising the risk of accidental or malign release.

A similar example of a social dilemma is where countries A and B are considering whether to eliminate their respective stockpiles of smallpox. The payoff matrix is shown in Table 11.1. The entries are

- R = reward
- T = temptation
- S = sucker
- P = punishment.

This matrix enables a definition of a social dilemma. A social dilemma exists if these four conditions hold

- $R > P$
- $R > S$
- $2R > T + S$
- $T > R$ or $P > S$.

Table 11.1 Pay-Off Matrix for a Two-Agent Game

		B	
		Eliminate	Retain
A	Eliminate	R_a, R_b	S_a, T_b
	Retain	T_a, S_b	P_a, P_b

Source: Author's representation.

There are three well-known dilemma structures, depending on how each country's decision-makers rank-order the consequences of the countries' joint decisions

- chicken: $T > R > S > P$
- prisoner: $T > R > P > S$
- trust: $R > T > P > S$.

We do not require quantification of the matrix entries; they only need to have a complete ordering for each player. We will denote the best outcome by 4 and the worst by 1. Of course, it is possible for the structure to differ between the two countries. In Table 11.2, the structure is 'Chicken' for country A and 'Prisoner' for country B.

Table 11.2 Chicken and Prisoner's Dilemma Combination

		B	
		Eliminate	Retain
A	Eliminate	3, 3	2, 4
	Retain	4, 1	1, 2

Source: Author's representation.

Table 11.2 makes it easy to see the roles played by greed and fear in a social dilemma. Each country can obtain its best outcome (rated 4) by retaining their supply if the other country eliminates theirs. Country B's worst outcome (rated 1) and country A's second-worst result are if each eliminates supply while the other retains theirs. If both act on fear and/or greed and retain their supplies then the joint outcome is the worst of all four (rated 1 for A and 2 for B).

Different structures yield distinct pressures for and against eliminating smallpox stockpiles. A 'cooperation index' is

$$K = \frac{R - P}{T - S}$$

which provides an overall indication of motivation for elimination. All else being equal, Prisoner will have a smaller value for K than Chicken or Trust. The cooperation index, in turn, may be decomposed into a Fear and Greed component

$$K = 1 - \left(K_f + K_g \right), \text{ where}$$

$$K_f = \frac{P - S}{T - S} \text{ and}$$

$$K_g = \frac{T - R}{T - S}.$$

Thus, in Trust and Prisoner $K_f > 0$ whereas in Chicken $K_f < 0$, while in Chicken and Prisoner $K_g > 0$, whereas in Trust $K_g < 0$. In Chicken, Greed is the component detracting from motivation to eliminate stockpiles, and in Trust, Fear is the detractor. Prisoner is the only dilemma in which both the Fear and the Greed components exceed zero, so that both detract from motivation to eliminate. This is why K generally is lowest for Prisoner.

Another crucial characteristic of a dilemma is the 'public' versus 'private' nature of the consequences. This strongly influences whether institutional solutions such as privatisation are potential solutions for social dilemmas. A good is subtractable if its use by one agent decreases the potential for its use by another. Attention is subtractable (devoting attention to one thing decreases the attention that can be given to others) whereas information is non-subtractable (simply acquiring information does not decrease its availability to others). A good is excludable if access to it can be restricted. Secrets and legally proprietary information are fairly excludable, whereas unsecured information on the Internet is not. Goods are privatisable insofar as they are excludable and subtractable.[2]

Public goods (and bads) are strongly non-subtractable and non-excludable. The open-access and communalistic norms of scientists render research outputs a public good. A virulent, easily transmissible pathogen quickly can become a public bad. Common-pool resources, on the other hand, are goods that are subtractable but non-excludable. Air or water quality is an example of a common-pool resource, and the diffusion of nanomaterials could threaten either of these. Toll goods are those, like proprietary information, that are excludable but non-subtractable. And finally, truly private goods are those, like well-guarded smallpox supplies, that are both excludable and subtractable.

The temporal dimension also can play an important role in dilemmas.[3] A large literature indicates that repeated dilemmas are more easily solved than one-shot dilemmas.[4] Repeated dilemmas permit agents to learn, build trust or negotiate and verify compacts, whereas these are considerably more difficult in one-shot dilemmas. Consider 'cat out of the bag' (COB) consequences: it takes only one instance of the research to yield the potential for misuse or accident; subsequent research replication usually does not increase those risks. The COB risk associated with a particular research project can be the basis of a one-shot dilemma; however, if we consider the potential for multiple research efforts to throw up COBs then we have the makings of a repeated dilemma. Packaging one-shot dilemmas into a common category reframes them as repeated dilemmas, enhancing the chances of solving them.

Finally, it should be noted in passing that we have implicitly assumed that both agents know not only their own outcome preferences but each other's as well. Of course, it is also crucial to take into account each agent's perception of the other's pay-offs, because those determine what each agent believes the other's (rational) motives and best moves will be. Referring back to Table 11.2, if country A's intelligence is that country B's pay-off matrix is identical to A's (that is, Chicken instead of Prisoner) then A will underestimate B's motivation

2 Ostrom, E., Gardner, R. and Walker, J. 1994, *Rules, Games and Common Pool Resources*, University of Michigan Press, Ann Arbor.

3 For example, Smithson, M. 1999, 'Taking exogenous dynamics seriously in public goods and resource dilemmas', in M. Foddy, M. Smithson, S. Schneider and M. Hogg (eds), *Resolving Social Dilemmas: Dynamic, Structural, and Intergroup Aspects*, Psychology Press, Philadelphia, pp. 17–32.

4 Danielson, P. 1992, *Artificial Morality: Virtuous Robots for Virtual Games*, Routledge, London.

for retaining smallpox supplies. This is because B's actual outcome ranks for retention are {4, 2} and for elimination they are {3, 1}, whereas A will believe they are {4, 1} and {3, 2} respectively.

Obviously there is much more to determining the nature of a dual-use dilemma's structure than has been dealt with in this section. The intention here is merely to provide a starting point by posing the question of whether a structure constitutes a social dilemma and, if so, what kind of social dilemma the structure corresponds to. The crucial difference between a social dilemma and a trade-off is that a social dilemma entails a conflict between individual and collective interests that does not appear in trade-offs. It is plausible, therefore, that the policies and procedures for dealing with dual-use dilemmas also will need to distinguish between the two.

Partition indeterminacy

Nearly all formal decision-making frameworks, including SEU, assume that all possible options and outcomes are known. In other words, the state space is predetermined. The nature of innovative research implies that in at least some dual-use dilemmas that assumption is untenable on three counts. First, the potential outcomes of research often are not completely known. The accidental creation of a mousepox 'super strain'[5] is a case in point, as is the current state of ignorance regarding potential consequences of third or fourth-generation nanomaterials. Second, the uses of research outputs also sometimes are unanticipated. Witness the applications in cryptography of number theory, a sub-discipline that once was held up as the epitome of pure mathematics beyond reach of any applicability. Third, the variety of responses to the threat of research misuse is not predetermined. The first two sources of state space indeterminacy are matters to be taken into account by those who make judgments and decisions, and this is the topic of the next subsection. The third, however, can be a matter of choice, and this is discussed in the subsection thereafter.

Unknown outcomes and consequences

In most standard probability theories, on the grounds of insufficient reason, a probability of $1/J$ is assigned to J mutually exclusive possible events when nothing is known about the likelihood of those events. For example, in a race involving three greyhounds, an agent who knows nothing about any of the dogs would assign a value of $1/3$ to the probability of each greyhound winning. Moreover, even under alternative probability assignments the probabilities

---

5   Miller and Selgelid, op. cit., pp. 523–80.

of the J events must sum to 1, meaning that the entire probability mass is concentrated on that set of events. Thus, a more knowledgeable rational agent who has assigned a probability of 1/2 to the first dog winning and 1/4 to the second dog is compelled to assign the remaining 1/4 to the third.

The number of possible elementary events or states in a space is determined by the *partition* of that space. The greyhound race has been partitioned into three outcomes: dog one wins, dog two wins or dog three wins. Were we to allow ties, the partition would expand to $J = 7$. The ignorant agent now would assign a probability of 1/7 to each dog winning, and the more knowledgeable agent could distribute the remaining 1/4 probability across the remaining five events instead of having to allocate it all to the third dog winning. Thus, probability assignments are *partition dependent*.

When partitions are indeterminate, partition dependence poses a problem for subjective probability assignments. This is not the same problem as unknown probabilities over a unique and complete partition (for example, where we know that there are only red and black marbles in a bag but do not know how many of each).[6] It is more profound. In the absence of a uniquely privileged partition, there is no defensible prior probability distribution to be constructed.

Two separable problems for partitions may arise. One is an incomplete account of possible events. A unique and complete partition might be attainable in principle, but we lack the necessary information. The other problem is the absence of a privileged partition even when one has a complete account of those possibilities. Shafer[7] presented an example of this problem as a motivation for the belief functions framework. He asked whether the probability of life existing in a newly discovered solar system should be partitioned into {life, no life} or {life, planets without life, no planets}. This issue arises naturally when a decision must be made that involves a threshold or an interval on a continuum. We shall revisit this particular problem in the next subsection.

Returning to the first problem, the most common situation confronting judges or decision-makers is partial knowledge of the possible outcomes. We may know some of the potential uses and misuses of a new biotechnology but not all of them. We might even be willing to assign a subjective probability that party X will misuse this technology in ways we can anticipate. But what probability should we assign to X misusing the technology in ways we haven't anticipated? Likewise, what probability should we assign to party Y finding a new way to use the technology for good?

6   Smithson, M. 2009, 'How many alternatives? Partitions pose problems for predictions and diagnoses', *Social Epistemology*, vol. 23, pp. 347–60.
7   Shafer, G. 1976, *A Mathematical Theory of Evidence*, Princeton University Press, Princeton, NJ.

Smithson[8] presents strategies for dealing with partition dependence, distinguishing those that apply when a privileged or at least agreed-upon partition is attainable from those that apply when it is not.

1. Where a privileged or agreed-upon partition is attainable
   a. de-biasing strategies
   b. establishing criteria for choosing partitions.

2. Where there is no privileged or agreed-upon partition
   a. using diverse partitions
   b. modelling partition-dependence effects
   c. using (non-standard) probabilistic frameworks that avoid partition dependence.

De-biasing strategies are needed because human judges are strongly influenced by partitions in their subjective probability assignments. Two important manifestations of partition dependence are distorted judgments of likelihoods of compound events and anchoring on an ignorance prior. A classic study[9] concerning people's assignments of probabilities to possible causes of a given outcome (for example, an automobile that will not start) revealed that possible causes that were explicitly listed received higher probabilities than when the same causes were implicitly incorporated into a 'catch-all' category of additional causes. The effect has since been referred to as the 'catch-all underestimation bias' and also sometimes the 'pruning bias'.[10]

Likewise, it has been empirically demonstrated[11] that subjective probability judgments are typically biased towards the ignorance prior determined by the partition salient to the judge. That is, people anchor on a uniform distribution of $1/J$ across all J possible events, even when taking into account prior evidence of how likely each event is. Because those adjustments typically are insufficient,[12] judges' intuitive probability assignments are biased towards probabilities of $1/J$.

Criteria for choosing partitions and methods for exploring diverse partitions are not well established. One recently proposed set of criteria will be elaborated in

---

8   Smithson, 2009, op. cit., pp. 347–60.
9   Fischhoff, B., Slovic, P. and Lichtenstein, S. 1978, 'Fault trees: sensitivity of estimated failure probabilities to problem representation', *Journal of Experimental Psychology: Human Perception Performance*, vol. 4, pp. 330–44.
10   Russo, J. E. and Kolzow, K. J. 1994, 'Where is the fault in fault trees?' *Journal of Experimental Psychology: Human Perception Performance*, vol. 20, pp. 17–32.
11   Fox, C. R. and Rottenstreich, Y. 2003, 'Partition priming in judgment under uncertainty', *Psychological Science*, vol. 13, pp. 195–200.
12   Tversky, A. and Kahneman, D. 1974, 'Judgment under uncertainty: heuristics and biases', *Science*, vol. 185, pp. 1124–31.

the next subsection,[13] but these have limited scope. Other criteria could be linked with strategies for manipulating and exploring judgment biases in informative ways. As a simple example, expert judges estimating probabilities of adverse consequences arising from the revival of an extinct pathogen could be randomly assigned to one of two conditions: a twofold partition (consequence versus no consequence) or a J-fold partition (a list of anticipated consequences plus a catch-all category for unanticipated ones). Partition dependence would predict that the average probability of an adverse consequence in the first condition should be less than the average sum of the probabilities across the J consequence categories in the second condition. The results would yield fairly defensible lower and upper expert estimates of the probability of adverse consequences. More sophisticated experimental designs would enable the construction and estimation of relevant partition-dependence effects.

Finally, let us briefly consider non-standard probability frameworks that are not partition dependent. These have appeared in the growing literature on generalised probability theories, and also in behavioural economics.[14] Walley[15] argues on normative grounds that imprecise probability frameworks can avoid partition dependence entirely. He proposes that when judges are permitted to provide a lower and an upper probability judgment (that is, imprecise probabilities) every ignorance prior should consist of vacuous probabilities $\{0, 1\}$. In the greyhound race example, the ignorant agent could assign a lower probability of 0 and an upper probability of 1 to every event regardless of whether the partition is threefold or sevenfold. The lower and upper probabilities of the first dog winning would be 0 and 1 regardless of the partition, thereby avoiding partition dependence. Walley developed an updating method (the imprecise dirichlet model) that is partition independent and has generated interest within the community of imprecise-probability theorists.

That said, recent studies[16] experimentally demonstrated that naive judges are just as strongly influenced by partitions when making imprecise probability judgments as they are when making precise probability judgments. Moreover, they demonstrated that many judges anchor on $1/J$ as the midpoint of their lower and upper probability judgments. No applicable de-biasing strategies have

13   Smithson, M. 2006, 'Scale construction from a decisional viewpoint', *Minds and Machines*, vol. 16, pp. 339–64; and Smithson, 2009, op. cit., pp. 347–60.

14   For example, Grant, S. and Quiggin, J. 2004, 'Conjectures, refutations and discoveries: incorporating new knowledge in models of belief and decision under uncertainty', Paper presented at the 11th International Conference on the Foundations and Applications of Utility, Risk and Decision Theory (FUR XI—Paris), under the joint auspices of the Ecole Nationale Supérieure d'Arts et Métiers (ENSAM) and the Ecole Spéciale des Travaux Publics (ESTP), Paris, 2 July.

15   Walley, P. 1991, *Statistical Reasoning with Imprecise Probabilities*, Chapman Hall, London; and Walley, P. 1996, 'Inferences from multinomial data: learning about a bag of marbles', *Journal of the Royal Statistical Society*, Series B, vol. 58, pp. 3–34.

16   Smithson, M. and Segale, C. 2009, 'Partition priming in judgments of imprecise probabilities', *Journal of Statistical Theory and Practice*, vol. 3, pp. 169–82.

yet been reported. Nevertheless, the possibility remains that allowing judges to express one kind of uncertainty (imprecision in their probability assignments) may militate against the impact of another kind (partition indeterminacy).

How many options?

Policies regulating responses to dual-use dilemmas could be limited to two options—for example: laissez faire and bans. But what about a third option, such as oversight by a regulatory body? Or more than two additional options? Are there criteria that could indicate how many options a rational agent should prefer? How would we know whether each option was worth retaining? This appears to be a relatively unexplored topic, but reasonably important given that this is one aspect of dual-use dilemmas where policy and decision-makers actually have choices. It is directly related to partition indeterminacy because we are constructing a partition of a space of possible acts.

In the context of legal standards of proof, a typical threshold probability of guilt associated with the phrase 'beyond reasonable doubt' is in the [0.9, 1] range.[17] For a logically consistent juror, a threshold probability of 0.9 implies the difference between the utility of acquitting versus convicting the innocent is nine times the difference in the utility of convicting versus acquitting the guilty.

Connolly demonstrated that the utility assignments to the four possible outcomes (convicting the guilty, acquitting the innocent, convicting the innocent, and acquitting the guilty) that are compatible with such a high threshold probability are counterintuitive. Specifically, if one does want to have a threshold of 0.9, 'one must be prepared to hold the acquittal of the guilty as highly desirable, at least in comparison to the other available outcomes'.[18] He also showed that more intuitively reasonable utilities lead to unacceptably low threshold probability values.

Smithson[19] showed that the incorporation of a third middle option (such as the Scottish not-proven verdict) with a suitable threshold can resolve this quandary, permitting a rational (subjective expected utility) agent to retain a high conviction threshold and still regard false acquittals as negatively as false convictions. The price paid for this solution is a more stringent standard of proof for outright acquittal.[20] The main point here is that a consideration of

---

17   Connolly, T. 1987, 'Decision theory, reasonable doubt, and the utility of erroneous acquittals', *Law and Human Behavior*, vol. 11, pp. 101–12.

18   Ibid., p. 111.

19   Smithson, 2006, op. cit., pp. 339–64.

20   For evidence that this also is what humans do, see Smithson, M., Gracik, L. and Deady, S. 2007, '"Guilty, not guilty, or …?" Multiple verdict options in jury verdict choices', *Journal of Behavioral Decision Making*, vol. 20, pp. 481–98.

preferences as expressed by the relative positions of utilities can aid in the choice of a partition of acts, due to the connection between these utilities and the threshold probabilities that determine when one act is chosen over another.

Applying Smithson's framework to dual-use dilemmas, consider the simplest set-up in which either some kind of misuse of a research output occurs or no misuse occurs. Suppose we must make a decision regarding the fate of a potential research project (for example, whether to prohibit it or allow it to proceed), and we wish to do so on the basis of an estimated probability that the research output could be misused. Let us assume that choices will affect the utility of the no-misuse outcome because of inhibited scientific progress and/or resource expenditure in security arrangements. Let us also assume that the utility of the misuse outcome also will be affected by choice because the same considerations will be combined with the consequences of misuse, even if they are dwarfed by the latter.

Suppose we have a J-fold partition of acts R_j, for $j = 0, 1, 2, \ldots, J\text{-}1$. There are two possible outcomes: no misuse and misuse. The act R_j has a utility H_j if there is no misuse and a utility G_j if there is misuse. We assume that the acts R_j are ordered so that $H_j > H_{j-1}$ and $G_{j-1} > G_j$ for any j. A straightforward argument shows that if the odds of no misuse exceed an odds threshold defined by

$$w_{j-1j} = \frac{G_{j-1} - G_j}{H_j - H_{j-1}}$$

then the decision-maker should prefer act R_j over R_{j-1}. The odds threshold W_{j-1j} therefore is determined by the ratio of utility differences.

Table 11.3 Twofold Partition of Acts

	R_1	R_0
	Laissez F.	Prohibit
No misuse	$H_1 = 1$	H_0
Misuse	$G_1 = 0$	G_0

Source: Author's representation.

The simplest set-up of this kind is shown in Table 11.3. There are two possible acts: prohibition or laissez faire. Without loss of generality we may assign $H_1 = 1$ (the best possible outcome) and $G_1 = 0$ (the worst). Therefore, the odds threshold is

$$w_{j-1j} = \frac{G_0}{1 - H_0}.$$

It immediately follows that if $G_0 < q_0$ for $0 < q_0 < 1$ then

$H_0 > 1 - q_0/w_{01}.$

Suppose we also wish to restrict $w_{01} > y_0 > 1$. This should seem reasonable, because we are merely restricting the odds-of-no-misuse threshold to be above 1. Then

$H_0 > (y_0 - q_0)/y_0.$

For example, if $q_0 = 0.1$ and $y_0 = 10$ then $H_0 > 0.99$; and in fact if $q_0 = 1$ and $y_0 = 100$ then we also have $H_0 > 0.99$. Thus, no misuse under prohibition has nearly as high a utility as no misuse under laissez faire, implying that prohibition hardly decreases utility at all. Moreover, in the special case where prohibition obviates misuse so that $G_0 = H_0$, a high odds threshold yields a correspondingly high value for G_0 and H_0. For instance, $y_0 = 10$ implies G_0 and H_0 both must exceed 10/11.

The problem is the inability to simultaneously have a high value of y_0, a low q_0 and a relatively low H_0. The chief result is that a high (and therefore cautious) odds-of-no-misuse threshold for invoking the prohibition of research requires a belief that prohibition results in only a very small decrease in utility relative to the improvement in the (dis)utility of misuse. As in the legal standard-of-proof case, this difficulty arises because we have only two possible acts. A way around this is to introduce a third act (middle option). Let us call it 'Regulate'. Table 11.4 shows the utility set-up for this threefold partition.

Table 11.4 Threefold Partition of Acts

	R_2	R_1	R_0
	Laissez F.	Regulate	Prohibit
No misuse	$H_2 = 1$	H_1	H_0
Misuse	$G_2 = 0$	G_1	G_0

Source: Author's representation.

The w_{01} threshold now determines when the Regulate option is chosen over Prohibit, and a new threshold, w_{12}, determines when Laissez-Faire is chosen over Regulate. Now, $H_1 > (y_1 - q_1)/y_1$ implies

$$w_{12} < \frac{G_1 - q_1}{(y_1 - q_1)/y_1 - H_1}$$

which in turn implies

$H_0 > 1 - (G_1 - q_1)/w_{12} - q_1/y_1.$

Setting $w_{12} = 5$ and $G_0 = 0.5$, for instance, and using the settings $q_1 = 0.1$ and $y_1 = 10$, gives

$$H_0 > 1 - (0.5 - 0.1)/5 - 0.1/10 = 0.91$$

If we are willing to lower the threshold to $w_{12} = 2$ and increase G_0 to 0.68 then

$$H_0 > 1 - (0.68 - 0.1)/2 - 0.1/10 = 0.7.$$

The threefold partition therefore can express a belief that outright prohibition could substantially negatively affect research (in this last example, a decline in utility from 1 to 0.7). Nevertheless, there are limits if we take certain additional constraints into account. It seems reasonable to stipulate that misuse cannot yield a greater utility than no misuse, so we impose the constraint $G_0 < H_0$. As mentioned earlier, the case where $G_0 = H_0$ corresponds to the situation where prohibition of research eliminates the possibility of misuse of its outputs, so that there is no difference between the 'no misuse' and 'misuse' states. The restriction $G_0 < H_0$ and the constraint $w_{01} = 1$ imply that $H_0 > 1/2$. Higher odds thresholds increase the lower bound on H_0. It is easy to prove that the general relationship is $w_{01} = x$ implies $H_0 > x/(x + 1) = p_{01}$, the corresponding probability threshold. In the two examples above, $w_{01} = 5$ implies $H_0 > 5/6$ and $w_{01} = 2$ imply $H_0 > 2/3$.

Thus, extreme cases where prohibition of further research would hardly alter the (dis)utility of misuse of an existing technology impose severe restrictions on the utility if there is no misuse. Table 11.5 shows a set-up like this, with similar low values of G_1 and G_2. We would be inclined to set the odds thresholds w_{01} and w_{12} to be very high—say, $w_{01} = 100$ and $w_{12} = 1000$. The result would be that H_1 and H_0 both would be very close to 1: $H_1 = 0.99999$ and $H_0 = 0.99989$. Therefore, a substantial difference between H_1 and H_0 (say, due to the inhibition of scientific progress) can only arise if there is a substantial difference between G_1 and G_0 and relatively low threshold odds of no misuse, w_{01}.

Table 11.5 Extreme Disutility of Misuse

	R_2	R_1	R_0
	Laissez F.	Regulate	Prohibit
No misuse	$H_2 = 1$	H_1	H_0
Misuse	$G_2 = 0$	$G_1 = 0.01$	$G_0 = 0.02$

Source: Author's representation.

Are there sets of utilities and threshold odds that could satisfy the intuition that some security measures should be in place when there is only a very small chance of misuse, but that severe restrictions on research will have a substantial impact on scientific progress? What would these look like? Table 11.6 illustrates a set-up similar to an earlier example that is compatible with these intuitions.

The bottom row shows the odds thresholds. Resetting w_{12} to values greater than 10 has relatively little impact on w_{01} (or alternatively on utilities G_0 and H_0 if we wish w_{01} to remain at 2) because H_1 is already close to 1 and G_1 is close to 0. And of course it is possible to solve for H_1 and G_1 such that w_{12} takes a specific value greater than 10 while w_{01} is unaffected and remains at 2. Thus, in the threefold partition of acts we are free to set w_{12} to very conservative (high) values while still retaining flexibility regarding w_{01} or the utilities it comprises.

Table 11.6: Extreme Disutility of Prohibition

	R_2	R_1	R_0
	Laissez F.	Regulate	Prohibit
No misuse	$H_2 = 1$	$H_1 = 0.99$	$H_0 = 0.6933$
Misuse	$G_2 = 0$	$G_1 = 0.1$	$G_0 = 0.6933$
Odds threshold	$w_{12} = 10$	$w_{01} = 2$	

Source: Author's representation.

The set-up is greatly affected, however, by changes in w_{01} because of the relationship described earlier between w_{12} and the lower bound on H_2. Increasing w_{01} from 2 to 5, as mentioned earlier, raises the lower bound on H_0 from 2/3 to 5/6. Preferences and intuitions regarding these effects will need to be guided by a sense of how harmful potential misuses are under prohibition versus regulation versus laissez faire in comparison with the loss of potential knowledge and benefits when research is prohibited versus regulated. These comparisons are admittedly not easy to make, let alone quantify. Nevertheless, decisional thresholds do need to be set, and setting them in a considered manner requires comparisons of this sort.

Therefore some considerations about utility scales are appropriate to conclude this subsection. The utility scales used here are not absolute, or even ratio-level. They have neither an absolute zero nor a fixed upper bound. At best, they are interval-level scales, meaning that the difference between two utility assignments (for example, $H_2 - H_1$) is a ratio-level scale. Recall that a ratio comparison of two such differences, $(G_{j-1} - G_j)/(H_j - H_{j-1})$, determines the odds threshold w_{j-1j}. Smithson[21] defines two kinds of risk-orientation bias in the utility differences when utilities are restricted to the [0, 1] interval. 'A-bias' is measured by the sum of the log of the odds thresholds and refers to greater risk aversion to one outcome than the other. In our examples thus far, all $w_{j-1j} > 1$, indicating greater risk sensitivity to misuse than to no misuse. 'R-bias', on the other hand, is measured by

---

21   Smithson, 2006, op. cit., pp. 339–64.

$$\sum_{j=1}^{J-1} \log\left(\frac{H_j - H_{j+1}}{H_{j-1} - H_j}\right) + \log\left(\frac{G_{j+1} - G_j}{G_j - G_{j-1}}\right)$$

and compares gains and losses in utility as the decision-maker moves from one act to another. A positive sum indicates greater risk sensitivity in choosing between acts for high js and a negative sum indicates greater risk sensitivity in choosing between acts with low js. In Table 11.6 these log ratios are 3.39 and 1.78, so there is greater risk sensitivity in choosing between Regulate and Prohibit than between Laissez Faire and Regulate. This is simply due to the greater changes in H and G utilities as we move from Regulate to Prohibit.

Finally, given that the utility scales have no absolute lower or upper bounds, a reasonable question to ask is whether some bounds are more useful or sensible than others. The [0, 1] interval probably is not well suited to human judgments because it lacks two features that have psychological significance: a reference point representing the status quo and a distinction between being better off or worse off than the status quo. A well-established empirical and theoretical literature[22] informs us that people judge the utility of future outcomes relative to a reference point (usually the status quo) instead of in absolute terms, and that they are more sensitive to losses than to gains.

Table 11.7 presents one way of rescaling Table 11.6 according to these considerations. Suppose we assign 0 to represent the status quo and represent the maximal loss by −100. Suppose also that we believe misuse of a research output under laissez faire would yield a loss that is 10 times the magnitude of the gains that could be realised if no misuse occurred. Then $G_2 = -100\$$ and $H_2 = 10$. The odds thresholds in Table 11.6 partially determine the remaining utility assignments. We require more than one constraint, so let us repeat the loss due to misuse being 10 times the gain with no misuse under Regulate. The end result reveals that we believe we will be worse off than the status quo under the Prohibit option whether there is misuse or not, but that will be our best option if the odds of misuse are shorter than 2 to 1.

Table 11.7 Rescaled Utilities from Table 11.6

	R_2	R_1	R_0
	Laissez F.	Regulate	Prohibit
No misuse	$H_2 = 10$	$H_1 = 9.9$	$H_0 = -26.4$
Misuse	$G_2 = -100$	$G_1 = -99$	$G_0 = -26.4$

Source: Author's representation.

---

22 Beginning with Kahneman, D. and Tversky, A. 1979, 'Prospect theory: an analysis of decision under risk', *Econometrica*, vol. 47, pp. 263–91.

Imprecision and bias in judgments

Probability and utility judgments regarding dual-use dilemmas ultimately must be made by human judges, and this last section discusses the most important issues regarding human judgments of this kind. We begin by considering issues of imprecision and conflict in judgments, and subsequently discuss relevant human tendencies towards overconfidence in predictions and confirmation bias.

Even when it is foreseeable, the probability of the misuse of a new technology or research output and the severity of its consequences almost never are known precisely, nor is there usually a consensus on their magnitudes. Imprecision and conflict are very likely to pervade judgments of probability and utility in dual-use dilemmas. These uncertainties must not be denied or ignored; decision-makers will treat falsely precise estimates as if they really are precise and decisions based on them will be far from robust. At the very least, decisions and their criteria should be subjected to sensitivity analyses to ascertain which components are the most affected by altering parameter values. In the preceding section, for instance, we saw that the three-option set-up in Table 11.6 was robust against changes in w_{12} but sensitive to changes in w_{01}.

I shall leave conflict aside as even a brief treatment of it is beyond the scope of this chapter, except to note in passing that some psychological investigations indicate that people prefer dealing with vague but consensual opinions to precise but disagreeing ones.[23] Thus, imprecision is viewed as a less severe kind of uncertainty than conflict.

Nevertheless, imprecision complicates decision-making. A precise probability assigned to the misuse of a technology either exceeds or fails to exceed a decisional threshold of the kind discussed in the preceding section, so the choice among alternatives is clear. Precise probabilities bring decisiveness with them; however, a probability interval may lie entirely below or above the threshold, or may include it. Standard decision frameworks for imprecise probabilities treat the lower bound as the probability to use in betting on misuse and the upper bound as the probability to use in betting against it. Therefore, these frameworks claim there is no basis in the probabilities themselves for preferring the alternative on either side of a decisional threshold if the probability interval straddles it.

Suppose, for instance, that the set-up in Table 11.6 is our decisional guide and we are confronted with a potential technological development for which

---

23 Smithson, M. 1999, 'Conflict aversion: preference for ambiguity vs. conflict in sources and evidence', *Organizational Behavior and Human Decision Processes*, vol. 79, pp. 179–98; Smithson, 2006, op. cit., pp. 339–64; and Cabantous, L. 2007, 'Ambiguity aversion in the field of insurance: insurers' attitude to imprecise and conflicting probability estimates', *Theory and Decision*, vol. 62, pp. 219–40.

experts estimate the odds of misuse to be somewhere between 5 and 50 to 1. This interval includes the threshold $w_{01} = 10$, so should we choose Regulate or Laissez Faire? If we can defer this decision pending more information, should we do so? This issue is an active topic of research and attempting a resolution of it is beyond the scope of this chapter, but the main purpose in raising it here is to point out that because imprecision really matters, decision-makers must work out how they will treat imprecise estimates differently from precise ones.

We now turn to probability judgments themselves. There is a large body of empirical and theoretical work on subjective probability judgments, but discussion will be restricted to just two judgment biases that are directly relevant. The first of these is probability weighting, which may be summarised by saying that people overweight small and underweight large probabilities. Note that this does not mean that people are necessarily under or overestimating the probabilities, but instead treating them in a distorted fashion when making decisions based on them. Rank-dependent expected utility theory[24] reconfigures the notion of a probability weighting function by applying it to a cumulative distribution whose ordering is determined by outcome preferences. Cumulative prospect theory[25] posits separate weighting functions for gains and losses.

Two explanations have been offered for the properties of probability weighting functions. The first[26] is 'diminishing sensitivity' to changes that occur further away from the reference points of 0 and 1. The second is that the magnitude of consequences affects both the location of the inflection point of the curve and its elevation. Large gains tend to move the inflection point downward and large losses move it upward.[27] Diminishing sensitivity has an implication for judgments and decisions based on imprecise probabilities as well as precise probabilities. A change from 0.01 to 0.05 is seen as more significant than a change from 0.51 to 0.55, but a change from 0.51 to 0.55 is viewed as less significant than a change from 0.95 to 0.99. An implication is that for decisional purposes people might view a probability interval [0.01, 0.05] as less precise than [0.51, 0.55], and so on. The prospect of large losses (as in the misuse of biotechnology) will exaggerate these effects for low probabilities. This issue is important for dual-use dilemmas because at least some of the possible outcomes under consideration will have extreme probabilities attached to them.

The second relevant bias concerns confidence judgments and the elicitation of prediction or confidence intervals from human judges. Numerous studies

24   For example, Quiggin, J. 1993, *Generalized Expected Utility Theory: The Rank Dependent Model*, Kluwer, Boston.

25   Tversky and Kahneman, op. cit., pp. 1124–31.

26   Camerer, C. F. and Ho, T. H. 1994, 'Violations of the betweenness axiom and nonlinearity in probability', *Journal of Risk and Uncertainty*, vol. 8, pp. 167–96.

27   For example, Etchart-Vincent, N. 2004, 'Is probability weighting sensitive to the magnitude of consequences? An experimental investigation on losses', *Journal of Risk and Uncertainty*, vol. 28, pp. 217–35.

demonstrate that both novices and experts tend to be overconfident in the sense that they construct prediction intervals that are much too narrow for their confidence criteria. A typical discrepancy is that when asked to construct an interval that has a 90 per cent probability of including the correct prediction the actual hit rate is less than 50 per cent.[28] Recent findings,[29] however, have suggested that when presented with prediction intervals, people do not overestimate their coverage rates. The take-home lesson from this literature is that asking experts to estimate how likely is the probability of, say, the theft of smallpox supplies from a particular source between two values will yield more well-calibrated results than asking the experts to construct, say, a 95 per cent confidence interval for that probability.

Finally, the catch-all underestimation bias described earlier is a special case of confirmation bias. This is a largely unconscious tendency in human information processing and judgment such that people seek out and pay more attention to information that confirms their beliefs than to disconfirming information. In the catch-all underestimation bias, confirmation bias manifests itself as a tendency to underestimate the likelihood of novel or unanticipated events. Unfortunately these are exactly the kinds of events that policy planners and decision-makers must be on the lookout for in dealing with dual-use dilemmas. There are few recommendations on record for militating against confirmation bias. One is to construct inclusive teams containing members with diverse backgrounds and viewpoints and ensure that decision-makers and planners listen attentively to those members with whom they disagree; however, this seemingly obvious strategy is complex and deceptively difficult to implement.[30] Another is the use of formal analyses, simulations and models to reveal consequences or possibilities that our preconceptions render invisible to us.[31]

Conclusions?

This chapter largely neglects ethical considerations, which may seem odd given the predominantly ethical nature of dual-use dilemmas. Ethical considerations have been set aside to enable a focus on some prerequisites for a 'fine-grained' analysis of dual-use dilemmas—namely, a systematic investigation of specific

---

28   For example, Russo and Kolzow, op. cit., pp. 17–32.
29   For example, Winman, A., Hansson, P. and Juslin, P. 2004, 'Subjective probability intervals: how to cure overconfidence by interval evaluation', *Journal of Experimental Psychology: Learning, Memory, and Cognition*, vol. 30, pp. 1167–75.
30   See, for example, Brown, V. A. 2010, 'Collective inquiry and its wicked problems', in V. A. Brown, J. Russell and J. Harris (eds), *Tackling Wicked Problems through the Transdisciplinary Imagination*, Earthscan, London, pp. 61–84.
31   See Lempert, R., Popper, S. and Bankes, S. 2002, 'Confronting surprise', *Social Science Computer Review*, vol. 20, pp. 420–40.

unknowns that such an analysis would have to contend with. My hope is that ethicists will find useful guidance in this investigation, avoiding some of the pitfalls and traps awaiting the unwary.

Some pertinent unknowns have not been dealt with here, so this chapter cannot be taken as anything like an exhaustive survey. Nevertheless, we have examined types of unknowns that are beyond the purview of standard decision theories, such as state space indeterminacy and imprecision. We have seen that there are genuinely different kinds of unknowns, not just different sources of the same kind, and that these play distinct roles. One of the key emergent points is that many of the unknowns in dual-use dilemmas (and in so-called 'wicked problems') are interconnected. They can be traded against one another, and how one unknown is dealt with has ramifications for other unknowns. Allowing imprecision in probability assignments, for instance, offers a way of handling state space indeterminacy. Conversely, choosing the 'right' number of options can rectify incompatibilities between preferences and decisional probability thresholds.

We may never be able to attain precise quantification of costs, benefits and probabilities of outcomes arising from dual-use dilemmas, so a fine-grained analysis in that sense also is unachievable. After all, accidental findings and consequences are legion in cutting-edge research and development, and so irreducible unknowns such as the catch-all underestimation problem are likely to dog policy formation and decision-making alike. Moreover, as I have argued elsewhere,[32] if we value creativity, discovery and/or entrepreneurship then we shall have to tolerate at least some irreducible unknowns.

Nevertheless, as Head[33] has pointed out, great uncertainty alone is not sufficient to render a problem 'wicked' in the sense used in most of the literature on that topic. Wickedness also requires complexity and divergent or contradictory viewpoints about the nature of the problem and preferences regarding alternative outcomes. I have tried to show here that even rather simple formal analyses in the form of thought experiments can frame and structure dual-use dilemmas in useful ways that avoid some aspects of wickedness, so that at least some of our psychological foibles can be taken into account and even overcome.

---

32 Smithson, M. 2008, 'The many faces and masks of uncertainty', in G. Bammer and M. Smithson (eds), *Uncertainty and Risk: Multidisciplinary Perspectives*, Earthscan, London, pp. 13–26.
33 Head, B. W. 2008, 'Wicked problems in public policy', *Public Policy*, vol. 3, pp. 101–18.

12. Moral Responsibility, Collective-Action Problems and the Dual-Use Dilemma in Science and Technology[1]

Seumas Miller

The dual-use dilemma

The so-called 'dual-use dilemma' arises in the context of research in the sciences as a consequence of one and the same discrete piece, or ongoing program of scientific research, intentionally undertaken for good ends having the potential to be intentionally used to cause great harm.[2] So there is a primary user who creates new knowledge or designs new technology for good—for example, discovers how to aerosolise chemicals for use in crop-dusting. But there is also a secondary user who uses the knowledge or technology for some harmful purpose—for example, uses the newly discovered process of aerosolisation to weaponise chemicals.

Many, if not most, so-called dual-use dilemmas are not really dilemmas in the narrow sense of being situations involving two options that are equally ethically problematic. In the first place, the dilemmas in question could be tri-lemmas; indeed, there could be four or five or some very large number of options, all of which are equally ethically problematic. In the second place, the options are not generally equally ethically problematic. Certainly, there are ethical considerations for and against each of the options, however, it may well be that, all things considered, one of the options is morally preferable to the others and that this is relatively obvious to any rational, morally sensitive person. The point is rather that there are at least some significant moral costs associated with each of the available options.

Naturally, many, if not most, scientific discoveries and especially new technologies have dual-use potential in the trivial sense that they could be used by someone for some malevolent purpose. Indeed, any newly designed object, such as the first baseball bat, has dual-use potential in this trivial sense. After all, baseball

---

1   This chapter is a condensed version of material provided by the author in Work Package 3 (WP3) of the European Union's Seventh Framework project, Synth-Ethics. (WP3 was jointly authored by Seumas Miller and Michael Selgelid.)
2   Miller, S. and Selgelid, M. 2008, *Ethical and Philosophical Consideration of the Dual Use Dilemma in the Biological Sciences*, Springer, London, ch. 1.

bats can be used to hit people over the head, as well as for the enjoyment of playing baseball; however, it is implicit in the use of the term 'dual use' in play in the academic literature that the potential harm in question is of a very great magnitude—for example, the potential of nuclear fusion to lead to the creation of the hydrogen bomb, and the potential of genetic engineering to lead to a super virus. Moreover, it is also implicit that the dual-use potential in question is not simply a repeat application of existing science or technology—as is the case with the construction of the first baseball bat—but genuinely new science or technology that can be used to provide a qualitatively or quantitatively new means of harming, for example: the invention of the first explosives, or creating a highly virulent form of an existing much less virulent pathogen. Hitting a person over the head with a solid object has been done since time immemorial, so hitting someone over the head with a baseball bat is hardly a novel means of harming. By contrast, blowing someone up with gunpowder was initially a novel means of harming. Moreover, unlike the use of a baseball bat, the use of an explosive device, such as a 10-tonne bomb, could harm on a very large scale.

Note that accidents involving science and technology, even accidents on a very large scale such as the Chernobyl disaster, are not dual use in my sense since there is no secondary evil user, although they may involve unethical behaviour such as negligence with respect to safety precautions. Nor are weapons designed as weapons—for example, guns—instances of dual-use science and/or technology. For even if their harmful use is intended to be ultimately for the good, such weapons are in the first instance designed to harm; their use to harm is not a secondary, but rather a primary, use.

One paradigmatic case of dual-use research was the biological research on the mousepox virus. The dilemma was as follows.

- Option 1: The scientists ought to conduct research on the mousepox virus and do so intending to develop a genetically engineered sterility treatment that combats periodic plagues of mice in Australia.
- Option 2: The scientists ought *not* to conduct the research since it might lead (and in point of fact did lead) to the creation of a highly virulent strain of mousepox and the consequent possibility of the creation—by, say, a terrorist group contemplating a biological terrorist attack—of a highly virulent strain of smallpox able to overcome available vaccines.

It is a dilemma since there are two options with good reasons in favour of both and it is an ethical dilemma since these options are morally significant. In essence the dilemma involves a choice between intentionally doing good and foreseeably providing others with the means to do evil.

A second and more recent paradigm of dual-use research is the biological research done on a deadly flu virus (A [H5N1]), which causes bird flu. Scientists in the United States and the Netherlands created a highly transmissible strain of this virus, albeit, as it emerged, a strain that is not as deadly as ordinary H5N1. Crucially, the work was done on ferrets, which are considered a very good model for predicting the likely effects on humans.

As with the mousepox case, there are two options, ethically speaking.

- Option 1: The scientists ought to conduct research on the bird flu virus and do so intending to develop vaccines against similar, but deadly, naturally occurring and artificially created strains of H5N1.
- Option 2: The scientists ought *not* to conduct the research since it will lead to the creation of a virus that is transmissible to humans, and is unlikely to lead to the development of vaccines against similar, but deadly, naturally occurring and artificially created strains of H5N1.

Notice here that the scientific claim in option two that the research is unlikely to lead to the development of vaccines contradicts the corresponding claim in option one. In this respect the 'dilemma' is unlike that posed by the mousepox research. Moreover, if this scientific claim made in option two is correct then the justification offered in option one for conducting the research collapses.

In such dual-use cases, the researchers—if they go ahead with the research—will have foreseeably provided the means for the harmful actions of others and, thereby, arguably infringed a moral principle (albeit their infringement might in some cases be morally justified). The principle in question is the principle of what we might refer to as the 'no means to harm principle'.[3] Roughly speaking, this is the principle that rules out providing malevolent persons with the means to do great harm—a principle that itself ultimately derives from the more basic principle: do no harm.

The 'no means to harm principle' (NMHP) is the principle that one should not foreseeably (whether intentionally or unintentionally) provide others with the means to intentionally do great harm and it assumes: 1) the means in question is a means to do great harm; and 2) the others in question will do great harm, given the chance.

As with most, if not all, moral principles, the NMHP is not an absolute principle and, therefore, it can be overridden under certain circumstances. For example, it is presumably morally permissible to provide guns to the police in order that they can defend themselves and others. Moreover, as is the case with most, if

---

3 This principle, or similar ones, is familiar in a variety of ethical contexts. See, for example, Scanlon, T. 1977, 'A theory of freedom of expression', in R. M. Dworkin (ed.), *The Philosophy of Law*, Oxford University Press, Oxford.

not all, moral principles, the application of the NMHP is very often a matter of judgment. In the case of NMHP the need for judgments depends in large part on the uncertainty of future harms.

The dual-use dilemma is a dilemma for researchers, governments, the community at large, and for the private and public institutions, including universities and commercial firms that fund or otherwise enable research to be undertaken. Moreover, in an increasingly interdependent set of nation-states—the so-called global community—the dual-use dilemma has become a dilemma for international bodies such as the United Nations. The dilemma is perhaps most acute in those areas of science and technology that operate on an 'engineering' or 'construction' model—for example, synthetic biology, nanotechnology (as opposed to a 'description' model, which restricts itself to the description of pre-existing entities and their causal and other relationships—for example, astronomy).

Scientific research and weapons of mass destruction

The history of science and technology is replete with examples of scientific research being used intentionally or unintentionally to create weapons, including weapons of mass destruction (WMD).

Scientists have developed biological, chemical and nuclear weapons. Such weapons include the following historical examples: the mustard gas used by German and British armies in World War I; the aerial spraying of plague-infested fleas by the Japanese military in World War II that killed thousands of Chinese civilians; the dropping of atomic bombs on Hiroshima and Nagasaki by the US Air Force in World War II; the large-scale biological weapons program in the Soviet Union from 1946 to 1992; the biological weapons program of the apartheid government in South Africa; and the use of chemical agents against Kurds by Saddam Hussein's Iraqi regime in 1988.

In recent years there have been a number of high-profile 'defections' of scientists from Western countries to authoritarian states with WMD programs. For example, Dr Abdul Qadeer Khan joined and in large part established Pakistan's nuclear weapons program after working for Urenco in the Netherlands, and Frans van Anraat (also from the Netherlands) went to Iraq to assist Saddam Hussein's WMD program in producing mustard gas.

Moreover, in recent years there have been a number of acts, or attempted acts, of bioterrorism—notably, by the Aum Shinrikyo in Japan (they attempted to acquire and use anthrax and botulinum toxin, and actually carried out a number of terrorist attacks using the chemical sarin gas), by al-Qaeda (they attempted to acquire and use anthrax) and by US Government employees in the so-called Amerithrax attacks (involving the actual use of anthrax).

Given that a small number of animal, human and plant pathogens are readily obtainable from nature, and that bioterrorists with some microbiological training could use these to inflict casualties or economic damage, and especially given the new possibilities provided by synthetic biology (involving the creation of pathogens de novo), evidently there is a non-negligible bioterrorist threat, and it is increasing.[4] This threat is perhaps greatest in unregulated environments, including in weak or failing states in which well-resourced international terrorist groups are allowed to flourish.

Moreover, it would be naive to assume that the scientific community can be entirely trusted to regulate itself in relation to dual-use problems. After all, thousands of scientists have worked in the abovementioned and other WMD— for example, biological weapons—programs, and in doing so have had as their institutional collective end the production of biological weapons. Accordingly, these scientists are directly collectively morally responsible for the existence of those weapons and, in the case of scientists working for authoritarian governments, for enabling authoritarian regimes to possess them. Moreover, on some occasions, as already noted, WMD have actually been used; accordingly, the scientists involved in the development of these WMD are morally implicated, even if only indirectly, in the harms caused by such use.

Scientific freedom and joint action

According to scientist cum philosopher Michael Polanyi:

> The existing practice of scientific life embodies the claim that freedom is an efficient form of organization. The opportunity granted to mature scientists to choose and pursue their own problems is supposed to result in the best utilization of the joint efforts of all scientists in a common task. In other words: if the scientists of the world are viewed as a team setting out to explore the existing openings for discovery, it is assumed that their efforts will be efficiently coordinated if only each is left to follow his own inclinations. It is claimed in fact that there is no other

---

4   Miller, S. 2009, *Terrorism and Counter-Terrorism: Ethics and Liberal Democracy*, Blackwell, London, ch. 7.

efficient way of organizing the team, and that any attempts to coordinate their efforts by directives of a superior authority would inevitably destroy the effectiveness of their cooperation.[5]

Polanyi's view is that each scientist acts freely but does so

1. on the basis of the work of past scientists

2. with constant reference and adjustment to the work of other contemporary scientists

3. in the overall service of a collective end of comprehensive knowledge (in the sense of understanding) of the scientific phenomena in question.

So his conception is one of individual scientific freedom in the overall context of intellectual interdependence in a joint *epistemic* project.

The epistemic character of scientific research is obvious: science aims at knowledge or, more broadly, at understanding. What is perhaps not quite so obvious, however, is that scientific work is, therefore, epistemic *action*. We often contrast the acquisition of beliefs, knowledge and other cognitive states with action—the notion of action being reserved for behaviour or bodily action. Action is intentional and voluntary; on the other hand, knowledge acquisition, it is often held, is unintentional and involuntary. Consider perceptual knowledge. If an object, say, a piece of litmus paper placed in a liquid solution in a test tube, turns blue before our eyes then we come to believe that it has turned blue; we have no choice in the matter, or so it seems.

This focus, however, on perceptual knowledge of ordinary middle-sized objects provides a distorted picture of the scientific *epistemic* project for reasons that will become clear in what follows. A further point to be made here picks up on the joint or cooperative nature of scientific work described above. If scientific research is epistemic action and, at the same time, it is joint action then scientific research is *joint epistemic action*.

Naturally, this being an epistemic academic project, the participants are governed by epistemic principles (for example, replication of experiments), peer-review processes to ensure quality control and uncensored publication.

The key notion in play here is what I am referring to as joint epistemic action. What is joint action?

Joint action consists of multiple individual actions performed by multiple agents and directed towards a collective end—for example, a team of workers building the Empire State Building; a team of terrorists destroying the Twin Towers,

---

5 Polanyi, M. 1951, *The Logic of Liberty: Reflections and Rejoinders*, Routledge & Kegan Paul, London, p. 34.

killing thousands; and a team of scientists discovering the cure for cancer.[6] A collective end is an individual end that each of the participating agents has, but it is an end that no one agent acting alone realises on his or her own. So each agent acts interdependently with the other agents in the service of the same, shared end: the collective end. Again, consider the collective end of a security organisation such as the Federal Bureau of Investigation (FBI) whose members may be jointly working to prevent harm—notably, great harms planned by criminal organisations such as terrorist groups; or consider a team of scientists working feverishly to develop an antidote for some infectious disease that is reaching epidemic proportions.

Joint actions exist on a spectrum. At one end of the spectrum there are joint actions undertaken by a small number of agents performing a one-off simple action at a moment in time—for example, two lab assistants lifting some equipment onto a bench. At the other end of the spectrum there are large numbers of institutionally structured agents undertaking complex and often repetitive tasks over very long stretches of history—for example, those who built the Great Wall of China, or biological scientists developing vaccines.

Joint activity within institutions typically also involves a degree of competition between the very same institutional actors who are cooperating in the joint activity. For example, life scientists within a university who are engaged in joint research and teaching activity in the service of the university's institutional goals are also competing for academic status within the relevant international scientific community, and competing in part on the basis of their individual contribution to those same institutional goals.

Moreover, in many institutional settings organisations compete with one another—for example, business organisations in market settings. Here there is joint activity at a number of levels. For one thing, each competing organisation (for example, a single corporation) comprises a 'team' of individual agents who are cooperating with one another and jointly working to secure the collective ends of the organisation (for example, a biotech company trying to maximise market share). For another thing, each 'team' (for example, each corporation) is engaged in *joint* compliance with the regulatory framework that governs their competitive market behaviour—that is, each complies with (say) the regulations of free and fair competition interdependently with the others doing so and in the service of ensuring the ongoing existence of the market in question. This is consistent with the existence of a regulator who applies sanctions to those organisations which breach the regulations, including safety regulations that might be regarded as a costly and unnecessary impost on business (for example, the 'select agent rule', some mandatory safety procedures in laboratories, and

---

6   Miller, S. 2001, *Social Action: A Teleological Account*, Cambridge University Press, Cambridge, ch. 1.

screening all orders of DNA sequences); the last compliance mechanism is an 'add-on' to the fundamental underlying structure of interdependence of action in the service of collective ends that is constitutive of market mechanisms.[7]

Let us consider further the notion of a joint action and the correlative notion of a collective end. As stated above, joint actions involve multiple agents with the same end—for example, to build a house (the collective end) or map the human genome. Note the following points.

- First, each agent's individual action is a (possibly small) causal contribution to the collective end—for example, building the Great Wall of China, or mapping the human genome.

- Second, each agent's individual action or omission is performed on condition that others perform their contributory actions/omissions; there is interdependence of action.

- Third, each has the collective end only on condition others have the collective end; there is interdependence of ends.

- Fourth, what the collective end is and that it is being pursued are matters of mutual true belief among the participants (A and B mutually truly believe that p if and only if A believes truly that p, B believes truly that p, A believes that B believes that p, and so on).[8]

- Fifth, collective ends are purely conative states; they are not affective states such as feelings or desires. Accordingly, we need to distinguish the mental states constitutive of joint actions (that is, intentions, ends and beliefs) from the mental states that might motivate some joint actions (for example, feelings and desires).

Joint actions can realise collective ends that are also goods—namely, collective goods. Examples of such collective goods are a law-abiding society and an economically viable biotechnology sector. At an organisational level, a collective good might be the realisation of a collective end that consists in harm minimisation or prevention. Thus a biotechnology firm might have as one of its collective ends to avoid any major industrial accidents or to prevent any serious security breaches. Note that in my sense of the term 'collective good', a collective good is simply a good that is produced by joint action directed to a collective end the realisation of which consists in the provision of that good. Such joint action includes action that consists in joint compliance with safety and security procedures that has as a collective end the collective good of prevention of harm.

---

7   Miller, S. 2010, *The Moral Foundations of Social Institutions: A Philosophical Study*, Cambridge University Press, Cambridge, pp. 50–2, and ch. 2.
8   Miller, 2001, op. cit., pp. 56–9.

Such collective goods are not necessarily reducible to an aggregate of individual benefits. Relational goods produced by joint activity—such as social harmony and mutual scientific knowledge (each knows that p and each knows that each knows that p, and so on)—are cases in point.

Some collective goods are intrinsic goods and are jointly pursued for their own sake—that is, they are not pursued merely as a means to an individual end. Various kinds of collective interest, such as the national interest or the interests of the scientific community, are examples of this.

Moreover, a belief in the value of collective goods can motivate action irrespective of individual self-interest—for example, a soldier giving his life in the national interest, or a whistleblower blowing the whistle on a secret biowarfare program.

It will be evident from the above that I am distinguishing between self-interested reasons (or motives) for individual and joint action—for example, so-called 'sticks' and 'carrots'—and moral reasons (or motives) for action (including joint action), and claiming that the latter are not reducible to the former (and vice versa).

It will be further evident that I hold that moral reasons—for example, a belief that biological warfare is wrong, or a belief in the common good—can motivate in and of themselves. So individual self-interest is not the only motive for action. Moreover, even when the motive of self-interest is present, which it obviously typically is, it is not necessarily the dominant motivation.

Armed with this general characterisation of joint action, collective goods and so on, let us now turn to joint epistemic action—the salient form of joint action involved in scientific activity.

Joint epistemic action

As we have seen, epistemic actions are actions of acquiring knowledge. Here we can distinguish between so-called 'knowledge-that' and 'knowledge-how', the former being propositional knowledge (knowledge of the truth of some proposition), the latter being practical knowledge (knowledge of how to undertake some activity or produce some artefact). The definition of propositional knowledge, in particular, is philosophically controversial, but let us assume for our purposes here that someone, A, has knowledge that p if and only if A has a true belief that p and A has a justification for believing that p, which does not rely on some other false belief.[9]

---

9   See, for example, Moser, P. 1989, *Knowledge and Evidence*, Cambridge University Press, Cambridge.

The methods of acquiring propositional knowledge are manifold but for scientific knowledge they include observation, calculation and testimony. Moreover, the acquisition of these methods is very often the acquisition of knowledge-how— for example, how to calculate, how to use a microscope, how to 'read' an x-ray chart.

In the case of the engineering sciences there is an even more obvious and intimate relationship between propositional and practical knowledge, since both are in the service of constructing or making things. Thus in order to build an aeroplane engineers have to have prior practical ('how-to') knowledge and that practical knowledge in part comprises propositional knowledge—for example, with respect to load-bearing capacity. Moreover, this engineering model has increasing applicability in new and emerging sciences such as synthetic biology and nanotechnology. In the case of synthetic biology, for example, scientists can develop new vaccines, enhance the virulence and transmissibility of existing pathogens and even create new pathogens (albeit, presumably using elements of existing pathogens as building blocks).

What counts as sufficient evidence for the possession of knowledge varies from one kind of investigation and one kind of investigative context to another. Thus a scientist seeking a cure for cancer would need his or her experimental results to be replicated by other scientists and the putative cure (say, a new drug) would need to be subjected to clinical trials before being made widely available. A detective investigating a series of murders—for example, by the Yorkshire Ripper—will be focused not only on physical evidence but also on motive (a mental state) and opportunity. Moreover, the evidential threshold for being found guilty is beyond reasonable doubt.

Whereas the acquisition of practical knowledge is readily seen as emanating from action and, indeed, as being a species of action ('knowledge-in-action'), the acquisition of propositional knowledge is a different matter; however, coming to truly believe that p on the basis of evidence—that is, propositional knowledge acquisition—is action in at least three respects.

First, the agent, A, makes a decision to investigate some matter with a view to finding out the truth; the action resulting from this decision is epistemic action. For example, a detective intentionally gathers evidence having as an end to know who the serial killer of prostitutes in Yorkshire is—that is, who the Yorkshire Ripper is. Thus the detective gathers physical evidence in relation to the precise cause and time of death of the Ripper's victims; the detective also interviews people who live in the vicinity of the attacks, and so on. Here A has decided that A will come to have a true belief with respect to some matter, as opposed to not having any belief with respect to that matter—for example, a true belief with respect to who the Yorkshire Ripper is. A's decision is between coming to have true belief and being in a state of ignorance, and in conducting the investigation A has decided in favour of the former. Similarly, a scientist seeking to discover the genetic structure of some organism makes a decision to come to have a true belief with respect to this matter rather than remaining in ignorance.

Second, the agent, A, intentionally makes inferences from A's pre-existing network of beliefs; these inferences to new beliefs are epistemic actions. For example, a forensic scientist might infer the time of death of a murder victim on the basis of her prior belief that rigor mortis sets in within 10 hours after death.

Third, in many cases A makes a judgment that p in the sense that when faced with a decision between believing that p and believing that not p, A decides in favour of p; again, A is performing an epistemic action. For example, our detective, A, intentionally makes an evidence-based judgment (mental act) that Sutcliffe is the Yorkshire Ripper (as opposed to that Sutcliffe is not the Yorkshire Ripper) and does so having as an end the truth of the matter. Here, A is deciding between believing that p and believing that not p; but A is still aiming at truth (not falsity). A is not deciding to believe what he thinks is false. Similarly, our forensic scientist makes an evidence-based judgment in relation to the cause of death of the victim having as an end the truth of the matter. Here the scientist is deciding between believing that the cause of death was x and believing that the cause of death was not x (but was, say, y).

As is the case with non-epistemic action, much epistemic action—whether it is propositional or practical epistemic action or, more likely, an integrated mix of both—is joint action, that is, joint epistemic action. Joint epistemic action is knowledge acquisition involving multiple epistemic agents seeking to realise a collective epistemic end. For example, a team of scientists seeking knowledge of the cure for cancer is engaged in joint epistemic action.[10]

In cases of joint epistemic action there is mutual true belief among the epistemic agents that each has the same collective epistemic end—for example, to discover the cure for cancer. Moreover, there is typically a division of epistemic labour. Thus in scientific cases some scientists are engaged in devising experiments, others in replicating experiments, and so on. So, as is the case with joint action more generally, joint epistemic action involves interdependence of individual action, albeit interdependence of individual epistemic action.

As we saw above, knowledge of the cure for cancer, for example, is joint epistemic action that involves a collective epistemic end and also involves a division of epistemic labour and interdependence of epistemic action. The further point to be made here is that there is interdependence in relation to such collective

---

10   See Miller, 2010, op. cit., ch. 11.

epistemic ends. This is because, given the need for replication of experiments by others, each can only know that p is the cure for cancer—to continue with our example—given that others also know this, that is, there is interdependence in relation to the collective end of knowledge.

A collective epistemic end can be both a collective intrinsic good—and thus an end in itself—and the means to further ends. Knowledge of the cure for cancer is a case in point. Such knowledge consists of propositional and practical knowledge: knowledge of the cure for cancer and knowledge of how to produce it. This knowledge, however, has as a further (collective) end: the actual production of the cure (say, a drug). And this end has in turn a still further end—namely, to save lives.

If knowledge of the cure for cancer is a collective end in itself then it is not simply a means to individual ends—namely, each having as an end that he/she knows the cure for cancer. Rather it is mutually believed that knowledge of the cure for cancer is a collective intrinsic good. In my view, however, moral beliefs can have motivational force.[11] In that case the mutual belief that knowledge of the cure for cancer is a collective intrinsic good can have motivational force.

It follows from this that—as we saw with joint action more generally—joint epistemic action can be collectively self-motivating and does not necessarily have to rely on prior affective states such as desires.

Layered structures of joint epistemic action

Organisational action typically consists of what elsewhere I have termed a *layered structure of joint actions*.[12] One illustration of the notion of a layered structure of joint actions is an armed force fighting a battle. Suppose at an organisational level a number of joint actions ('actions') are severally necessary and jointly sufficient to achieve some collective end. Consider an army fighting a battle. Here the 'action' of the mortar squad destroying enemy gun emplacements, the 'action' of the flight of military planes providing air cover and the 'action' of the infantry platoon taking and holding the ground might be severally necessary and jointly sufficient to achieve the collective end of defeating the enemy; as such, these 'actions' taken together constitute a joint action. Call each of these 'actions' level-two 'actions', and the joint action that they constitute a level-two joint action. From the perspective of the collective end of defeating the enemy, each of these level-two 'actions' is an individual action that is a component of a (level-two) joint action: the joint action directed to the collective end of defeating the enemy.

---

11 See, for example, Benn, S. 1988, *A Theory of Freedom*, Cambridge University Press, Cambridge, ch. 2.
12 Miller, 2001, op. cit., ch. 5.

Each of these level-two 'actions' is, however, already in itself a joint action with component individual actions; and these component individual actions are severally necessary and jointly sufficient for the performance of some collective end. Thus the individual members of the mortar squad jointly operate the mortar to realise the collective end of destroying enemy gun emplacements. Each pilot, jointly with the other pilots, strafes enemy soldiers to realise the collective end of providing air cover for their advancing foot soldiers. Further, the set of foot soldiers jointly advances to take and hold the ground vacated by the members of the retreating enemy force.

At level one there are individual actions directed to three distinct collective ends: the collective ends of (respectively) destroying gun emplacements, providing air cover, and taking and holding ground. So at level one there are three joint actions—namely, the members of the mortar squad destroying gun emplacements, the members of the flight of planes providing air cover, and the members of the infantry taking and holding ground. Taken together, however, these three joint actions constitute a single level-two joint action. The collective end of this level-two joint action is to defeat the enemy; and from the perspective of this level-two joint action, and its collective end, these constitutive actions are (level-two) individual actions.

Importantly for our purposes here there are layered structures of joint *epistemic* action. Consider a crime squad, comprising detectives, forensic scientists and so on, attempting to solve a crime.

At level one, a victim, A, communicates the occurrence of the crime (say, an assault) and description of the offender to a police officer, B. But A asserting that p to B is a joint epistemic action; it is a cooperative action governed by conventions—the convention that the speaker, A, tells the truth and the hearer trusts the speaker to tell the truth.[13]

Also at level one, a couple of detectives interviews the suspect to determine motive and opportunity; the detectives are cooperating with one another in the performance of a joint epistemic action the collective end of which is to discover motive and opportunity.

Finally, at level one, a team of forensic scientists analyses the available physical evidence—for example, the DNA of the blood samples of the offender found on the victim are matched to the suspect's DNA; the forensic scientists are engaged in joint epistemic action to determine whether there is or is not a DNA match.

These three level-one joint epistemic actions are constitutive of a level-two joint epistemic action—namely, the level-two joint epistemic action directed towards

---

13 Miller, 2010, op. cit., ch. 11.

the collective end of determining who committed the crime. Accordingly, when each of the level-one joint epistemic actions is successfully performed then the level-two joint epistemic action is successfully performed—that is, the crime squad solves a crime.

Now consider an example of a large scientific project conducted by a number of cooperating organisations and hundreds of scientists over many years—namely, the human genome project. The project involved multiple connected goals—collective ends—and multiple layered structures of joint action, including joint projects in publishing, undertaken to realise those goals.

In fact most organisations are hierarchical institutions comprising task-defined roles standing in authority relations to one another, and governed by a complex network of conventions, social norms, regulations and laws. Consider a science department in a university or the forensic laboratory in a police organisation: both comprise heads of department, scientists, laboratory assistants and so on, and the work of both is governed by scientific norms of observation, replication of experiments, and so on. So most layered structures of joint action, including joint epistemic action, are undertaken in institutional settings, and scientific joint epistemic action is not an exception.

Institutions have de facto purposes/strategic directions—that is, collective ends, such as to maximise shareholder profit (corporations), find a cure for cancer (university research team), or build an atomic bomb (military organisation). Institutions also have specific structures (hierarchical, collegial and so on) and they have specific cultures (for example, a competitive, status-driven ethos).[14] In this connection consider scientific activity—for example, biological research, undertaken in three different institutional settings: that of the university, the commercial firm and the military biodefence organisation. Some of the principal purposes/strategic directions (collective ends) of commercial firms—for example, to maximise shareholder profits—are quite different from, and possibly inconsistent with, those of universities (for example, scientific knowledge for its own sake), and quite different again from those of military research establishments (for example, to save the lives of *our* military personnel). Again, the hierarchical structures within a military research establishment are quite different from the more collegial structures prevailing in universities; and the structure of commercial firms is quite different again. The general point to be made here is that scientific activity is not only a form of complex joint activity (a layered structure of joint epistemic action); it is activity that is inevitably shaped by the non-scientific institutional setting in which it is conducted—that is, by the specific collective ends, structure and cultures of particular institutions.

---

14   Miller, 2010, op.cit., 'Introduction'.

Here we also need to stress the distinction between the de facto institutional collective end, structure and/or culture and what it *ought* to be; cultures, for example, can vary greatly from one organisational setting to another, notwithstanding that the type of institution in question is the same or very similar.

In the light of the above, we can distinguish the normative account of science as a joint intellectual activity—for example, aimed at knowledge for its own sake—from science as a means to broader social ends (for example, vaccines to save lives). Moreover, we can distinguish both from the normative account of specific institutions in which science exists principally as a means—for example, commercial firms (vaccines to make profit) and a military biodefence organisation (vaccines to save the lives of our military personnel). Importantly, in the context of discussion of dual-use concerns, we can distinguish within the normative account of science (both at the level of joint intellectual activity and at the level of specific institutions) between its positive ends (for example, knowledge for its own sake and knowledge as a means to combat disease) and negative ends (for example, harm prevention). Earlier we discussed the no means to harm principle (NMHP)—namely, the principle that one should not foreseeably (whether intentionally or unintentionally) provide others (for example, bioterrorists) with the means to do great harm. Clearly the means in question is essentially scientific or technological knowledge and this knowledge is the product of joint epistemic action (indeed, joint epistemic action undertaken in the context of multiple layered structures of joint epistemic action). Accordingly, harm prevention in relation to dual-use concerns is, or at least morally ought to be, a joint enterprise with respect to joint epistemic action; it is something that biological scientists as members of their scientific community (or communities) and specific institutions (for example, biology departments in universities, biotech companies) morally ought to jointly address including (presumably) by way of education and regulation of their potentially harmful joint epistemic action. But let us be clear on the moral responsibilities in play here.

Collective moral responsibility

As we have seen, dual-use issues in science are inherently ethical, and scientific activity is essentially joint epistemic activity undertaken in various institutions with possibly divergent normative underpinnings. Given the multilevel, cooperative character of scientific activity and given its ethical or moral significance, a key theoretical normative notion in play here is that of collective moral responsibility.

Let us distinguish between natural, institutional and moral responsibility and, in respect of responsibility, between individual and collective responsibility. I note that the notions of natural, institutional and moral responsibility are not mutually exclusive. I also note that agents responsible for some outcome might have both individual and collective responsibility for that outcome.

An agent, A, has *natural* responsibility for some action, x, if A intentionally did x for a reason and x was under A's control. Bench scientists engaging in routine scientific research—for example, replication of experiments—have natural responsibility for their actions. Moreover, such actions might not have any obvious moral implications.

Agent A has *institutional* responsibility for action x if A has an institutional role that has as one of its tasks x. Thus, for example, laboratory assistant, A, has the institutional responsibility to clean the test tubes; moreover, A has this responsibility even if A does not in fact do this.

What of moral responsibility? Roughly speaking, agents have moral responsibility for natural or institutional actions if those actions have moral significance. So if A is naturally or institutionally responsible for x (or for some foreseeable outcome of x, O) and x (or O) is morally significant then, other things being equal, A is morally responsible for x (or O) and, other things being equal, can be praised/blamed for x (or O).

Note that other things might not be equal if, for example, A is a psychopath (and, therefore, incapable of acting in a morally responsible fashion) or if A does something wrong but has a good excuse (and, therefore, ought not to be blamed).

Note also that if O involves some intervening agent, B, who directly causes O, then A may have diminished moral responsibility for O.

Let us now consider collective moral responsibility. In essence, the account of collective moral responsibility mirrors that of individual moral responsibility, the key difference being that the actions in question are joint actions, including joint epistemic actions.

Accordingly, if agents A, B, C and so on are naturally or institutionally responsible for a joint (including epistemic) activity, x (and/or some foreseeable outcome of x, O), and x (and/or O) is morally significant then, other things being equal, A, B, C and so on are collectively (that is, jointly) morally responsible for x (and/or O) and, other things being equal, can be praised or blamed for x (and/or O).

The 'other things being equal' clauses function here as they did in the above account of individual moral responsibility. Moreover, as was seen to be the case with individual moral responsibility, if there are additional intervening (individual or joint) actions then those jointly responsible for the joint action in question, and its outcome, may have diminished moral responsibility. Scientists who engage in dual-use research that is subsequently used in the construction of WMD may well have diminished responsibility for the harm caused by those WMD. Diminished responsibility is not, however, necessarily equivalent to no responsibility. Further points to be made here are as follows.

First, each agent may have full or partial moral responsibility for x jointly with others for the joint action x and/or its outcome. If, for example, five men each stab a sixth man once, killing him, each is held fully morally (and legally) responsible for the death even though no single act of stabbing was either necessary or sufficient for the death. In some cases each agent might have full moral responsibility (jointly with others) for some outcome, O—notwithstanding the fact that each only made a very small causal contribution to the outcome—in large part because each is held to have prior full institutional (including legal) responsibility (jointly with others) for O.

On the other hand, each agent might have partial and minimum moral responsibility jointly with others if each only makes a very small and incremental contribution as a member of a very large set of agents performing their actions over a long period—for example, the scientists who worked on the human genome project.

Second, we need to distinguish cases in which agents have collective moral responsibility for some joint action or its outcome from cases in which agents only have collective moral responsibility for failing to take adequate preventative measures against O taking place. Many untoward dual-use cases are of the latter kind.

On the other hand, agents may not have any collective moral responsibility with respect to some foreseeable morally significant outcome, O, if O has a low probability, takes place in the distant future and involves a large number of intervening agents.

The collective moral responsibilities of scientists are multiple. Scientists have a collective institutional (professional) and moral responsibility as scientists to acquire knowledge for its own sake.

Scientists functioning in universities also have a collective institutional and moral responsibility to acquire knowledge for the good of humanity—for example, vaccines for poverty-related diseases.

Scientists functioning in commercial firms might have a collective institutional and (contractually based) moral responsibility to acquire (say) knowledge of vaccines for rich people's diseases—since that is a commercial imperative of their employer and they are being paid to do just that.

Scientists functioning in biodefence organisations have a collective institutional (and moral?) responsibility to acquire knowledge of vaccine-resistant pathogens if this is a national security imperative of their employer—that is, the government.

As human beings, scientists have a collective moral responsibility not to provide the means for others to intentionally do great harm—for example, the means to allow others to drop atomic bombs on Hiroshima and Nagasaki or engage in biowarfare.

Moreover, these various collective institutional and moral responsibilities may be inconsistent with one another—notably, the collective moral responsibilities scientists have as human beings and the institutional responsibilities that they might have as members of military research organisations.

Regulatory frameworks

In the light of the dangers stemming from the dual uses of science and technology there is a need to consider a range of regulatory measures to reduce the risks. Such measures include the imposition of limits on dual-use experiments and on the dissemination of potentially dangerous information resulting from dual-use discoveries. These measures themselves exist on a spectrum ranging from the least intrusive/restrictive to the most intrusive/restrictive.[15]

Some specific regulatory measures that might be considered include the following ones.

1. Mandatory physical safety and security regulation: Should there be regulations providing for physical safety and security in relation to the storage of, the transport of and physical access to samples of pathogens, equipment, laboratories and so on? And should compliance with such regulations be mandatory? In both cases the answer is presumably in the affirmative.

2. Licensing of dual-use technologies/techniques: Should there be mandatory licensing of dual-use technologies/techniques/pathogen samples? Only certain laboratories in the public sector and the private sector might be licensed to

---

15 For a detailed treatment of these and other options, see Miller and Selgelid, 2008, op. cit.

engage in research involving the use of certain dual-use technologies and licences for DNA synthesisers might be required.

3. Mandatory education and training: Given the potential harms arising from, for example, 'experiments of concern', it is clear that some process of education and/or training for relevant researchers and other personnel is called for.

4. Mandatory personnel security regulation: Doubtless it is prudent, indeed it is a moral requirement, that access to virulent pathogens be disallowed to a researcher diagnosed as a psychopath or to a known member of a terrorist organisation.

5. Censorship/constraint of dissemination: The question of whether research findings ought to be freely disseminated, censored or their dissemination in some lesser way restricted is an extremely difficult issue and it is by no means obvious who the ultimate decision-maker ought to be. A relevant important distinction here is that made above between first-tier and second-tier dual-use research. For example, first-tier research findings might need to be disseminated in such a way that anyone being informed of these findings would not be able to replicate the experiments that enabled the results reported in the findings. Indeed, this is precisely what was recommended at one point (for example, by the US National Science Advisory Board for Biosecurity) in relation to the dual-use bird flu research mentioned at the outset of this chapter.

Let us assume that a range of regulatory options ought to be pursued, at the institutional (university, commercial firm, government research laboratory), national and international levels (for example, Biological Weapons Convention verification processes). There remain some in-principle obstacles to the establishment of adequate measures to deal with the dual-use problem in science and technology and a number of these stem from various perverse incentive structures that derive from collective-action problems. In the final section of this chapter, I will briefly discuss some of the more salient of these.

Collective-action problems

Thus far we have characterised the scientific enterprise as essentially a joint epistemic one: the emphasis has been on intellectual cooperation to achieve common scientific (epistemic) goals in an institutional context of minimal restrictions on the freedom of individual scientists. This picture, however, while acceptable as far as it goes, is an oversimplification. Specifically, it obscures the competitive dimension of scientific activity and, in particular, it masks various collective-action problems arising from such competition. This is important for our purposes here, not least because it casts the dual-use problem in a somewhat different light.

On the purist (as we might call it) model of scientific activity as joint epistemic action performed under conditions of scientific freedom, the dual-use problem arises only because scientific research undertaken for the benefit of humankind can be misused by others for harmful purposes. Accordingly, there is a need to monitor dual-use research and erect safeguards against misuse by malevolent individuals and groups—for example, 'lone wolf' malcontents, nihilistic terrorist groups, 'rogue states', and so on.

Notice, first, that there is here an implicit additional assumption—namely, that scientific activity will be undertaken in the first place in order to benefit humankind. This is, as we have seen in relation to WMD programs, not necessarily the case. On the other hand, WMD research is not dual use in our sense (unless military research undertaken for protective purposes yet having the potential, as is probable, to be misused for aggressive purposes is to be regarded as dual use).

Notice, second, that much scientific work, including in the biological sciences, is not undertaken under conditions of scientific freedom. Consider research undertaken in the private sector or for various government laboratories in, for example, authoritarian states such as China. Which research is undertaken, and whether or not it is published, is not necessarily or even typically a decision made by individual scientists or, indeed, by groups of scientists. Rather these are commercial decisions made by managers or they are decisions made by government officials in the national interest (presumably). Accordingly, it is simply not true that scientific work, including scientific work in the biological sciences, is necessarily, or even typically, conducted under conditions of scientific freedom.

But to return to the main point at issue—namely, competition and, relatedly, collective-action problems: in the biological sciences, as elsewhere, there is competition between individual scientists, between scientific institutions and between nation-states.

It is self-evident that there is competition between, for example, biotechnology companies in the private sector. Moreover, governments compete insofar as they have an interest in promoting their own biotech industries and, more generally, insofar as they want to ensure that they do not fall behind in research and development (R&D) in the biological sciences (not least for military reasons). Further, even in the case of scientific work undertaken under conditions of scientific freedom (for example, in universities), there are important elements of competition—for example, between rival teams of scientists in competition for status and (relatedly) for scarce funding.

As suggested above, competition in these various sectors gives rise to a variety of collective-action problems that have important implications for the dual-use issue.

First, in the private sector there are collective-action problems arising from commercial competition. As already noted, many scientists work in commercial firms in which there is an imperative to maximise profit. In such a context of fierce commercial competition restrictions on dual-use research may handicap an organisation. This is a collective-action problem insofar as an organisation—all things considered and given appropriate weight—ought to choose *not* to perform a particular dual-use experiment on the grounds that the potential harm to humankind resulting from this kind of experiment might outweigh the potential benefits to humankind. All things might not be considered, however, or given appropriate weight. Specifically, the firm might give excessive weight to its commercial interests, especially if it believes that some other competing firm is likely to be less scrupulous and go ahead with the experiments in question. In short, in dual-use cases where discretionary judgment is called for, the judgment might be skewed by considerations of commercial self-interest in a fiercely competitive commercial environment. I note that commercial self-interest may well be dominant in such cases, notwithstanding the commitment of individual scientists to the NMHP. For one thing, it is not necessarily the decision of the scientists, who are, after all, mere employees; and for another, their self-interest as employees might align them with the firm's commercial interest, especially given the relative ignorance of scientists of security issues.

Second, in the university sector there are collective-action problems arising from competition for status. As already noted, many scientists working in the university sector are engaged in a competition for status (and for scarce funds to undertake projects by means of which they can achieve status), both for themselves as individuals and on behalf of the institutions they work for. Accordingly, there is an analogue of the above-described collective-action problem in the private sector. In dual-use cases where discretionary judgment is called for, the judgment might be skewed by considerations of individual or institutional self-interest in a competitive environment, albeit the competition in question is primarily for status (and scarce funds to achieve status).

Third, in the government sector there are collective-action problems arising from competition among nation-states. As noted, in the past and, indeed, in the present there has been a variety of arms races—for example, the nuclear arms race—in which scientists played a central role. The problem here is that national self-interest is pitted against humanity's collective interest in a context in which there is no enforceable international law; evidently nation-states cannot effectively collectively self-regulate—hence the WMD programs of the United States, Iran, North Korea and so on. The inability or unwillingness to collectively self-regulate exists, however, at least potentially, in relation to the dual-use problem and does so independently of any inherent desire on the part of nation-states to maintain WMD programs. We saw above that the self-interest of individual scientists and the institutions in which they work (for example, commercial firms and universities) can under conditions of fierce competition lead to collective-action problems in relation to dual-use research. Nation-states are, however, themselves in competition with one another, and it is in the economic, military and so on interests of nation-states to support their own R&D in science and technology—that is, to support the work of their commercial firms and universities, and to do so in the face of 'foreign' competition. Accordingly, we cannot necessarily look to individual governments to regulate adequately the scientific research in their own institutions, at least if 'adequately' in this context refers to an all-things-considered, morally informed decision made in the long-term interests of humankind—as opposed to a decision made in the (possibly short-term) national interest.

A fourth collective-action problem arising from competition is of a somewhat different kind: it is a species of the generic problem of freeriding. It is the possibility of the untoward consequences of scientific freeriding, so to speak. Let us assume that Polanyi's scientific freedom model—for example, no censorship—is in fact the best model to acquire new knowledge; those operating entirely outside the model cannot compete. Accordingly, so the argument runs, the 'good guys' (for example, the scientists making vaccines within the framework of scientific freedom) stay ahead of the 'bad guys' (for example, the scientists weaponising pathogens outside the framework of scientific freedom); the 'bad guys' are always playing catch-up. Contrary to this argument, however, it might be claimed that a well-qualified national cohort of 'bad guys' can always free ride but then get ahead of 'good guys'—for example, scientists in an authoritarian state with biodefence projects benefit from the work of those in the scientific freedom model but don't share their own work.

13. Biosecurity and the Just-War Tradition[1]

Koos van der Bruggen

Biosecurity and the just-war tradition occupy separate worlds. In debates and discussions about biosecurity and dual use, no references are made to just-war criteria. And in textbooks on just-war tradition, biosecurity and dual use hardly get any attention. This chapter will deal with the question of whether this separation is justified. First, an oversight will be provided of the just-war tradition and more especially of recent developments within it. Attention will be given to which questions this tradition deals with, why and how. The answers it offers will then be applied to biosecurity and dual-use issues.

Just-war tradition: Between past and present[2]

Thinking about what counts as a 'just war' can truly be called a tradition. It has its roots in authors such as Aristotle, Cicero, Augustine, Thomas Aquinas, Grotius, as well as Spanish theologians such as Vitoria and Suarez, and the Swiss diplomat de Vattel. They and many other philosophers, theologians and lawyers have contributed to debate about how to approach the notion of 'just' in the context of conflict.[3] In recent years, debate has continued through authors such as Paul Ramsey, James T. Johnson, Michael Walzer, Jeff McMahan and—again—many others.[4]

Just-war *tradition* as a name for this normative approach to conflict is preferable to just-war *theory*. The use of the word theory implies a framework was developed at a certain time and has remained more or less static. In fact, just-war thinking has evolved and has been influenced by political, social, technological and military developments. The political situation of the Roman Empire cannot be compared with that of the Italian city-states or the Spanish colonial expeditions.

---

1 The views expressed in this chapter are those of the author and do not represent those of the Royal Netherlands Academy of Arts and Sciences.
2 See also: van der Bruggen, K. 2009, 'Other wars, other norms?' in G. Molier and E. Nieuwenhuys (eds), *Peace, Security and Development in an Era of Globalization. The Integrated Security Approach Viewed from a Multidisciplinary Perspective*, Martinus Nijhof, Dordrecht/Leiden/Boston, pp. 355–76.
3 An anthology of relevant texts is published by: Reichberg, G. M., Syse, H. and Begby, E. (eds) 2006, *The Ethics of War. Classic and Contemporary Readings*, Blackwell, Malden, Mass.
4 Classic studies are: Walzer, M. 1977, *Just and Unjust Wars. A Moral Argument with Historical Illustrations*, Basic Books, New York; Johnson, J. T. 1981, *Just War Tradition and the Restraint of War*, Princeton University Press, Princeton, NJ; Johnson, J. T. 1984, *Can Modern War Be Just?* Yale University Press, New Haven, Conn.

Also technical developments have impacted thinking. A well-known example involves a high-tech weapon of the Middle Ages: the crossbow. During the Second Lateran Council, its use was prohibited, at least against Christians.[5] This ban can be seen as an early application of the principle of proportionality, at least for Christians. The example shows that debates about the moral acceptibility of the use of weapons are longstanding. This certainly holds true for biological and chemical weapons. May and Crookston write that 'for 2,500 years the use of poisons during battles has been forbidden'.[6] The moral and legal rejections of poisons sometimes were even more rigorous than those of other means. From Roman authors such as the historian Valerius Maximus, we know the saying *'armis bella non venenis geri'* (wars are fought with weapons, not with poison).[7] In other words: poison was not seen as a legitimate weapon. This prohibition did not mean that no poisons have been used in times of war though. History provides many examples of poisoning:

> Evidence can be found for the existence of forms of chemical and biological warfare in ancient and classical times. The evidence for chemical and toxin warfare is the clearest. Solon of Athens is said to have used hellebore roots (a purgative) to poison the water in an aqueduct leading from the Pleistrus River around 590 B.C. during the siege of Cirrha. Writings of the Mohist sect in China dating from the Fourth Century B.C. tell of the use of ox-hide bellows to pump smoke from furnaces in which balls of mustard and other toxic vegetable matter were being burnt into tunnels being dug by a besieging army to discourage the diggers. The use of a toxic cacodyl (arsenic trioxide) smoke is also mentioned in early Chinese manuscripts. Sparta used the toxic smoke generated by burning wood dipped in a mixture of tar and sulfur during one of its periodic wars with Athens.[8]

This overview can be extended with examples from the Middle Ages until the Iran–Iraq war in the 1980s.[9] In spite of the use of poisons and other biological agents during wars, the saying of Valerius Maximus has always kept a kind of validity. Biological (and chemical) weapons never reached the status of 'normal' weapons, comparable with swords, spears, crossbows and later firearms. Using biological 'weapons' always had the smell of doing something foul. An example is the praise of Cicero for a Roman general who refused the offer of a deserter to

---

5   <http://www.nuclearmuseum.org/online-museum/article/waging-peace-the-challenge-of-nuclear-stewardship/> (viewed 2 January 2011).
6   May, L. and Crookston, E. 2008, 'Introduction', in L. May (ed.), *War. Essays in Political Philosophy*, Cambridge University Press, Cambridge, p. 2.
7   Valerius Maximus, *Factorum Et Dictorum Memorabilium Libri Nouem*, <http://www.thelatinlibrary.com/valmax6.html> (viewed 11 January 2011).
8   <http://www.cbwinfo.com/History/History.html#0001> (viewed 2 January 2011).
9   Ibid.

poison King Pyrrhus.[10] And Samuel von Pufendorff (1632–94) wrote that 'the more civilized nations condemn certain ways of inflicting harm on an enemy: for instance the use of poison'.[11]

Back to just-war thinking in general. New times and new situations ask for revisiting what moral and legal norms should be weighed and how in seeking to answer the question if (and under what conditions) the use of violence is justified. Sometimes developments lead to more than marginal adjustments of assessments. So after World War II, the issue was posed of whether the invention of nuclear weapons led to novel considerations in relation not only to the possible *use* of these weapons, but also to *deterrence*. Is it, for instance, morally justified to threaten their use if this would go against all prevailing criteria regarding what is appropriate in terms of violence? What if deterrence is the only way to prevent the actual use of nuclear weapons? Such complex questions raised by this case could lead to the conclusion that the just-war tradition has lost its relevance and significance.[12] But such a view is too rash given it is based on only one—however important and shocking—development in the waging of war. In the heydays of the Cold War, (too) many so-called conventional wars were fought in which just-war criteria could and should have provided good guidance. More than this: just-war thinking could provide criteria for moral argument on the paradoxes of nuclear deterrence, as is shown in the development of a theory of justified deterrence.[13]

Just-war criteria

In the just-war tradition, a distinction is made between the so-called *jus ad bellum* and the *jus in bello*. The *jus ad bellum* indicates if and when it can be justified to start a war. The criteria one to five below refer to this *jus ad bellum*. The sixth one belongs to both domains and the remaining two criteria deal with the question of the *jus in bello*: what behaviour is permitted or forbidden during a war. More recently a third domain has been added: the *jus post bellum*. There is much debate about whether this new distinction is useful and necessary for judging modern wars. Elsewhere I have questioned the utility for a *jus post*

---

10   Cicero, 'On duties', in Reichberg et al., op. cit., p. 53.

11   von Pufendorf, S. 2006, 'On the duty of man and citizen', in Reichberg et al., op. cit., p. 459.

12   A common way of thinking during the Cold War was that a nuclear war never could be a just war and that—by consequence—just-war tradition had lost its significance. Often this view was promoted by pacifists, who had already disputed just-war thinking by definition. But another group was the so-called nuclear pacifists, who limited their view to the conclusion that a just nuclear war was a contradiction in terms.

13   van der Bruggen, K. 1986, *Verzekerde Vrede of Verzekerde Vernietiging. Ontwikkeling van een Theorie vanGerechtvaardigde Afschrikking* [*Assured Peace or Assured Destruction. Development of a Theory of Justified Deterrence*], Kok, Kampen.

bellum.[14] Setting aside this wider question though, the *jus post bellum* will not be considered in this chapter, because it is not very relevant for biosecurity issues.

Jus ad bellum

1. The war must be waged by a legitimate government

This criterion expresses the assumption that only a legitimate government has and should have the monopoly on violence. In a sovereign state the government and the government alone is allowed to use violence. This may seem obvious, but in practice things often are not so obvious, because the legitimacy of a government is not always undisputed. Much violence (for example, in civil wars or anticolonial wars) stems from conflicts over who owns the legitimate authority. Often it works out that legitimacy can be decided only post hoc: the winner takes all. In addition, the notion of legitimate authority had varying meanings over time. Thomas Aquinas had a much different concept than Hugo Grotius, and also in our days definitions are changing. But the fact remains that the notion of 'legitimate authority' has a certain core holding. By convention—actually since the Peace of Westphalia in 1648—sovereign states hold the monopoly on violence.[15] States and government embody the public domain and because of that they differ from all other (private) groups and entities. Given all disputes and discussions about sovereignty, this first just-war criterion certainly is not ideal, but something better is not available. If the public domain collapses, the war of all against all remains a possibility, even today. And in such a situation sovereign authority will be sought. Recent examples can be found in so-called failed states, where in practice the central government has virtually disappeared. Many people in that situation eventually accept the authority of terrorist movements, because—in the view of powerless groups—their violent and arbitrary exercise of power is always better than the absence of any authority. To give an example: it is no wonder that in Afghanistan many people accept (which is not the same as support) the authority of the Taliban instead of what they see as a condition of anarchy.

2. The war must be waged for a just cause

Fighting war for a just cause is the core of the *jus ad bellum*. Influenced by the Charter of the United Nations (1945), the interpretation of what counts under this category has become more and more restricted in recent times. Only a reaction against aggression or an intervention that is sanctioned by the UN

---

14   van der Bruggen, 2009, op. cit.
15   Schrijver, N. 1998, 'Begrensde soevereiniteit—350 jaar na de Vrede van Munster [Limited sovereignty—350 years after the Peace of Münster]', *Transaktie*, vol. 27, no. 2, pp. 141–74.

Security Council is taken as a legally acceptable base for waging war. This means that—to give some examples—retribution or recoveries of previously suffered injustices are not acceptable justifications.

In a sense the interpretation has also been extended. Since the end of the Cold War the idea has become widespread that the international community has the right—if not the duty—to use military force to intervene in cases of gross violations of human rights. The intervention in Kosovo in 1999 serves as one example, even though a UN Security Council authorisation was missing. A further extension came after 11 September 2001, when the 'war against terror' was added to the list of just wars by many people. The invasion of Afghanistan to expel the Taliban regime was not explicitly mandated by the UN Security Council, but the American action was widely seen as a legitimate form of self-defence supported by resolutions that were adopted after 9/11.[16] In relation to biosecurity the question arises if and under which conditions the development or possession of biological weapons can be a reason to start a war. This is not a purely academic question as is shown below in the heated debates about the American–British intervention in Iraq.

3. The war must be waged with a right intention

This criterion is in line with the previous one, but it goes further. It says that it is not justified to have secondary intentions when fighting a war for a just cause. The only intention should be that just cause (as resisting the aggressor). If the defender succeeds in that objective, he must stop fighting. He is not allowed to aim at military or economic destruction of the opponent. According to the classical tradition, feelings of revenge also should play no role. Most recent wars have shown that standard proves unrealistic. In fact, in most wars, intentions other than the 'official' ones can be discerned. Economic (oil), political or personal secondary intentions are never far away. If and how these secondary intentions played a part in the Iraq war—formally intended to remove biological and chemical weapons—will be illustrated below.

4. All other means of conflict resolution must be exhausted; war is the last resort

This criterion is designed to prevent governments from taking up arms too quickly. The question that arises is: when is 'too quickly' and who is to decide? Parties that are contemplating force would likely argue that it is the only solution. The relevance of this criterion is that it is forcing the 'aggressor' to make the case for why aggression is necessary. Is a military intervention the

---

16 For example, United Nations Security Council (UNSC) Resolution 1373 (2001): <http://daccess-dds-ny.un.org/doc/UNDOC/GEN/N01/557/43/PDF/N0155743.pdf?OpenElement> (viewed 11 January 2011).

last remaining option? Are all other means really exhausted? These kinds of questions have to be asked and answered. The problem is that the answers to these and similar questions can never be given with mathematical certainty.

Fortunately, history also shows many examples where war could be prevented. Humankind has repeatedly escaped the battlefield.[17] Of course we owe this to several causes, but certainly the message of the just-war tradition—and more specifically of this criterion—played a role in the idea that starting a war is a decision that has to be avoided if possible. This can be illustrated by the way political leaders try to persuade others that their war is justified and that they do not have any other possibility. One need not believe what is said to still note the importance of making a justification. Peace is the rule; war (unfortunately all too often) the exception.

5. There must be a reasonable chance that the intended purposes are reached through the war

The purpose of this criterion is to prevent a government from starting a war if it is clear beforehand that the intended targets cannot be reached. At least one objection to this criterion rises immediately: it seems to support the side of the strongest party. Yet military superiority is not always a guarantee for victory. The most famous example in recent history is the Vietnam War. Despite overwhelming military superiority, the Americans did not manage to reach their goals. Especially in a humanitarian intervention, the militarily stronger party does not always have the greatest interest in going on with fighting. The tragedy of Srebrenica seems an example. If the UN Protection Force (UNPROFOR, the UN troops in Bosnia) had used all available resources, the fall of Srebrenica could have been prevented. Among other factors, the fact that the self-interest of the Netherlands and the Dutch military was not at stake certainly played a role. The professional soldiers of Dutchbat had signed up for the army and realised that an ultimate consequence of this was to die, '[b]ut dying for Srebrenica. It was not worth it.'[18]

6. The objectives and means of war must be in a reasonable relationship to each other. This is the principle of proportionality

The principle of proportionality can relate to both the *jus ad bellum* and the *jus in bello*. In the first case, the principle refers to waging the war: is there a proportionate relationship between its (expected) cost and the aim to be reached? During the war the principle can be applied to judge if a concrete action or the

---

17   Gerrits, A. and de Wilde, J. 2000, *Aan het slagveld ontsnapt. Over oorlogen die niet plaatsvonden*, Waburg Pers, Zutphen. This book is about wars that did not take place.
18   Westerman, F. and Rijs, B. 1997, *Srebrenica. Het zwartste scenario* [*Srebrenica. The Blackest Scenario*], Atlas, Amsterdam/Antwerpen, p. 11. Quote from a Dutchbat soldier.

use of a certain kind of weapon is proportionate in relation to the goals to be reached. In both cases the discussion will concentrate on the issue of weighing and judging the factors that determine the aims and the means. Who is to decide on this and on the basis for weighing factors? This problem becomes even more complicated because it is almost impossible to know beforehand what the costs of a war or even a concrete action during a war will be. Who would have predicted that hundreds of thousands of French, British and German soldiers would be killed in the trenches of World War I? When these soldiers marched into the battlefields in August 1914, they expected to be home at Christmas in the same year, after having won a *'frischen und fröhlichen Kriege'*. Proportionality can become a salami-like criterion: the lives already lost become an argument for going on, because not doing so would mean that these victims lost their lives in vain.[19] So the limits of what still is seen as proportional shift.

7. The use of force should be limited to the minimum that is necessary in order to resist the aggressor

This criterion has a certain affinity with the just mentioned proportionality principle. But it goes further by suggesting even less than proportionally acceptable violence has to be used, if that is enough for achieving the aims of war. Again the same questions return as in the preceding two criteria: what is minimal violence and who is to decide? Once started wars have their own momentum that often leads to more escalation. Von Clausewitz, the famous Prussian theorist of war, assumed that wars were by definition absolute and that they should be.[20] Only politics could prevent that absolute character.

Indeed, the influence of politics in the recent wars of intervention has been great, but that has not always led to restraint. Regarding the Kosovo war there has been much criticism of the decision to carry out aerial bombardment for strategic reasons on some civilian targets in Yugoslavia. First there was the question of whether it was necessary from a military perspective. The attack on the TV studio in Belgrade is one of the most infamous examples. This brings us to the final criterion of the just-war tradition to discuss here: the principle of noncombatants.

---

19   This was the reason that some parents of Dutch soldiers who had been killed in Afghanistan were among the people who resisted the return of Dutch troops. The same argument can be read in the memoir of George W. Bush, where he describes his meetings with parents of killed soldiers: George W. Bush 2010, *Decision Points*, Crown, New York, ch. 12, pp. 355–94.

20   von Clausewitz, C. 1982, *Over de oorlog*, Het Wereldvenster, Bussum, p. 15.

8. A distinction should be made between military and civilians; the latter group may not be involved in the fight. This is the noncombatants principle

The noncombatants principle has always played an important role in the just-war tradition and it still does. This principle has been 'translated' in international treaties like the Geneva conventions. What is important is that not only citizens belong to the category of noncombatants, but also soldiers who are not actively involved in the fight (for instance, those wounded, and prisoners). Developments in modern technology have, however, led to wars that are conducted in ways different than in the time when the noncombatants principle was developed. Massive battles seem at least in the Western world to belong to the past—and whoever remembers the images of the trenches of World War I can of course only be happy with it. But it begs the question of the actual meaning of the noncombatants principle.

Today's wars by major military powers are often wars at a distance. There is hardly direct physical contact with the combatants of the counterparty and most governments are reluctant to create situations in which such a direct confrontation is provided. The discussion about whether or not to proceed with a ground war during the intervention in Kosovo (1999) is such an example. At the other end of the spectrum—that of the noncombatants—there have been developments also. To give only one instance: the distinction between combatants and noncombatants coincided for centuries with the difference between civilians and soldiers. That is not true anymore. Most terrorists do not belong to the military, but they surely are to be categorised as combatants.

Despite this noncombatant criterion, the general trend is that attacks in which civilians are deliberately targeted have become the 'normal' picture of modern wars. Names of cities such as Guernica, Dresden and Hiroshima suffice as an illustration. And even when citizens are not deliberately chosen targets, questions can be asked—for instance, regarding the concept of collateral damage. A question in this context is whether this so-called collateral damage is too easily accepted. How unintended is unintended if you know that the chosen method of attack will lead almost by definition to civilian casualties or to victims because of destroyed infrastructure? Walzer stated in his famous Just and Unjust Wars that collateral damage in such a case cannot be morally justified.[21] More recently, however, he was less convinced of this judgment. In a lecture in Amsterdam in 2007, Walzer states that the distribution of responsibility is the ultimate moral factor in judging if the killing of civilians can be justified. As an example he gives his view on the Lebanon war (2006):

---

21   Walzer, 1977, op. cit., pp. 151–4.

In the Lebanon war, the Israeli army caused most of the civilian deaths, but some (or many) of the villages it attacked were being used by Hezbollah as bases for rocket attacks on Israeli cities, so the greater part of the responsibility for civilian deaths in those villages lay with Hezbollah—as did the greater part of the responsibility for the war itself, which began with a rocket barrage and a Hezbollah raid across the international frontier. Israel is responsible for deaths caused by unjustifiable bombings like the Qana raid or by the cluster bombs used late in the war. But (again) it shouldn't be the proportionality argument that guides our judgment of those deaths; they were wrong whether or not they were disproportionate. In Vietnam, Kosovo, and Lebanon, it is the balance of responsibility that is morally determinative.[22]

Just-war tradition and biological weapons: The Iraq war of 2003

How can a link be made between just-war principles and the problems of biological weapons, or—more generally—biosecurity? At first sight the *jus ad bellum* is not directly linked to questions of biosecurity. Seen in the widest sense of the term though, some recent international conflicts, and more especially the Iraq war (2003), do make such a link. Albeit this is a link that evokes many questions!

Officially, one of the main arguments the United States and the United Kingdom used to justify their attack on Iraq was the consideration that Saddam Hussein had weapons of mass destruction (WMD), and as part of this arsenal, biological weapons. In his oral evidence before the Chilcot Inquiry Committee in 2010, former UK prime minister Tony Blair confirmed that he still maintains the possession of WMD by Iraq as a justification for the war against Saddam.[23]

Is the development or possession of biological weapons by a country a 'just cause' for starting a war? Here it will be argued that this is not the case. The development of WMD—objectionable as it may be on legal or moral grounds—is not equal to aggression as defined in the UN Charter. This means that the only justification for an intervention in Iraq had to be found in a resolution under Chapter 7 of the UN Charter. This is what the United States and the United Kingdom tried to realise in the months before they invaded Iraq. When such a resolution appeared to be unattainable, however, they decided to start the war

---

22 <http://www.ru.nl/soeterbeeckprogramma/terugblik/terugblik-2007/teksten-2007/thomas_more_lezing_0/> (viewed 12 November 2010).

23 <http://www.iraqinquiry.org.uk/media/45139/20100129-blair-final.pdf> (viewed 4 October 2010).

without a (new) resolution. Attempts were made to justify this by an appeal to the so-called 'revival argument'. This contended that the whole range of UN resolutions against Saddam since 1991 could be legitimating for the war. This argument was defended by Peter Goldsmith, the then attorney-general for England and Wales.[24] Because the Dutch Government took his arguments on, the Dutch Committee of Inquiry on the War in Iraq analysed this argument in detail. They concluded that it was not a valid way of reasoning on the basis of public international law.[25]

Of course this argument is based on legal judgments. A case on purely moral grounds could lead to a different judgment. This is acknowledged by the Dutch Committee of Inquiry:

> [S]ome defend the position that a basis in international law alone cannot be the deciding factor for the justification of international action by states. The observance of international law rules is very important, so runs the argument, but cannot always be decisive. Sometimes, in an international conflict, values of such importance are at stake that states can feel compelled to act even when this may not be according to the prevailing international law. However, these are exceptional situations in which very compelling moral imperatives apply. The legal maxim, 'need before law', and the international law concept 'state of necessity' also offer safety valves within the law for finding a way out of this situation. It is striking that the debate on the possible justification for the Iraq war never went down this path.[26]

In other words: none of the members of the 'coalition of the willing' used this 'need for law' argument. Such an argument usually is used as a reason for human intervention in case of genocide. It has never been used in the case of the (suspected or expected) development of weapons of mass destruction.[27]

In the debates about the Iraq war the question has often been asked whether considerations or intentions other than the official ones played a part in the decision for intervention. And, if so, can these intentions be seen as proper? Many commentators have pointed to other motivations: the importance of oil interests, the—hardly hidden—wish for regime change and last but perhaps

---

24   'Iraq resolution 1441 advice—original memo', *BBC News*, 7 March 2003, <http://news.bbc.co.uk/1/shared/bsp/hi/pdfs/28_04_05_attorney_general.pdf> (viewed 4 October 2010).
25   Dutch Committee of Inquiry on the War in Iraq 2010, 'Report of Dutch Committee of Inquiry on the War in Iraq, Chapter 8: The basis in international law for the military intervention in Iraq', *Netherlands International Law Review*, vol. LVII, no. 1, pp. 126–8.
26   Ibid., pp. 87–90.
27   Although not all (applications of) biological weapons lead to mass destruction, I will follow the usual terminology for chemical, biological, radiological and nuclear (CBRN) weapons as weapons of mass destruction.

not least the desire of President George W. Bush to finish the job of his father. Moreover, it is well known that Tony Blair had an almost religious zeal to restore human rights in Iraq.

Were all other means of conflict resolution exhausted and was war indeed the last resort?

This third criterion of the just-war tradition played an important role in the discussions about the Iraq war. Opponents did not stop to mention the efforts of the International Atomic Energy Agency (IAEA) and the UN Monitoring, Verification and Inspection Commission (UNMOVIC) in the search for possible WMD. Both organisations had confidence in the success of their efforts. Because of that they were opposed to the strategy of Bush and Blair, which was directed at starting a conflict—almost independently of the results of the IAEA and UNMOVIC. The United States and the United Kingdom had, however, developed their 'plan of attack' and they were not willing to make that plan dependent on the possible results of the inspections. The planning of the invading countries was related more to practical circumstances in and around Iraq.[28]

Looking at the Iraq war, it seems that biological weapons hardly played any role in the final decision. But in the declaratory policy to justify the war as well as in the long lead-up to the war, the issues certainly were important. It helped to make war acceptable: if a country possesses and even uses chemical or biological weapons, it must be a very despicable regime and thus war is an acceptable way to get rid of that regime. In other words: the (supposed) possession of WMD as such is seen a *casus belli*.

The next criterion (number five) that should be taken into account is that there must be a reasonable chance that the intended purposes are reached by waging war. If the purpose of a war is eliminating possible stocks of WMD or dual-use capacities, the feasibility of this purpose is not only dependent on the outcome of the war. Of course it is helpful and perhaps even necessary to have military superiority in those areas where the WMD or other materials are stockpiled, but in addition an intensive scientific survey has to be set up. Anyone who takes a look at the complexity of the activities of the UN Special Commission (UNSCOM), UNMOVIC, the IAEA and—after the war—the Iraq Survey Group gets an impression of the work that has to be done to find, identify and eventually eliminate WMD materials. The inspectors Hans Blix (UNMOVIC) and Mohamed El Baradei (IAEA) had repeatedly declared that they would be able to finish their complex job without military intervention. This makes the argument that war is necessary to reach the goals of eliminating biological (and nuclear or chemical) weapons at least implausible. Assuming the 'hidden' goals or intentions of the

---

28   Woodward, R. 2004, *Plan of Attack*, Simon & Schuster, New York.

Iraq war, it can be argued that the goal of regime change indeed was reached. Within two months the regime of Saddam Hussein collapsed. But, as we know now, this was not the end of the fighting.

What can be said of proportionality (criterion six) as an argument for whether or not to start a war? Proportionality as a criterion of the *jus ad bellum* becomes relevant if—according to the other *ad bellum* criteria—starting a war can be justified. Proportionality in that case is an added criterion to judge if the relation between purposes and means is balanced. There is no need to consider the proportionality criterion if it has already been determined that the other *ad bellum* arguments define a war as unjust. Given the fact that despite all counterarguments the Iraq war took place, the *jus in bello* context of the proportionality criterion is applicable. But how to apply it to the specific aspect of biosecurity—or broader biological weapons—is not very obvious. The Iraq war—once started—led to many actions that were and could not be foreseen. This war is no exception to the rule that a war creates its own judgments on proportionality: the longer a war lasts, the less some actions are appreciated as disproportionate. The formal link with the search for biological weapons drifted out of sight and out of mind.

Finally, both parties have violated the noncombatant principle many, many times during the Iraq war. Many civilians were killed by Iraqi militants as well as by coalition troops.[29] But it is not possible to link these violations and the biosecurity issue. The only thing that can be said is that this war—at least partially—was based on handling the problem of weapons of mass destruction; however, this was a very exceptional situation. This evokes the question if and how just-war criteria are relevant for 'everyday' biosecurity policy. This question will be addressed by looking for elements of this policy to which just-war criteria could or should be applied.

Biosecurity, dual use and just-war tradition

The Iraq war was an exceptional event in the way in which the (alleged) development and possession of WMD led to an armed conflict. In this section, attention will be paid to some more mundane aspects of biosecurity, where there is no imminent threat of war. Is an appeal to criteria of the just-war tradition in these circumstances helpful?

---

29   The number of civilian deaths is registered at <www.iraqbodycount.org> (viewed 12 January 2011). See also: Hsiao-Rei Hicks, M., Dardagan, H., Guerrero Serdán, G., Bagnall, P. M., Sloboda, J. A. and Spagat, M. 2009, 'The weapons that kill civilians—deaths of children and noncombatants in Iraq, 2003–2008', *New England Journal of Medicine*, vol. 360, pp. 1585–8.

Suspicion of development of biological weapons

The cases of Iraq and also—more recently—Iran, North Korea and Syria are about states that are involved in or at least suspected of developing nuclear, biological or chemical weapons. What if the main actors threatening to develop WMD are non-state actors? Is it allowable to attack a state if inhabitants of this state are suspected of acting in such a way? Is there a right for third parties to start a war and, if so, under what conditions? At first sight, this right does not exist. Two scenarios can be discerned. First, it is possible that the government of a country itself is developing, producing, stockpiling or otherwise acquiring biological weapons. If this country is a state party of the BWC, it should act in accordance with Article 1 of the convention.

The second possibility is that the development, production, stockpiling or acquisition of biological weapons is taking place within the territory of a state or under its jurisdiction or control. In that case Article 4 of the BWC is violated. It seems evident that the first party that is to act in such a case is the involved national state. Each government has the duty to prevent the misuse of biological agents. The state should do all that is possible to put an end to this situation. This has been arranged in the BWC and in many more treaties and agreements, such as UN Security Council Resolution 1540.

The BWC also indicates what has to be done if there is a breach of an obligation of the convention. This is described in Articles 6 and 7 of the convention:

Article VI

(1) Any State Party to this convention which finds that any other State Party is acting in breach of obligations deriving from the provisions of the Convention may lodge a complaint with the Security Council of the United Nations. Such a complaint should include all possible evidence confirming its validity, as well as a request for its consideration by the Security Council.

(2) Each State Party to this Convention undertakes to cooperate in carrying out any investigation which the Security Council may initiate, in accordance with the provisions of the Charter of the United Nations, on the basis of the complaint received by the Council. The Security Council shall inform the States Parties to the Convention of the results of the investigation.

Article VII

Each State Party to this Convention undertakes to provide or support assistance, in accordance with the United Nations Charter, to any Party to the Convention which so requests, if the Security Council decides that such Party has been exposed to danger as a result of violation of the Convention.

Articles VI and VII determine that it is the UN Security Council which has to decide if any measures will be taken. This means that any attack by other states to end such a violation is not justified, unless there are other reasons that justify such an attack (such as an imminent threat or a resolution of the UN Security Council that justifies the use of 'any other means').

Of course there are other measures that can be taken to prevent a violation of Articles I and IV. Most important is to look for peaceful solutions: cooperation, helping to counter terrorist threats, education and training, and so on. These measures are especially fitting and relevant in cases where states are not the ones violating the BWC, but groups or organisations within a country. The importance of these measures has been stressed by Resolution 1540 of the Security Council on the nonproliferation of weapons of mass destruction. In this resolution attention is paid explicitly to the possibility that states need help in realising the goals of this resolution. The council

> [r]ecognizes that some States may require assistance in implementing the provisions of this resolution within their territories and invites States in a position to do so to offer assistance as appropriate in response to specific requests to the States lacking the legal and regulatory infrastructure, implementation experience and/or resources for fulfilling the above provisions.

This resolution has been adopted under Chapter VII of the UN Charter, so it has an obligatory character, but it does not entail any direct or indirect legitimation for using violence.

Biodefence as bio-offence

According to Article I, BWC state parties are allowed to undertake activities for prophylactic, protective or other peaceful purposes. This implies that biodefence is allowed as far as this is limited to protective or peaceful purposes. But who is to decide what purposes are protective and peaceful? In practice biodefence can coincide with bio-offence. And even this could be defended with an argument that is derived from the debate on nuclear deterrence. The argument could be that having the *ability* to produce biological weapons will deter another state or non-state actor from using their biological weapons because of fear of retaliation. This of course was the logic of the nuclear-deterrence policy during the Cold War. In those days many debates were devoted to the question of whether nuclear deterrence was justified from a moral point of view: was it allowable to threaten using a weapon that clearly should lead to a violation of the proportionality and the noncombatant principles? Nothing like consensus was reached in this debate between people who were of the opinion that it could never be right to threaten with these WMD and others, who defended their view that deterrence was the only way to prevent the use of nuclear weapons.[30]

If this latter view could be defended from a moral point of view—and there are some convincing arguments for it—then that was only possible in very specific circumstances. This includes the bilateral relationship between the United States and the Soviet Union during the Cold War in combination with the strategic importance of nuclear weapons. Biological weapons never had the same strategic importance in practice. During the Cold War there was never a situation that gave reason for a deterrence strategy with biological (or chemical) weapons. In fact, the United States and the Soviet Union even agreed on the BWC during the heyday of the Cold War, although the Soviet Union for at least some years still went ahead with expanding its program—in part because they thought the United States was doing the same.

If biological deterrence was not defendable during the Cold War, the same holds today. There can be no strategic or political argument that overrules the moral inhibitions of the proportionality principle and noncombatants principle. Moreover, there is reason to believe that such a policy of biological deterrence would undermine the BWC. A possible reasoning that biological deterrence is not directed at other states (let alone state parties of the BWC), but at non-state actors, is untenable. Terrorist groups are almost certainly not deterred by fear of retaliation. Besides, it is almost impossible to react with a targeted action. Terrorists are often not directly linked with a specific area or state. Of course it is conceivable, as happened after the 9/11 attacks, to attack the (presumed) host state of the terrorists; however, the case of the military actions against Afghanistan and al-Qaeda could be spoken of as proportionate and targeted actions, but using biological weapons is almost by definition disproportionate and untargeted.

In summary: biodefence may be allowed according to the BWC, but the margins of this research are limited. Caution is required, especially since the risk of dual use of biodefence is not at all imaginary. The anthrax letters of 2001 (which allegedly came from a biodefence laboratory) are an already classic example. And of course there is the risk of accidents. The Sverdlovsk accident (1979)

---

30 In 1996 the International Court of Justice gave this judgment on nuclear weapons: 'the threat or use of nuclear weapons would generally be contrary to the rules of international law applicable in armed conflict, and in particular the principles and rules of humanitarian law. However, in view of the current state of international law, and of the elements of fact at its disposal, the Court cannot conclude definitively whether the threat or use of nuclear weapons would be lawful or unlawful in an extreme circumstance of self-defence, in which the very survival of a State would be at stake.' International Court of Justice 1996, *Legality of the Threat or Use of Nuclear Weapons*, Advisory Opinion of 8 July 1996.

suffices to draw attention to these kinds of risk. Although such accidents do not take place during a war, the 'collateral' damage that almost by definition will be caused by biological weapons can be seen as a violation of the noncombatants principle of the just-war tradition.

Conclusion

This chapter started with the observation that biosecurity and the just-war tradition occupy separate worlds. This observation has been confirmed. There are not many overlaps between both. But although the overlaps are few, some clear lines can be drawn between the two.

Biological weapons are in the category of weapons of mass destruction, but biological weapons have much less military value today in deterrence and in practice than nuclear weapons. The most important example of a war that was waged for reasons that were linked to biological weapons was the Iraq war of 2003. But this link existed more on paper and in the declaratory policy than in reality. As far as the argument was used, it cannot be accepted as a just cause for the war.

This does not mean that developing and storing biological weapons are justified from a just-war perspective. Most of these weapons (certainly the ones with contagious agents) are by definition indiscriminate, and using them would be a violation of the noncombatants principle. The same argument also applies to a possible bioterrorist use of biological agents. And for the category of non-indiscriminate biological weapons, the argument against their use can be found in the inhumane character of these means. From a more military point of view, the argument of Valerius Maximus ('*armis bella non venenis geri*': wars are fought with weapons, not with poison) still can be seen as valid.

The consideration that possession of biological weapons could be legitimated for reasons of deterrence is refuted by the current political and military situation. Because of the BWC, which became possible because of the limited military value of biological weapons, there is no credible reasoning that these weapons have a deterring function.

14. The Precautionary Principle and the Dual-Use Dilemma

Steve Clarke

Three precautionary principles

The precautionary principle (PP) is a conceptual tool used to guide decision-making in the management of risk.[1] It has been widely taken up in environmental law, and is now being applied in a variety of contexts, including the regulation of potentially dangerous technologies. It was first developed in Sweden and the former West Germany in the late 1960s,[2] was explicitly used in West German environmental law by the 1980s[3] and has become increasingly influential in many countries since then, particularly in Europe.[4] That the PP is usually referred to as *the* PP might seem to suggest that there is a canonical formulation of this principle. But this is not the case. There are at least 20 different versions of the PP[5] and new ones appear on a regular basis. The fact that there are many different versions of an abstract principle is perhaps not surprising in and of itself. What is somewhat surprising is that these different principles do not appear to be variants of a more general principle. Instead, they are only loosely associated with one another. What they have in common is a shared history and the fact that they all advance precaution in some way. As we will see, they involve at least three very different approaches to the advancement of precaution, which can be understood in terms of their differing relationships with cost–benefit analysis (CBA) (also known as benefit–cost analysis).

Development of the PP was initially motivated by dissatisfaction with CBA, which was the dominant conceptual tool used in risk management until the rise of the PP, and which continues to be very widely applied. The application of

---

1   The PP is also sometimes understood, more broadly, as a set of guidelines for structuring the deliberative processes involved in risk management. See, for example, Rappert, B. and Moyes, R. 2010, 'Enhancing the protection of civilians from armed conflict: precautionary lessons', *Medicine, Conflict and Survival*, vol. 26, no. 1, pp. 24–47.
2   See Sunstein, C. 2005, *Laws of Fear: Beyond the Precautionary Principle*, Cambridge, Cambridge University Press, p. 16.
3   Majone, G. 2002, 'What price safety? The precautionary principle and its policy implications', *Journal of Common Market Studies*, vol. 40, pp. 89–109.
4   The PP was referred to in 27 resolutions of the European Parliament between 1992 and 1999, is referred to in the 1992 Maastricht Treaty on the European Union, and appeared in a draft constitution for the European Union. See Sunstein, op. cit., p. 17.
5   See ibid., p. 18.

CBA involves attempting to determine the probability of benefits occurring, and the probability of costs being incurred, as well as determining the relative sizes of the benefits and costs of a particular course of action and balancing these. This calculation is compared with the relative balance of costs and benefits for alternative courses of action from which the option with the best overall balance (adjusting for the probability of these occurring) is selected.[6] From the 1960s, environmental lawyers, policymakers and activists became increasingly dissatisfied with many of the decisions that were made with the use of CBA.

One source of dissatisfaction was that, in many actual applications, only potential costs that were established with 'full scientific certainty' were considered. A precautionary corrective to this tendency was the development of versions of the PP that were intended to guide the use of CBA so as to ensure that potential costs other than those established with 'full scientific certainty' were given due consideration.[7] A good example of this form of the PP is Principle 15 of the *Rio Declaration on Environment and Development*: 'In order to protect the environment, the precautionary approach shall be widely applied by States according to their capabilities. Where there are threats of serious or irreversible damage, lack of full scientific certainty shall not be used as a reason for postponing cost-effective measures to prevent environmental degradation.'[8]

A second complaint about decisions guided by CBA was that in some of these it was implicitly assumed that the onus of proof of the existence of a cost fell on critics of an activity. A precautionary response to this tendency was to supplement CBA with a second step specifying conditions under which the 'onus of proof' lay with proponents, rather than critics, of an activity. The well-known *Wingspread Statement* is a good example of this sort of PP: 'Where an activity raises threats of harm to the environment or human health, precautionary measures should be taken even if some cause and effect relationships are not fully established scientifically. In this context the proponent of the activity, rather than the public, should bear the burden of proof.'[9]

Neither insistence on full scientific certainty nor placement of the onus of proof on critics is intrinsic to CBA; so the above versions of the PP supplement CBA,

---

6   This decision-making process is philosophically problematic in various ways, most obviously because benefits often appear to be incommensurable with risks. For a discussion of further philosophical problems with CBA, see Hansson, S. O. 2007, 'Philosophical problems in cost–benefit analysis', *Economics and Philosophy*, vol. 23, pp. 163–83. For a defence of CBA against various criticisms, see Schmidtz, D. 2001, 'A place for cost–benefit analysis', *Philosophical Issues (A Supplement to Nous)*, vol. 11, pp. 148–71.

7   For more on the history of the development of this type of precautionary principle, see Magnus, D. 2008, 'Risk management versus the precautionary principle: agnotology as a strategy in the debate over genetically modified organisms', in R. N. Proctor and L. Schiebinger (eds), *Agnotology: The Making and Unmaking of Ignorance*, Stanford University Press, Stanford, Calif., pp. 250–65.

8   United Nations Environment Programme, 1992.

9   *Wingspread Statement on the Precautionary Principle 1998*, <www.gdrc.org/u-gov/precaution-3.html> (viewed 1 November 2010).

rather than replacing it with an alternative. What have come to be known as strong versions of the PP (sPP), however, have been developed with the intention of replacing CBA, at least under certain circumstances, with an altogether different approach to risk management that does not involve weighing the costs and benefits of a particular policy and comparing these weightings with those of possible alternative policies. Instead, it involves an exclusive focus on the potential costs of a particular policy. An example of sPP is the *Final Declaration of the First European 'Seas at Risk' Conference*, 1994: 'If the "worst case scenario" for a certain activity is serious enough then even a small amount of doubt as to the safety of that activity is sufficient to stop it taking place.'[10]

Here we are instructed to focus our attention exclusively on costs and ignore the potential benefits of particular activities when making policy, no matter how significant these potential benefits might be.

We have focused on the above three examples of the PP because these are clear examples of different types of PP that are intended to advance precaution in particular ways. Not all versions of the PP are as clear as these. Indeed, some are couched in highly nebulous language that makes it hard to see what their creators are trying to achieve, beyond conveying enthusiasm for precaution. We will go on to examine an example of the PP, created especially for application in dual-use contexts,[11] which suffers from this problem. Jordan and O'Riordan[12] embrace the vagueness of (many versions of) the PP, which they see as a virtue that enables those who employ it to be more politically effective. This may be true if the PP is understood, as Jordan and O'Riordan understand it, as an intellectual tool of protest movements; however, its vagueness becomes a vice when the PP is used to try to steer policy.[13] Indeed, scholars who are concerned about effective precautionary regulation sometimes complain that while adoption of the PP changes conceptualisations of risks, it does not appear to have had a clear effect on regulatory practice.[14]

I have analysed the PP in terms of its different possible relationships to CBA. Some commentators, such as Sandin,[15] might want to object to the direct

---

10 Cited in Sunstein, op. cit., p. 29.

11 Kuhlau, F., Hoglund, A. T., Evers, K. and Eriksson, S. 2011, 'A precautionary principle for dual use research in the life sciences', *Bioethics*, vol. 25, no. 1, pp. 1–8.

12 Jordan, A. and O'Riordan, T. 1999, 'The precautionary principle in contemporary environmental policy and politics', in C. Raffensperger and J. A. Tickner (eds), *Protecting Public Health and the Environment: Implementing the Precautionary Principle*, Island Press, Washington, DC, pp. 15–35.

13 Clarke, S. 2005, 'Future technologies, dystopic futures and the precautionary principle', *Ethics and Information Technology*, vol. 7, pp. 121–6.

14 O'Riordan, T. and Cameron, J. (eds) 1994, *Interpreting the Precautionary Principle*, Earthscan, London; and Eckley, N. and Selin, H. 2004, 'All talk, little action: precaution and European chemicals regulation', *Journal of European Public Policy*, vol. 11, pp. 78–105.

15 Sandin, P. 1999, 'Dimensions of the precautionary principle', *Human and Ecological Risk Assessment*, vol. 5, pp. 889–907.

comparison of any form of the PP with CBA, taking the view that, while CBA has been developed to be applicable in circumstances where we need to make policy decisions under risk, the PP has been developed to address circumstances of uncertainty. The distinction between risk and uncertainty goes back to Knight.[16] In his terminology, situations of risk are circumstances where the probabilities of possible outcomes can be specified, on the basis of reliable evidence, and situations of uncertainty are circumstances where the probability of possible outcomes cannot be specified, on the basis of reliable evidence. Flipping a normal coin creates an instance of risk without uncertainty as we know all the possible outcomes of a coin flip (heads and tails) and we are warranted in specifying particular probabilities for these outcomes. Speculation about the details of a possible afterlife is a case of uncertainty without risk. There are many possible accounts of the ways in which an afterlife might be experienced, but we appear to lack good grounds for assigning probabilities to any of them.

While coin flips are pure instances of risk and speculations about the afterlife seem to be pure instances of uncertainty, these are exceptional cases. The majority of real-world cases where decisions involving probabilities need to be made involve a mixture of risk and uncertainty. Consider a couple of real-world decisions: 1) an insurer who writes home insurance policies will try to price those policies on the basis of assessments of the risks to which particular houses are exposed. It is not possible, however, to anticipate all such risks and work out exactly how likely these are, so there is inevitably an element of uncertainty in such assessments. Exact probabilities can, of course, be assigned to possible outcomes, but such assignments will involve a degree of stipulation. 2) A patient who is contemplating an operation will want to know what the risks involved are and the surgeon who is set to conduct the operation will generally try to transmit this information to the best of her ability. But while it is possible to anticipate many of the risks involved in a complex operation, estimates of how likely these are to occur in particular contexts will inevitably involve a degree of speculation. In many real-world circumstances, we attempt to turn uncertainties into risks, however, this typically involves some speculation and so an element of uncertainty remains.[17] Most real-world circumstances involve a mix of risk and uncertainty and the PP and CBA can both be applied to these. So it is appropriate to make direct comparisons of CBA and the PP in dealing with most real-world cases.

---

16   Knight, F. 1921, *Risk, Uncertainty, and Profit*, Hart, Schaffner & Marx, Boston.
17   Knight's view of risk is based on the (I think plausible) presupposition that we try to assign objective probabilities to the world, and that when we are unable to do so, we are left with uncertainties: LeRoy, S. F. and Singell, L. D. 1987, 'Knight on risk and uncertainty', *The Journal of Political Economy*, vol. 95, pp. 394–406. Some commentators, such as Friedman, who take the view that we only ever assign subjective probabilities to the world, hold that we never need to have recourse to the Knightian concept of uncertainty at all. Friedman, M. 1962, *Price Theory: A Provisional Text*, Aldine, Chicago.

Precaution, paradox and bias

Strong versions of the PP (sPP) have attracted much controversy and have been subjected to an apparently devastating form of criticism, which is that they typically lead to contradictory policy recommendations, if applied consistently, and are therefore paralysing.[18] This is because viable alternatives to a policy typically have risks associated with them and if all of these are given due consideration under sPP then policy paralysis will result. Consider, for example, the application of sPP to the possible development of nuclear power plants, intended to address an impending shortage of energy in a particular country. Clearly there are risks associated with building and running nuclear power plants, so an application of sPP would lead to the recommendation that we do not develop nuclear power plants. But what will happen if we fail to develop nuclear power in this context? Viable alternatives seem also to involve risks. If we develop new sources of power that are based on fossil fuels then we increase the chance and severity of climate change. So sPP precludes the development of new fossil fuel-based power sources. Another alternative is not to provide additional power but to try to convince people to use less energy. But this alternative involves the risk of social instability, so sPP precludes the policy of not developing new sources of power. If the three options considered above are our only viable alternatives then the consistent application of sPP necessarily provides incoherent policy recommendations, precluding all viable courses of action, and also precluding inaction.

One way that advocates of sPP have attempted to avoid paralysing paradoxical outcomes is by insisting that only risks over a certain threshold of significance (understood in terms of probability of occurrence and/or seriousness of costs involved) be considered in formulations of the PP.[19] Indeed our exemplar of sPP, the 'Final Declaration', mentions risks being 'serious enough' to warrant consideration (although it does not specify how likely it is that these will be realised), so it might be supposed that it is an instance of sPP that can avoid paradoxical outcomes. The advocate of this strategy to defend sPP from the change of paradox needs to make a choice about where to set the levels of significance and/or likelihood of occurrence of risks. If these are set too high then sPP will not be applicable to many of the risks that advocates of precaution have wished to apply it to. On the other hand, if they are set too low then the problem of paradox won't be avoided and sPP will continue to issue contradictory advice.[20] In effect this is a 'Goldilocks strategy': it might work

---

18 Manson, N. A. 2002, 'Formulating the precautionary principle', *Environmental Ethics*, vol. 24, pp. 263–74; and Sunstein, op. cit.

19 Sandin, P., Peterson, M., Hannson, S. O., Rudén, C. and Juthe, A. 2002, 'Five charges against the precautionary principle', *Journal of Risk Research*, vol. 5, pp. 287–99.

20 See Clarke, op. cit., p. 126.

if we know in advance where to set levels of significance, but because we are dealing with the management of uncertain outcomes we usually don't have any reliable way of knowing where to set such thresholds, so there is not usually any reason to be confident that our strategy has succeeded. There have been other attempts to evade or resolve the paradox of sPP,[21] but they do not appear to be successful either.[22]

If sPP leads to incoherent policy recommendations and causes policy paralysis then it seems odd that this is not often noticed by those who attempt to apply sPP, such as those who attempt to apply the 'Final Declaration'. The reason this is not often noticed is that when sPP is applied it is usually applied selectively. We are able to apply sPP when we focus on one possible policy outcome and fail to consider alternatives; and it seems that we have a propensity to do this. This propensity is encouraged by the structure of the PP. Whereas CBA involves an explicit consideration of alternative policies, most versions of the PP are designed to be applied to one policy option at a time, and do not involve explicit comparison with alternatives. According to Sunstein,[23] there is also a deeper reason we tend to consider risks selectively when applying the PP, which is that we are hostage to a variety of significant cognitive biases that have the effect of blinding us to the fact that we are typically faced with risks however we decide to act (and even if we don't act). Sunstein mentions various sources of cognitive bias that have this effect including probability neglect, a belief in the benevolence of nature, loss aversion and systems neglect;[24] however, the factor he considers the most significant in encouraging a selective approach towards risk is the widespread use of the availability heuristic.[25]

The availability heuristic is a rule of thumb that people intuitively use to estimate the magnitude of particular risks. If I am asked how likely an earthquake, a flood or a bushfire is in Oxford, I will tend to make intuitive assessments of the likelihood of these events by seeing how readily I can bring to mind episodes in which such events have taken place in Oxford and will adjust my estimates in accordance with the 'availability' to me of instances of such events. Availability

---

21   Gardiner, S. M. 2006, 'A core precautionary principle', *Journal of Political Philosophy*, vol. 14, pp. 33–60; and Weckert, J. and Moor, J. 2006, 'The precautionary principle in nanotechnology', *International Journal of Applied Philosophy*, vol. 20, pp. 191–204.

22   Clarke, S. 2009, 'New technologies, common sense and the paradoxical precautionary principle', in P. Sollie and M. Duwell (eds), *Evaluating New Technologies: Methodological Problems for the Ethical Assessment of Technological Developments*, Springer, Dordrecht, pp. 159–73.

23   Sunstein, op. cit.

24   Ibid., p. 37.

25   See ibid., p. 5. For more on the availability heuristic, see Tversky, A. and Kahneman, D. 1982, 'Judgment under uncertainty: heuristics and biases', in D. Kahneman, P. Slovic and A. Tversky (eds), *Judgment under Uncertainty: Heuristics and Biases*, Cambridge University Press, Cambridge, pp. 3–21; and Kahneman, D. and Frederick, S. 2002, 'Representativeness revisited: attribute substitution in intuitive judgment', in T. Gilovich, D. Griffin and D. Kahneman (eds), *Heuristics and Biases: The Psychology of Intuitive Judgment*, Cambridge University Press, Cambridge, pp. 49–81.

is influenced by both familiarity and salience.[26] The more familiar I am with a class of events, all things being equal, the greater will be my intuitive assessment of its likelihood. If a class of events is particularly salient to me—if, say, I have personal experience of a flood or have just watched a documentary about the danger of flooding in the Thames Valley, where Oxford is located—then that will also increase my intuitive estimate of the likelihood of flooding in the area.[27]

When particular risks are highly available to us they tend to 'crowd out' other risks that we would otherwise be inclined to consider. After the events of 11 September 2001, ordinary estimates of the dangers associated with air travel rose dramatically, particularly in America, and this led many people to alter their travel plans and drive or take a train rather than fly. It seems that such decisions, made at that time, were a response to the high availability of the dangers of air travel, which crowded out consideration of the risks involved with other forms of transport. But the risks of alternative travel choices were significant. It is estimated than an extra 350 road fatalities occurred in America in the final three months of 2001 as a result of people avoiding air travel in the aftermath of the events of 11 September. This is a higher number of deaths than the number of deaths of passengers and crew in all four of the crashed flights of 11 September combined (266 deaths).[28]

Dual use and lessons from debates about the precautionary principle

A dual-use dilemma arises when a piece of research—typically in the life sciences—has potential benefits as well as the potential to cause harm.[29] If we go ahead and conduct this research then we may enjoy the benefits that might result from it, but only if we are willing to bear the risk of potential harms. If we do not conduct such research then potential harms are avoided but benefits are also forgone. How are we to decide what to do? The most obvious way to attempt to make such decisions is by applying CBA, under which we attempt to weigh the potential benefits of use of the potential piece of research in question against the potential harms. It has recently been suggested, however, that the PP

---

26   See Sunstein, op. cit., p. 37.
27   The passage of time will tend to decrease both familiarity and salience, and thereby lead to a decrease in intuitive estimates of likelihood.
28   Gigerenzer, G. 2004, 'Dread risk, September 11 and fatal traffic accidents', *Psychological Science*, vol. 15, pp. 286–7.
29   Miller, S. and Selgelid, M. 2007, 'Ethical and philosophical consideration of the dual-use dilemma in the biological sciences', *Science and Engineering Ethics*, vol. 13, p. 524.

could be usefully applied in dual-use contexts.[30] Even more recently, a specific formulation of the PP for application in 'dual-use life-science research' has been developed, which we will go on to consider.

If a PP is to be applied to dual-use dilemmas then I would urge those who are attempting to apply it to be as clear as they can about what role the PP is supposed to play in the resolution of dual-use dilemmas. As we have seen, it might be expected to play (at least) three very different roles in decision-making. First, the PP might be used to ensure that some harms, which might not be properly considered, are properly considered in applications of CBA in dual-use contexts. Second, it might be invoked in an attempt to specify where the 'onus of proof' lies in respect of either the significance of particular harms or the viability of particular remedies for those harms. Third, it might be invoked as an alternative to CBA. There is no one way to be cautious, and it is important that we try to be clear about how we intend to be cautious, if we are to do so in dual-use contexts. It is also important that we explain why we need to be cautious in this or that particular way. What is the particular failing of current approaches to dual-use dilemmas that the PP is supposed to help address; and are there any other possible responses to this problem that we might also consider?

As we have seen, one of the key problems with some versions of the PP (sPP) is that application of these depends on us adopting a selective approach to risk. We can adopt a selective approach to risk and be oblivious to the fact that we are doing so because of widespread cognitive bias that affects ordinary assessments of risk. We should ask ourselves whether framing a problem as a dual-use dilemma makes us more or less susceptible to adopting selective attitudes towards risk. We should also ask ourselves if applying sPP (or other versions of the PP) adds to such problems or reduces these. When we are considering whether to develop a new drug, framing our choice as a dual-use dilemma encourages us to focus our attention on a comparison of the consequences of going ahead and developing the drug, which will involve potential benefits as well as potential costs, with the consequences of not developing the drug. We may be in a position, however, to develop a variety of different drugs that provide overlapping benefits and involve overlapping risks. Rather than considering choices about which of these to develop and which not to develop as a series of isolated dilemmas, it may be more sensible to try to make a comparison of the overall benefits and costs of each, before deciding which options to pursue.

Framing a problem as a dual-use dilemma may well increase our propensity to adopt selective approaches to risk. When we frame a potential development as a dual-use dilemma (or any other sort of dilemma), we are encouraging a focus on

---

30 See Rappert, B. 2008, 'The benefits, risks and threats of biotechnology', *Science and Public Policy*, vol. 35, p. 40.

a choice between exactly two alternatives; however, as is the case with the above example, there may be contexts in which we may be able to make comparative choices that are more complicated than simple dilemmas, and such possibilities may be obscured from us by framing our choice as a dilemma.[31]

Framing a problem as a dual-use dilemma encourages us to consider a choice between two alternatives in isolation from other possible choices. And at least some versions of the PP (sPP) can only be applied if we make a decision about risks in isolation from consideration of alternatives. Other versions of the PP, including the *Wingspread Statement* and the Rio declaration, are also structured around consideration of the risks that are associated with a particular activity and do not encourage us to consider risks involved with alternatives to that activity. The concern here is that if the PP is applied to the dual-use dilemma then the tendency of both of these conceptual structures to focus our attention on a particular policy option, to the exclusion of consideration of other options, may reinforce one another. So we have a reason to be especially wary about applying PP when that is combined with the framing of options as dilemmas.

A suggested precautionary principle for dual-use contexts

Kuhlau et al. suggest the following formulation of the PP for application in 'dual-use life-science research': 'When and where serious and credible concern exists that legitimately intended biological material, technology or knowledge in the life sciences pose threats of harm to human health and security, the scientific community is obliged to develop, implement and adhere to precautious measures to meet the concern.'[32]

This is a very vague formulation of the PP and a key problem here is to understand what sort of PP Kuhlau et al.[33] intend us to apply. The most straightforward reading is that they are pointing out that when the risk of serious costs is present, serious remedies will be required if these are to be addressed. The problem with this suggestion is that it is hard to see what work is being done by the PP, as an implication of ordinary CBA is that when the risk of serious costs is at stake serious remedies will be required if these are to be addressed.

---

31 This problem may become resolved as language evolves. The word 'dilemma' is sometimes used these days in a way that is meant to be inclusive of tri-lemmas, quadri-lemmas, and so on. If this usage becomes sufficiently common then framing a decision as a dilemma will no longer encourage a forced choice between exactly two alternatives.

32 Kuhlau et al., op. cit., p. 6.

33 Ibid., pp. 1–8.

But perhaps Kuhlau et al. mean to offer the scientific community a reminder that there are threats to human health and security that can arise from life-science research and they are insisting that the scientific community considers these when deciding whether or not to conduct particular research that may lead to harms. On this second reading, their version of the PP is supposed to function like the Rio declaration and is designed to ensure that, when costs and benefits are weighed up, the potential costs of certain risks are not excluded from consideration. If this is the sort of reading that is intended then it would be useful to know what motivates Kuhlau et al. to suppose that there is a danger of these concerns not being considered, when consideration of them is expected under the application of ordinary CBA. The Rio declaration was developed in response to a history of significant risks being ignored on the (fallacious) grounds that if these had not been established with 'full scientific certainty' then they should not be considered at all. But is there a reason to think that threats to human health and security are liable to be ignored by the scientific community?[34] If there is no particular reason to believe that this may happen then it seems that Kuhlau et al.[35] have created a gratuitous version of the PP that performs no needed function.

On a third possible reading, Kuhlau et al. intend a strong version of the PP. Their wording is ambiguous, but they may intend that the scientific community only allows research in the life sciences to go ahead once all concerns regarding possible threats to human health and security have been met, no matter how potentially beneficial such research may be. If they do intend this strong reading then it would be useful to know how the following two questions can be answered. First, it would be useful to know why they think we should accept this strong precautionary approach. If the benefits of some piece of research are judged to outweigh the risks, and if all significant possible costs and benefits have been considered, why shouldn't we accept the risks of conducting such research, in order to try to reap potential benefits? Second, we need to be told how the problem of paradox is to be avoided.[36] In many situations there will be risks involved in not conducting research in the life sciences. If we do not take the risks involved in developing new vaccines for currently lethal diseases, for example, we implicitly accept the risks of people continuing to die as a result of

34 Tom Douglas suggests that there is a tradition in the scientific community of denying that scientists should be held responsible for the ways in which scientific knowledge is employed and that Kuhlau et al. may be intending to issue a special reminder to scientists, as a way of trying to overcome the influence of this tradition. Also Selgelid suggests that biological scientists and bioethicists have a history of ignoring the dual-use potential of genetics. See Selgelid, M. 2010, 'Ethics engagement of the dual-use dilemma: progress and potential', in B. Rappert (ed.), *Education and Ethics in the Life Sciences*, ANU E Press, Canberra, pp. 23–34.
35 Kuhlau et al., op. cit.
36 Kuhlau et al. consider some objections to the PP, including the objections that it stifles scientific development, it lacks practical applicability and that it is poorly defined and vague. They do not consider the applicability of these charges to specific versions of the PP and they do not consider the charge that (strong versions of) the PP leads to contradictory policy recommendations.

these diseases. It looks like Kuhlau et al.'s version of the PP (on a strong reading) is susceptible to the problem of there being 'risks on all sides' and so it looks like it leads to paradoxical recommendations, if applied consistently.[37]

The PP can be useful in particular contexts when it has specific uses. There may be particularly good uses for a PP developed to suit dual-use contexts. Unfortunately, Kuhlau et al.[38] are not clear about the uses to which they wish to put their version of the PP and it looks like their version is not designed to suit any specific use. If others wish to develop new versions of the PP for dual-use contexts then I would urge that they use precise language and specify what their version of the PP is intended to achieve.

---

37   Thanks to Linsey McGoey, Brian Rappert and Tom Douglas for some very helpful comments on an earlier version of this chapter.
38   Kuhlau et al., op. cit.

Part III: Ethical Practices

15. Scientific Control Over Dual-Use Research: Prospects for Self-Regulation

David B. Resnik

Introduction

In the past decade, scientists, policymakers, ethicists and citizens have become increasingly aware that scientific research that promotes public health and safety has the potential to be used for terrorist, criminal or other malevolent purposes.[1] The phrase 'dual-use research' refers to research that may have beneficial as well as detrimental consequences. The National Science Advisory Board for Biosecurity (NSABB), a US Government committee that provides advice to researchers and federal agencies, has defined 'dual-use research of concern' as 'research that, based on current understanding, can be reasonably anticipated to provide knowledge, products, or technologies that could be directly misapplied by others to pose a threat to public health and safety, agricultural crops and other plants, animals, the environment or materiel'.[2] Examples of published research that has raised dual-use issues include research on a mousepox virus that could be used to enhance the virulence of the human smallpox virus,[3] a study showing how to manufacture a polio virus from available sequence data

1   Atlas, R. 2002, 'National security and the biological research community', *Science*, vol. 298, pp. 753–4; Atlas, R. and Dando, M. 2006, 'The dual-use dilemma for the life sciences: perspectives, conundrums, and global solutions', *Biosecurity and Bioterrorism*, vol. 4, pp. 276–86; National Science Advisory Board for Biosecurity (NSABB) 2007, *Proposed Framework for the Oversight of Dual Use Life Sciences Research: Strategies for Minimizing the Potential Misuse of Research Information*, National Science Advisory Board for Biosecurity, Bethesda, Md, <http://oba.od.nih.gov/biosecurity/pdf/Framework%20for%20transmittal%200807_Sept07.pdf> (viewed 30 December 2011); Selgelid, M. 2007, 'A tale of two studies; ethics, bioterrorism, and the censorship of science', *Hastings Center Report*, vol. 37, no. 3, pp. 35–43; Resnik, D. B. and Shamoo, A. S. 2005, 'Bioterrorism and the responsible conduct of biomedical research', *Drug Development Research*, vol. 63, pp. 121–33; Selgelid, M. 2009, 'Governance of dual use research: an ethical dilemma', *Bulletin of the World Health Organization*, vol. 87, pp. 720–3; National Research Council 2004, *Biotechnology Research in the Age of Terrorism*, The National Academies Press, Washington, DC; National Research Council 2006, *Globalization, Biosecurity and the Future of the Life Sciences*, The National Academies Press, Washington, DC; Resnik, D. B. 2010, 'Can scientists regulate the publication of dual use research?' *Studies in Ethics, Law, and Technology*, vol. 4, no. 1, article 6, <http://www.bepress.com/selt/vol4/iss1/art6> (viewed 30 December 2011).
2   NSABB, op. cit.
3   Jackson, R., Ramsay, A., Christensen, C., Beaton, S. and Hall, D. 2001, 'Expression of mouse interleukin-4 by a recombinant ectromelia virus suppresses cytolytic lymphocyte responses and overcomes genetic resistance to mousepox', *Journal of Virology*, vol. 75, pp. 1205–10.

and mail-order supplies,[4] a study on the genetics of human smallpox virus that could be used to develop a strain of Vaccinia (the virus used in smallpox vaccine) that would overcome the immune system's natural defences,[5] a paper describing how a terrorist could poison the US milk supply with botulinum toxin,[6] and a paper demonstrating how to reconstruct the 1918 Spanish influenza virus from published sequence data.[7]

Preventing the use of scientific research for malevolent purposes raises dilemmas for ethics and public policy. Although most of these issues have existed in some form or another since the dawn of science, the globalisation of scientific research, advances in biotechnology and the spectre of bioterrorism create problems for bioscientists and other researchers that are complex, difficult and urgent.[8] The fundamental dilemma related to dual-use research is how to protect society from harm without unduly hampering the advancement of science.[9] While some restrictions on research are widely recognised as necessary to prevent the misuse of science, administrative, legal and bureaucratic oversight of research can hinder collaboration, the sharing of data and materials, and publication.[10]

Policymakers have explored two forms of oversight of dual-use research: governmental control and self-regulation.[11] Government oversight mechanisms that have been used or proposed include[12]

- regulations, such as the *Patriot Act*, that control access to and transfer and storage of dangerous biological, chemical or radiological materials[13]
- laws related to the transfer of technology across national borders[14]

---

4   Cello, J., Paul, A. and Wimmer, E. 2002, 'Chemical synthesis of poliovirus cDNA: generation of infectious virus in the absence of natural template', *Science*, vol. 297, pp. 1016–18.

5   Rosengard, A, Liu, Y., Nie, Z. and Jimenez, R. 2002, 'Variola virus immune evasion design: expression of a highly efficient inhibitor of human complement', *Proceedings of the National Academy of Sciences*, vol. 99, pp. 8808–13.

6   Wein, L. and Liu, Y. 2005, 'Analyzing a bioterror attack on the food supply: the case of botulinum toxin in milk', *Proceedings of the National Academy of Sciences*, vol. 102, pp. 9984–9.

7   Tumpey, T. M., Basler, C. F., Aguilar, P. V., Zeng, H., Solórzano, A., Swayne, D. E., Cox, N. J., Katz, J. M., Taubenberger, J. K., Palese, P. and García-Sastre, A. 2005, 'Characterization of the reconstructed 1918 Spanish influenza pandemic virus', *Science*, vol. 310, pp. 77–80.

8   Atlas and Dando, op. cit.

9   Miller, S. and Selgelid, M. 2007, 'Ethical and philosophical consideration of the dual-use dilemma in the biological sciences', *Science and Engineering Ethics*, vol. 13, pp. 523–80.

10   Ibid.; Atlas and Dando, op. cit.

11   Miller and Selgelid, op. cit.; Atlas and Dando, op. cit.

12   Miller and Selgelid, op. cit.

13   Malakoff, D. 2002, 'Biological agents: new U.S. rules set the stage for tighter security, oversight', *Science*, vol. 298, p. 2304; Bhattacharjee, Y. 2010, 'Biosecurity: new biosecurity rules to target the riskiest pathogens', *Science*, vol. 329, pp. 264–5.

14   National Research Council, 2006, op. cit.

- classification of government-sponsored research that could threaten national security if disclosed[15]
- restrictions on government funding of research with potential dual-use implications[16]
- restrictions on immigration of scientists with questionable backgrounds
- censorship of non-classified research that poses a threat to national security.[17]

Self-regulation by the scientific community might include[18]

- control over the dissemination of materials and technology by researchers and their institutions[19]
- review of research by committees charged with overseeing research[20]
- education and training in responsibilities related to dual-use research[21]
- development of professional ethics codes and institutional policies that address dual-use research[22]
- journal review of dual-use research.[23]

There is currently a vigorous debate over whether government oversight, self-regulation by the scientific community or some combination is the best way to control dual-use research.[24] The main argument for government regulation is that self-regulation by the scientific community has significant limitations and is often ineffective.[25] There is some justification for this view. First, scientists (and non-governmental research organisations) lack the legal authority to adequately control some types of research. For example, scientists do not have the authority to classify information, restrict access to dangerous materials or prevent the transfer of technologies across national borders.[26] Second, scientists often lack the expertise or resources needed to make decisions pertaining to dual-use research.[27] For example, scientists reviewing dual-use research for

---

15   Resnik, D. B. 2009, *Playing Politics with Science*, Oxford University Press, New York; National Research Council, 2004, op. cit.

16   Resnik and Shamoo, op. cit.

17   van Aken, J. 2006, 'When risk outweighs benefit: dual-use research needs a scientifically sound risk–benefit analysis and legally binding biosecurity measures', *EMBO Reports*, vol. 7(SI), pp. S10–13; Miller and Selgelid, op. cit.

18   NSABB, op. cit.

19   Atlas and Dando, op. cit.

20   NSABB, op. cit.

21   Somerville, M. and Atlas, R. 2005, 'Ethics: a weapon to counter bioterrorism', *Science*, vol. 307, pp. 1881–2.

22   Rappert, B. 2004, 'Responsibility in the life sciences: assessing the role of professional codes', *Biosecurity and Bioterrorism*, vol. 2, pp. 164–74.

23   van Aken, J. and Hunger, I. 2009, 'Biosecurity policies at international life science journals', *Biosecurity and Bioterrorism*, vol. 7, pp. 61–71.

24   Miller and Selgelid, op. cit.; van Aken, op. cit.; Resnik, 2010, op. cit.; Atlas and Dando, op. cit.

25   Miller and Selgelid, op. cit.; Selgelid, op. cit.; van Aken, op. cit.

26   National Research Council, 2006, op. cit.

27   Miller and Selgelid, op. cit.

an institutional biosafety committee (IBC) may not have access to information related to national security interests or threats. Third, scientists have a vested interest in promoting research that may conflict with their social responsibility to prevent harmful uses of research.[28] Journals may be more interested in publishing research that will generate interest and discussion than in preventing the dissemination of information that could threaten society.

The argument against government regulation is that it may pose a significant threat to the advancement of science. Freedom of inquiry, association and expression are essential to scientific research.[29] Government restrictions on science and technology can undermine collaboration, creativity, criticism, publication and sharing of data, materials and methods. Especially worrisome are restrictions on publication—the lifeblood of science.[30] Since the 1980s, the official policy of many governments, including the US Government, has been not to interfere with the publication of non-classified scientific research;[31] however, concerns about dual-use research have led some commentators to question the wisdom of this policy.[32] The NSABB has reviewed papers with dual-use implications, but its opinions are purely advisory and do not determine publication decisions.[33] Government restrictions on access to research materials can also have negative impacts on science, since they can interfere with collaborations among researchers, especially international ones. In the United States, scientists must submit to background checks in order to access dangerous biological materials classified as select agents,[34] which has generated concerns about the potential for racial or ethnic discrimination.[35] Any type of government restriction on scientific research also raises the possibility of politically motivated interference in the scientific process.[36]

I will not discuss or assess government oversight of dual-use research in depth, since the aim of this chapter is to examine the prospects for self-regulation of dual-use research; however, the conclusions reached in this chapter may have implications for government oversight, since one of the main arguments for government involvement is that attempts at self-regulation are likely to be ineffective. This chapter will describe the different mechanisms that scientists might use to control dual-use research and assess the potential effectiveness

---

28   Ibid.
29   Resnik, 2009, op. cit.
30   Journal Editors and Authors Group 2003, 'Uncensored exchange of scientific results', *Proceedings of the National Academy of Sciences*, vol. 100, p. 1464.
31   Resnik, 2009, op. cit.
32   Miller and Selgelid, op. cit.
33   NSABB, op. cit.
34   Bhattacharjee, op. cit.
35   Resnik and Shamoo, op. cit.
36   Resnik, 2009, op. cit.

of these mechanisms, including: control over transfer of materials, review by institutional committees, education and training, professional codes, and journal policies.

Control over the transfer of materials

Every day, scientists send materials and technologies, such as reagents, cell lines, pathogens and laboratory animals, to other scientists. Most institutions sign a material transfer agreement (MTA) when they transfer materials. MTAs state the terms and conditions for using such materials, and can be enforced through contract law. Although laws and regulations govern the transfer of some types of materials, such as select biological agents and radioactive substances, they do not cover all the types of materials that might be used for harmful purposes, as some materials might not be deemed dangerous enough to warrant regulations, and other dangerous materials might not yet have been identified (as such). For example, when investigators published research on the 1918 Spanish influenza virus, this pathogen was not on the list of select agents, and it was not until later that the list was amended to cover this material.[37]

Scientists can play an important role in preventing the transfer of dangerous materials to individuals or organisations with reprehensible motives by attempting to determine whether the request for materials comes from a legitimate, respected researcher or whether it comes from a suspicious individual or organisation. They can ask for curriculum vitae, research protocol and other information necessary to verify the requestor's expertise, background and affiliations. If they have any doubts about the legitimacy of the request, they can delay the transfer until they are satisfied that the requestor is a responsible scientist, or they can consult institutional officials on how to proceed.

Whether scientific control over the transfer of materials and technologies will be an effective mechanism for preventing the use of these items for malevolent purposes remains to be seen. For this mechanism to be effective, investigators must first be aware of the dual-use implications of transferring materials and technologies. A survey of attitudes of US life scientists concerning dual-use research indicates that most are aware of biosecurity issues, although a majority considers the risk of a bioterror attack to be small. A small percentage of those surveyed indicated that they had refused to collaborate with some individuals due to biosecurity concerns. [38] Although many scientists are aware of dual-use

---

37 National Select Agent Registry 2010, <http://www.selectagents.gov/Select%20Agents%20and%20 Toxins%20List.html> (viewed 30 December 2011).
38 Committee on Assessing Fundamental Attitudes of Life Scientists as Basis for Biosecurity Education 2009, *A Survey of Attitudes and Actions on Dual Use Research in the Life Sciences: A Collaborative Effort of*

issues, education (discussed below) can help to raise awareness. Investigators must also be willing to invest some of their time and effort into verifying the legitimacy of individuals or organisations requesting materials or technologies. Investigators who have extensive responsibilities related to research, teaching or administration may feel that they do not have enough time to devote such attention to requests for materials and technologies. Educational sessions may be able to convince investigators of the importance of this issue.

Review by institutional committees

Several different institutional committees may have the opportunity to review research with dual-use implications. These committees can decide whether the research should be approved, disapproved or approved with additional oversight or restrictions to prevent harmful consequences. Most of the discussion thus far has focused on the role of IBCs in addressing the dual-use implications of research. IBCs are responsible for reviewing research involving hazardous biological materials, such as pathogens and toxins, as well as recombinant DNA research, including experiments to transfer genes to plants, animals and humans.[39] IBCs are in charge of protecting researchers and the public from risks related to research on hazardous biological materials. IBCs include experts in microbiology, genetics, pathology, immunology and other disciplines with the requisite expertise. IBCs are a natural place for the review of research with dual-use implications, given their function and expertise.[40] Institutional animal care and use committees (IACUCs) review research involving research on animals. The main function of IACUCs is to protect the welfare of laboratory animals, and to ensure that investigators are using methods that minimise pain, suffering and distress.[41] Though IACUCS lack the mandate and expertise to review dual-use issues, they could refer any research that raises concerns to an IBC. Institutional review boards (IRBs) review research involving human subjects, and their main function is to protect the rights and welfare of individuals participating in research. IRBs, like IACUCs, lack the mandate or expertise to review dual-use issues, but they could make a referral to an IBC.[42]

the National Research Council and the American Association for the Advancement of Science, The National Academies Press, Washington, DC, <http://books.google.com/books/about/A_Survey_of_Attitudes_and_Actions_on_Dua.html?id=2-MLru0bHkQC> (viewed 30 December 2011).

39 Cornell University 2007, Institutional Biosafety Committee, <http://www.ibc.cornell.edu/responsibilities> (viewed 30 December 2011).

40 NSABB, op. cit.

41 Shamoo, A. S. and Resnik, D. B. 2009, Responsible Conduct of Research, 2nd edn, Oxford University Press, New York.

42 Resnik, D. B. 2010, 'Dual-use review and the IRB', Journal of Clinical Research Best Practices, vol. 6, no. 1, <http://www.firstclinical.com> (viewed 30 December 2011).

Although these three committees have the ability to identify scientific research with dual-use implications, inevitably some research may fall through the cracks, such as biomedical research that does not involve dangerous pathogens or animal or human subjects; chemical research; and engineering research. For example, the study of how to infect the US milk supply with botulinum toxin would not have been reviewed by any IBC, IACUC or IRB.[43] Because these committees may lack the expertise to assess the dual-use implications of research that falls within their purview, the NSABB has recommended that institutions consider appointing committees that focus on dual-use research.[44] It is not known how many institutions have created such committees, although the National Institutes of Health (NIH) intramural program has a dual-use committee.[45]

Review of research by institutional committees is likely to be one of the most effective ways of preventing harms related to dual-use research. First, most institutions already have IBCs, IACUCs and IRBs or equivalent committees. Second, since these committees are accustomed to dealing with a variety of ethical, legal and social issues related to research, it will not come as a surprise to most committee members that they should be mindful of dual-use concerns. Third, these committees are likely to know how to access the appropriate institutional officials (for example, vice-president for research, compliance officer) for dealing with dual-use issues. Although institutional committees are well positioned to deal with dual-uses issues, it is not known how often these committees encounter research that raises dual-use concerns, whether they recognise the dual-use implications, or how they typically respond to these situations. More research is needed to get a better understanding of the effectiveness of institutional committees of dealing with dual-use issues.

Education and training

The NSABB[46] and other commentators[47] have recommended that investigators, students and trainees receive education concerning their responsibilities related to dual-use research. Education should cover topics pertaining to dual use, such as select agents, control of dangerous materials and the potential consequences of research and publication. Many funding organisations and institutions already require or support some type of education in responsible conduct of research (RCR) to promote scientific integrity. In the United States, the NIH and the National Science Foundation (NSF) mandate research ethics training

---

43   Wein and Lu, op. cit.
44   NSABB, op. cit.
45   J. Schwartz, Personal Communication, 15 December 2009.
46   NSABB, op. cit.
47   Resnik and Shamoo, op. cit.; Sommerville and Atlas, op. cit.

requirements for students supported with grants funds.[48] Education in RCR varies considerably. Some institutions require students to complete online modules in RCR, while others require attendance at classes, workshops or seminars.[49] RCR education typically addresses fabrication, falsification, plagiarism, data management, conflict of interest, authorship, publication, mentoring and peer review, but this list could be expanded to include dual-use issues. It is not known how many institutions have conducted education and training in dual-use research, and more research is needed on this topic. In 2009, the NIH required that all intramural researchers and trainees receive education on their social responsibilities related to dual-use research. The educational materials developed by the NIH included an overview of dual-use research as well as several case studies.[50]

While education in dual-use issues is likely to play an important role in preventing the misuse of scientific research or materials, its effectiveness has been questioned. According to some studies, RCR education does not decrease the prevalence of negative behaviour, such as data falsification or plagiarism.[51] Other studies suggest that RCR education can promote awareness of ethical issues and knowledge of ethical concepts, but does not influence attitudes or behaviour.[52] Although there have been no published studies on the effectiveness of dual-use education and training, it is plausible to hypothesise that these efforts may have mixed results, given the scientific community's experiences with RCR education and training. A survey of life scientists indicates they support education and training initiatives that cover dual-use research;[53] however, more research is needed on the effectiveness of dual-use education, and the effectiveness of different pedagogical techniques—for example, seminars, online training, and so on.

---

48   Shamoo and Resnik, op. cit.
49   Ibid.
50   Schwartz, op. cit.
51   Anderson, M. S., Horn, A. S., Risbey, K. R., Ronning, E. A., de Vries, R. and Martinson, B. C. 2007, 'What do mentoring and training in the responsible conduct of research have to do with scientists' misbehavior? Findings from a national survey of NIH-funded scientists', *Academic Medicine*, vol. 82, pp. 853–60.
52   Schmaling, K. B. and Blume, A. W. 2009, 'Ethics instruction increases graduate students' responsible conduct of research knowledge but not moral reasoning', *Accountability in Research*, vol. 16, pp. 268–83; Plemmons, D., Brody, S. and Kalichman, M. 2006, 'Student perceptions of the effectiveness of education in the responsible conduct of research', *Science and Engineering Ethics*, vol. 12, pp. 571–82; Powell, S., Allison, M. and Kalichman, M. 2007, 'Effectiveness of a responsible conduct of research course: a preliminary study', *Science and Engineering Ethics*, vol. 13, pp. 249–64.
53   Committee on Assessing Fundamental Attitudes of Life Scientists as Basis for Biosecurity Education, op. cit.

Professional codes

The NSABB[54] and other commentators[55] have recommended that professional associations should develop codes of ethics (or conduct) to promote self-regulation of dual-use research. In response to the dual-use concerns that arose in 2001, the American Society for Microbiology (ASM) revised its code of ethics, which now includes the following paragraph:

> ASM members are obligated to discourage any use of microbiology contrary to the welfare of humankind, including the use of microbes as biological weapons. Bioterrorism violates the fundamental principles upon which the Society was founded and is abhorrent to the ASM and its members. ASM members will call to the attention of the public or the appropriate authorities misuses of microbiology or of information derived from microbiology.[56]

Other professions with codes that address dual-use issues include the International Union of Biochemistry and Molecular Biology (IUBMB)[57] and the American Medical Association (AMA).[58] It is not known how many other professions have such codes but development thus far has been sparse.[59] The NSABB has published some considerations for developing a code of conduct related to dual-use research in the life sciences, which address such topics as proposing, managing and conducting research; collaborations; public communications; and mentoring.[60]

The effectiveness of professional codes at promoting ethical conduct has been the subject of much debate.[61] Some argue that professional codes are merely symbolic statements of values that have little effect on behaviour. Others argue that codes provide useful guidance for members of the profession and help promote public trust by defining standards of conduct;[62] however, most professional codes lack enforcement mechanisms. Those that are the most effective are linked to professional licensure or certification, which can result

---

54 NSABB, op. cit.
55 Rappert, op. cit.
56 American Society for Microbiology (ASM) 2005, *Code of Ethics*, <http://www.asm.org/ccLibraryFiles/FILENAME/000000001596/ASMCodeofEthics05.pdf> (viewed 30 December 2011).
57 International Union of Biochemistry and Molecular Biology (IUBMB) 2005, *Code of Ethics*, <http://www.babonline.org/bab/babcethics.pdf> (viewed 30 December 2011).
58 American Medical Association (AMA) 2004, *Code of Medical Ethics, Opinion 2.078—Guideline to Prevent Malevolent Use of Biomedical Research*, <http://www.ama-assn.org/ama/pub/physician-resources/medical-ethics/code-medical-ethics/opinion2078.shtml> (viewed 30 December 2011).
59 Rappert, B. 2007, 'Codes of conduct and biological weapons: an in-process assessment', *Biosecurity and Bioterrorism*, vol. 5, pp. 1–10.
60 NSABB, op. cit.
61 Rappert, 2007, op. cit.
62 Bayles, M. 1988, *Professional Ethics*, 2nd edn, Wadsworth, Belmont, Calif.

in suspension of one's licence or disbarment for misbehaviour. For example, lawyers who violate ethical standards set by their bar association can have their licence to practise law revoked.[63] Since licensure by a professional association is not necessary to conduct many types of scientific research, scientific codes of ethics usually lack the coercive power to control behaviour. A microbiologist who violates the ASM code of ethics can still practise microbiology even if he is sanctioned by the organisation. Despite these limitations, many argue that scientific professional codes can still play an important role in guiding conduct, even if they lack coercive power.[64]

Institutional policies

The NSABB has also recommended that institutions develop policies addressing dual-use issues, complimenting existing RCR policies.[65] Many scholars and commentators have argued that institutional policies, combined with education and training, can help promote ethical conduct in research;[66] however, it is difficult to assess the effectiveness of institutional policies due to social, economic and cultural differences among institutions. Studies have shown that there is considerable variation in the content of some institutional ethics policies, such as conflict-of-interest and misconduct rules.[67] In the United States, the content of institutional policies is largely driven by government mandates, such as NIH or NSF rules that require institutions to establish conflict-of-interest, misconduct, animal research and human subjects research policies as a condition of obtaining funding.[68] Federal agencies could promote institutional policy development by requiring that funding recipients have dual-use policies; however, so far no federal agency has done so.

Institutional dual-use policies could play a key role in preventing the misuse of scientific work for malevolent purposes. Policies could address controlling, securing, transferring and accessing hazardous materials, as well as publishing research that could be readily used to cause significant harm to

---

63   Rotunda, R. 2007, *Legal Ethics*, 3rd edn, West Publishing, St Paul, Minn.
64   Shamoo and Resnik, op. cit.; Rappert, 2007, op. cit.
65   NSABB, op. cit.
66   Institute of Medicine 2002, *Integrity in Scientific Research: Creating An Environment that Promotes Responsible Conduct*, The National Academies Press, Washington, DC; Geller, G., Boyce, A., Ford, D. E. and Sugarman, J. 2010, 'Beyond "compliance": the role of institutional culture in promoting research integrity', *Academic Medicine*, vol. 85, pp. 1296–302.
67   Cho, M. K., Shohara, R., Schissel, A. and Rennie, D. 2000, 'Policies on faculty conflicts of interest at US universities', *Journal of the American Medical Association*, vol. 284, pp. 2203–28; McCrary, S. V., Anderson, C. B., Jakovljevic, J., Khan, T., McCullough, L. B., Wray, N. P. and Brody, B. A. 2000, 'A national survey of policies on disclosure of conflicts of interest in biomedical research', *New England Journal of Medicine*, vol. 343, pp. 1621–6; Lind, R. A. 2005, 'Evaluating research misconduct policies at major research universities: a pilot study', *Accountability in Research*, vol. 12, pp. 241–62.
68   Shamoo and Resnik, op. cit.

public health, society, security or the environment. In theory, institutional policies could be more effective than professional codes, because institutional policies are enforceable. Scientists or students who violate university policies can be suspended, demoted, fired or dismissed. It is not known how many institutions have developed dual-use policies, but probably very few have, since universities usually do not develop ethical oversight mechanisms in the absence of legal or financial threats.[69] Government action could spur the development of institutional policies.

Journal policies

Scientific journal dual-use review policies are another form of self-governance. Journals have been at the centre of the dual-use controversy, as several papers published by prominent journals have been controversial. For example, several members of the US Congress criticised journal editors for publishing the paper on how to manufacture a polio virus mentioned earlier. The US Department of Health and Human Services (DHHS) asked the editors of the *Proceedings of the National Academy of Sciences* (*PNAS*) not to publish the paper on contaminating the US milk supply, also mentioned earlier.[70] The editors of *PNAS* met with DHHS representatives prior to publication to discuss the benefits and risks of public dissemination of the findings. The Department of Homeland Security (DHS) asked the NSABB to review the paper on reconstructing the 1918 pandemic influenza virus prior to publication in *Science*. Although the NSABB voted unanimously in favour of publication, the editor of *Science* publicly stated he would have ignored the NSABB's recommendations if had been against publication.[71]

Recently, the NSABB recommended that journals omit important details from research that used genetic-engineering techniques to create a mutated form of an avian flu virus, A(H5N1), which could be transmitted between mammalian species, including humans. Key details would be available only to responsible scientists. Currently, the virus can only be transmitted from birds to other mammalian species, not between members of mammalian species. Six hundred people have contracted the virus since 1997, and more than half of them have died. The journals reviewing the research, *Science* and *Nature*, had not made a decision concerning publication as of the writing of this chapter.[72]

---

69   Ibid.
70   Resnik and Shamoo, op. cit.
71   Kennedy, D. 2005, 'Better never than late', *Science*, vol. 310, p. 195.
72   Grady, D. and Broad, W. 2011, 'Seeing terror risk, U.S. asks journals to cut flu study facts', *The New York Times*, 20 December 2011, p. A1.

Journals began developing dual-use review policies in the early part of the twenty-first century, following the publication of several controversial papers.[73] Some of the leading journals with dual-use review policies include *Science*, journals published by the Nature Publishing Group (NPG)—that is, *Nature, Nature Biotechnology*, and so on—and journals published by ASM.[74] Review policies include an additional level of review for papers that raise dual-use concerns. Outside reviewers with expertise in national security, terrorism or other relevant subjects may be asked to assist with the review. The outcome of the review could include a recommendation to publish, to not publish or to publish with appropriate revisions, such as restrictions on access to key information.[75] The NPG's dual-use policy provides that:

> Nature journal editors may seek advice about submitted papers not only from technical reviewers but also on any aspect of a paper that raises concerns … As in all publishing decisions, the ultimate decision whether to publish is the responsibility of the editor of the Nature journal concerned. The threat posed by bioweapons raises the unusual need to assess the balance of risk and benefit in publication. Editors are not necessarily well qualified to make such judgements unassisted, and so we reserve the right to take expert advice in cases where we believe that concerns may arise. We recognize the widespread view that openness in science helps to alert society to potential threats and to defend against them, and we anticipate that only very rarely (if at all) will the risks be perceived as outweighing the benefits of publishing a paper that has otherwise been deemed appropriate for a Nature journal. Nevertheless, we think it appropriate to consider such risks and to have a formal policy for dealing with them if need arises. The editorial staff of Nature journals maintains a network of advisers on biosecurity issues. All concerns on that score, including the commissioning of external advice, will be shared within an editorial monitoring group consisting of the Editor-in-Chief of Nature publications, the Executive Editor of the Nature research journals, the Chief Biological Sciences Editor of Nature, and the chief editor of the journal concerned. Once a decision has been reached, authors will be informed if biosecurity advice has informed that decision.[76]

---

73   Resnik, 2010, op. cit.
74   van Aken and Hunger 2009, op. cit.
75   *Report of the NSABB Working Group on Journal Review Policies 2010*, NSABB Board Meeting, 19 October 2010,   <http://oba.od.nih.gov/oba/biosecurity/meetings/Oct2010/Journal_Review_Policies_Slides.pdf> (viewed 30 December 2011).
76   Nature Publishing Group (NPG) 2010, *Biosecurity Policy*, <http://www.nature.com/authors/editorial_policies/biosecurity.html> (viewed 30 December 2011).

There have been three published studies to date on the dual-use review policies of scientific journals. In 2009, van Aken and Hunger published a survey of 28 major life-science journals that regularly publish research that may raise biosecurity issues. They found that 25 per cent of these journals had dual-use review policies.[77] In October 2010, a working group from the NSABB reported their findings at an NSABB board meeting. Members of the working group reported on their discussions with the editors of 18 high-impact life-science journals that had published dual-use review policies online. Members of the working group found that different models of dual-use review were in place, and that editors felt they needed additional guidance.[78]

Though these two studies provide some useful information about the dual-use policies of scientific journals, they do not provide systematic information about journal policy development because the sampling was focused (not random) and the sample sizes were small. To overcome these limitations, my colleagues and I conducted a larger survey of biomedical journals. We drew a random sample of 400 journals from the *ISI Web of Knowledge Journal Citation Reports 2009 Edition*. We eliminated journals that had little chance of reviewing dual-use research. We sent out an email survey to the editors and reminders after seven and 14 days if we did not receive a response. Of these, 155 journals responded to our survey (response rate 39 per cent). Only 7.7 per cent said they had a formal (written) dual-use policy; 72.8 per cent said they had had no experience with reviewing dual-use research in the past five years, 5.8 per cent indicated they had some experience, and 21.8 per cent did not give a definite answer to this question. We attempted to determine whether several variables were associated with having a dual-use policy. Belonging to the NPG was the most significant predictor of having a policy (positive association). Having experience with reviewing manuscripts with dual-use implications was also positively associated with having a policy.[79]

Our research indicates that less than 8 per cent of biomedical journals have a dual-use policy. This is much lower than the percentage reported by van Aken and Hunger. This is an important finding, since it shows that the scientific community has not made much progress in an important area of self-regulation. Another important finding is that most journals have not had any experience with reviewing research with dual-use implications. Indeed, some editors said they had not developed a policy because they saw no need for one, and others had not even heard of the term 'dual use'. Some editors said they were planning to develop a policy after they learned about our study.[80]

---

77   van Aken and Hunger, op. cit.
78   *Report of the NSABB Working Group on Journal Review Policies*, op. cit.
79   Resnik, D. B., Barner, D. D. and Dinse, G. E. 2011, 'Dual use policies of biomedical journals', *Biosecurity and Bioterrorism*, vol. 9, pp. 49–54.
80   Ibid.

Conclusion: Prospects for self-regulation

So what are the prospects for self-regulation of dual-use research? It is difficult to say at this point, because there has been very little empirical research that addresses this question. While a survey of life scientists indicates that most are aware of biosecurity issues, research on biomedical journals indicates that only a small percentage has developed dual-use review policies. Also, very few professional associations have developed dual-use policies. We lack systematic data on other important aspects of self-regulation, such as institutional policies and oversight, and education and training efforts. Clearly, more research is needed to better understand how the scientific community is responding to the dual-use dilemma, and any conclusions drawn at this point about prospects for self-regulation should be viewed as tentative.

Looking to the past as a predictor of the prospects for self-regulation, we find that scientists have had some successes and some failures. One of the best examples of successful self-regulation would be the management of the risks related to recombinant DNA. In the early 1970s, scientists began to conduct experiments involving the transfer of DNA to micro-organisms. The public became gravely concerned about the threat of 'superbugs' escaping from the laboratory and causing a modern plague. The NIH responded to the public's anxiety by forming the Recombinant DNA Advisory Committee (RAC) in 1974 to oversee NIH-funded recombinant DNA research. Scientists who were at the forefront of recombinant DNA research met in Asilomar, California, in February 1975 to discuss the risks of their work and make some recommendations to protect laboratory workers, the public and the environment from biohazards. The recommendations included a number of different safety procedures, such as only using organisms that are unable to survive outside the laboratory, physical containment protocols and following good laboratory practices. The scientists also recommended a voluntary moratorium on experiments that were deemed too risky and said that work should move forward cautiously.[81] In 1976, the RAC published its first set of guidelines for NIH-funded recombinant DNA research. Among the RAC's most important recommendations was that institutions establish biosafety committees to oversee recombinant DNA research. Although the RAC guidelines apply only to NIH research, scientists who are not funded by the NIH have also often followed the RAC's recommendations.[82] While the government has played an important role in helping to promote responsible

---

81 Berg, P., Baltimore, D., Brenner, S., Roblin III, R. O. and Singer, M. F. 1975, 'Summary statement of the Asilomar Conference on Recombinant DNA Molecules', *Proceedings of the National Academy of Sciences*, vol. 72, pp. 1981–4.
82 Kimmelman, J. 2009, *Gene Transfer and the Ethics of First-in-Human Research: Lost in Translation*, Cambridge University Press, Cambridge.

recombinant DNA research, scientists have taken it upon themselves to establish and follow safety standards. Although public anxieties about genetic engineering persist, the safety record of recombinant DNA research is excellent.[83]

An example of not-so-successful self-regulation would be research involving human subjects. Prior to World War II, there were no widely recognised ethical standards for research involving human subjects. Scientists and physicians appealed to their own sense of right and wrong in treating human beings used in research.[84] In most cases, research was ethical and humane; however, there were many abuses involving risky procedures, experimentation without consent and the use of vulnerable populations, such as children, prisoners and mentally disabled adults, as research subjects. The worst of these abuses occurred in Germany under the Nazi regime, when scientists and physicians used concentration camp prisoners as research subjects. These experiments often produced severe pain and suffering and often resulted in death. Informed consent was out of the question.[85] After the war ended, German physicians and scientists were tried for war crimes related to human experimentation. In 1947, the Nuremberg War Tribunals adopted the Nuremberg Code, the world's first ethical guidelines for research with human subjects, to use as a basis for trying physicians and scientists accused of war crimes.[86] While the Nuremberg Code provided important guidance for researchers, it did not constitute government oversight because it established an ethical standard, not a legal one. The first steps towards government oversight of research with human subjects in the United States occurred in 1966, when the NIH adopted rules for research that it sponsors. In 1971, the Food and Drug Administration (FDA) adopted human subjects research rules.[87]

The US Government started to play a much more prominent role in the oversight of research with human subjects in the 1970s, when Congress held hearings on the Tuskegee syphilis study and other ethical abuses in research involving human subjects. The Tuskegee syphilis study, in which researchers investigated the natural course of untreated syphilis in black men, was funded by the Public Health Service and ran from 1932 to 1972. It included 399 patients with untreated syphilis and 400 controls. The subjects were not told that they were participating in a research study. They were told only that they were getting medical treatment for 'bad blood'. The investigators also took steps to prevent the syphilitic patients from receiving penicillin, an effective treatment, when it

---

83   Ibid.
84   Shamoo and Resnik, op. cit.
85   Ibid.
86   Nuremberg Code 1947, *Directives for Human Experimentation*, <http://ohsr.od.nih.gov/guidelines/nuremberg.html> (viewed 30 December 2011).
87   Shamoo and Resnik, op. cit.

became available in the 1940s.[88] In 1973, Congress responded to the abuses of human research subjects by passing the *National Research Act* (*NRA*), which established the National Commission for the Protection of Human Subjects in Biomedical and Behavioral Research to study the ethics of research with human subjects and authorised federal agencies to draft regulations. The national commission published its findings, known as the Belmont Report, in 1979, which provided the conceptual underpinning for a major revision of the federal research regulations in 1981, which have changed little in the past 30 years.[89] Many other governments, following the United States' lead, regulate research involving human subjects. Some states, such as California, have also adopted their own laws dealing with human experimentation.[90]

While recombinant DNA research has avoided a great deal of government regulation, research with human subjects has evolved from self-regulation to robust government regulation. Why have these two areas of research taken different trajectories? One reason human subjects research has been highly regulated is that it directly impacts human rights and wellbeing. When people die or become gravely ill as a result of an experiment, the public becomes upset and laws are passed. While the risks of recombinant DNA research to laboratory workers are concrete and realistic, other risks are abstract and theoretical. The public may be less apt to react to risks that are yet to be realised. Another reason human subjects research is highly regulated is that scientists have not done a good job developing and following their own ethical standards. Laws related to research with human subjects have been passed in response to highly publicised scandals when investigators failed to follow ethical rules.[91] Not surprisingly, investigators who work with human subjects often complain that regulations and rules create needless administrative work that impedes research.[92]

Will dual-use research become externally regulated or will self-regulation prevail? Much depends on how scientists respond to opportunities for self-regulation presented by dual-use dilemmas. If scientists take these issues seriously and develop professional codes, journal policies, institutional oversight mechanisms, educational and training programs, and other forms of self-regulation, they may be able to avoid a high degree of government oversight. If, however, scientists fail to take appropriate steps towards self-control, a high degree of external control may result. A tragic event related to dual-use research, such as transfer of a dangerous pathogen to a terrorist who uses it to develop and deploy a lethal weapon, could motivate governments to impose stronger controls on dual-

---

88   Jones, J. H. 1981, *Bad Blood*, Free Press, London.
89   Shamoo and Resnik, op. cit.
90   Ibid.
91   Ibid.
92   Gunsalus, C. K. 2004, 'The nanny state meets the inner lawyer: overregulating while underprotecting human participants in research', *Ethics and Behavior*, vol. 14, pp. 369–82.

use research. Hopefully, scientists will be motivated to take appropriate steps towards self-regulation before an incident occurs that compels governments to impose extensive restrictions.

Acknowledgments

This research was sponsored by the National Institute of Environmental Health Sciences (NIEHS), National Institutes of Health (NIH). It does not represent the views of the NIEHS, NIH or US Government.

16. Contrasting Dual-Use Issues in Biology and Nuclear Science

Nicholas G. Evans

Dual use and nuclear science

The so-called 'dual-use dilemma'—which arises when scientific research, materials or technologies can be used to both benefit and harm humanity[1]—is not an altogether new phenomenon. Early in the twentieth century, nuclear science—here, the study of the properties of atomic nuclei—raised similar concerns about materials, technologies and knowledge that had the potential to be used to benefit humanity, but also posed serious risks of misuse. Comparing the experiences of the nuclear sciences with those of biology in regulating dual-use materials, knowledge and technology will be the focus of this chapter.

Nuclear science raises a number of dual-use issues. Nuclear science has many beneficial uses, the best known being nuclear-fission (the process by which a nucleus is split into two or more smaller parts) reactors. These reactors are useful in power generation, and are also used to create radioisotopes—elements that are subject to radioactive decay that have a variety of uses described below. Nuclear fusion (the process of 'fusing' two lighter nuclei into a heavier nucleus) is currently being explored as a cleaner, more plentiful fuel source that could become an important source of energy in years to come.

Non-fissile radionuclides—radioactive substances not capable of energetic fission—have a range of useful and beneficial medical, agricultural, construction, military and aviation applications, to name a few. Radioisotopes have therapeutic applications in medicine, as their emissions can be used to bombard and kill cancerous cells with radiation. Radioisotopes are also used in diagnostic testing, for example, where the emissions of ingested radioisotopes are plotted to gain information about physiological processes. In agriculture, radionuclides are used to create blight or weather-resistant crops, and kill parasites in food storage. In construction, radioisotopes are used to verify the

---

1 This corresponds with a common definition of dual use, such as that found in Miller and Selgelid. It, as others have discussed previously, does not deal with all definitions of the dual-use dilemma, but for the purposes of this chapter I restrict my analysis to the explicitly normative definition of dual use. See Miller, S. and Selgelid, M. J. 2008, *Ethical and Philosophical Considerations of the Dual-Use Dilemma in the Biological Sciences*, Springer, Dordrecht; Selgelid, M. J. 2010, 'Ethics engagement of the dual-use dilemma: progress and potential', in B. Rappert (ed.), *Education and Ethics in the Life Sciences*, ANU E Press, Canberra, pp. 23–34.

integrity of metals by detecting the scattering of emitted particles to determine the internal structure of metal or concrete under load. Depleted uranium—a byproduct of the enrichment process where the uranium-235 used in nuclear weapons and reactors is separated from non-fissile uranium-238—is incredibly dense and malleable, making it an ideal form of armor plating in tanks. In aviation, depleted uranium is used to weight propeller rotors. Depleted uranium also provides an ideal form of radiation shielding in reactors due to its high rate of capture of alpha, beta and gamma radiation, as well as neutrons. A layer of depleted uranium in a reactor can function as reactor shielding, and, as the depleted uranium captures neutrons and is converted into plutonium, can be used as reactor fuel.

The study of the atom in nuclear science and its connection to quantum mechanics and materials science have also led to the development of the materials and electronics that presently dominate the technological landscape. Many scientific innovations that are now ubiquitous in contemporary society (for example, the personal computer) owe their genesis to the pioneers of nuclear physics.

Nuclear science also has well-known harmful uses that pose a risk of being used by malevolent actors. Nuclear fission is the driving mechanism behind the atomic bombs dropped on Hiroshima and Nagasaki. Nuclear fusion is the mechanism that, combined with fission, drives the 'super' or H-bomb. Contemporary research into either nuclear fission or nuclear fusion can enable the creation and development of these weapons.

Many radioisotopes are highly toxic and can cause radiation sickness if ingested by or embedded in a person for an extended period. Radioisotopes packed around conventional explosives, known as 'dirty bombs', can cover a large area in radioactive ash. Survivors of such a blast would have to deal with radioactive particulates embedded in their bodies as a result of the explosion, infrastructure affected would need decontamination or have to be destroyed and removed, and the treatment requirements and potential for contamination in caring for the injured would increase significantly.[2] Radionuclides can cause radiation poisoning or other health problems if released—accidentally or intentionally—into water supplies, or as a gas or powder. The murder of former KGB agent and defector Alexander Litvinenko, who died of radiation poisoning after ingesting significant amounts of polonium-210, is an example of malevolent uses of radioisotopes.[3]

---

2   For example, Zimmerman, P. D. and Loeb, C. 2004, 'Dirty bombs: the threat revisited', *Defense Horizons*, vol. 38, pp. 1–11.
3   Stoll, W. 2007, '210–polonium', *Atw. Internationale Zeitschrift für Kernenergie*, vol. 52, no. 1, pp. 39–41.

Wither nuclear science?

Despite the dual-use potential of nuclear science, there has been little examination of the parallels between nuclear science and biology in current debates about the dual-use dilemma.[4] One important reason is that, from the outset, discourse has neglected or downplayed the relevance of the nuclear sciences to the debate in the life sciences. In 2004, a canonical report by the National Research Council chaired by Gerald Fink titled *Biotechnology Research in An Age of Terrorism* (the 'Fink Report') stated:

> The nature of the biotechnology problem—indeed the nature of the biological research enterprise—is vastly different from that of theoretical and applied nuclear physics in the late 1930s. The contrast between what is a legitimate, perhaps compelling subject for research and what might justifiably be prohibited or tightly controlled cannot be made a priori, stated in categorical terms, nor confirmed by remote observation.[5]

The Fink Report argued that the primary mechanism of arms control in the case of nuclear weapons, materials control, is not applicable to the life sciences.[6] The report highlighted differences between the properties of nuclear fissile materials and biological pathogens, which it claimed rendered materials control impractical as a regulatory solution (Table 16.1).

A follow-up report in 2006, *Globalization, Biosecurity, and the Future of the Life Sciences* (the 'Lemon-Relman Report'), gave a similar argument to the Fink Report regarding dual use:

> Clear thinking on the issue must proceed from an understanding of the significant differences among [biological and nuclear] weapons. Although there are lessons to be learned from the history of and our experience with nuclear weapons technology, many of the differences between the nuclear and biological realms are too great to adopt a similar mix of nonproliferation, deterrence, and defense.[7]

---

4   Notable exceptions include Selgelid, M. J. 2007, 'A tale of two studies: ethics, bioterrorism, and the censorship of science', *The Hastings Centre Report*, vol. 37, no. 3, pp. 35–48; Finney, J. L. 2007, 'Dual use: can we learn from the physicists' experience?' in B. Rappert (ed.), *A Web of Prevention: Biological Weapons, Life Sciences and the Governance of Research*, Earthscan, London, pp. 67–76; Forge, J. 2008, *The Responsible Scientist*, University of Pittsburg Press, Pittsburg; Miller and Selgelid, op. cit.
5   National Academies of Science 2004, *Biotechnology Research in An Age of Terrorism*, The National Academies Press, Washington, DC, p. 23.
6   Ibid., pp. 81–3.
7   National Academies of Science 2006, *Globalization, Biosecurity, and the Future of the Life Sciences*, The National Academies Press, Washington, DC, p. 46.

Table 16.1 Characteristics of Fissile Materials and Pathogens

Fissile materials	Biological pathogens
Do not exist in nature	Generally found in nature
Non-living, synthetic	Living, replicative
Difficult and costly to produce	Easy and cheap to produce
Not diverse: plutonium and highly enriched uranium are the only fissile materials used in nuclear weapons	Highly diverse: more than 20 pathogens are suitable for biological warfare
Can be inventoried and tracked in a quantitative manner	Because pathogens reproduce, inventory control is unreliable
Can be detected at a distance from the emission of ionising radiation	Cannot be detected at a distance with available technologies
Weapons-grade fissile materials are stored at a limited number of military nuclear sites	Pathogens are present in many types of facilities and at multiple locations within a facility
Few non-military applications (such as research reactors, thermo-electric generators, and production of radioisotopes)	Many legitimate applications in biomedical research and the pharmaceutical/ biotechnology industry

Source: National Academies of Science 2004, *Biotechnology Research in An Age of Terrorism*, The National Academies Press, Washington, DC.

The above conclusion—echoing the Fink Report—indicates that, due to the marked differences between the materials in question, methods of prevention used in the nuclear sciences are unsuitable to regulate biological science and technology. Moreover, the current culture of biology is not one that is conducive to these regulations.

Yet, I wish to contend, the lack of immediate compatibility between regulatory measures does not entail that the differences between fields are too great for anything valuable to be learned from nuclear history when reflecting upon how dual-use biology should be governed. It is doubtful that the regulatory framework used in nuclear sciences would provide a good model to use even if biology and nuclear sciences were more similar. No-one wants another arms race. That said, even just the lessons learned from the regulatory and policy mistakes in the nuclear sciences would be useful—but I hope to show that we can learn far more than that.

The purpose of this chapter is to compare and contrast regulation of the dual-use dilemma in the nuclear and biological sciences. This will not be a comprehensive comparison—the history of the nuclear sciences from its beginning in the early twentieth century through to the political implications of nuclear science in the twenty-first century is too large for a single chapter. Rather, I will provide a series of examples from the nuclear sciences that invites exploration of different ethical issues of dual use in the life sciences. I also view this chapter as a call for further investigation of the nuclear sciences and their history to understand the pitfalls of regulating science and technology, and apply lessons learnt to dual use in the life sciences.

Finally, it is not my intention to claim that biology and the nuclear sciences are the same regarding the dual-use dilemma. I will show where the issues presented by dual-use materials, knowledge, techniques or technologies in each field overlap, and where they come apart. The differences between nuclear science and biology do not preclude an analysis of their similarities; likewise, those differences may be important in and of themselves.

Second, the examples I have chosen serve to highlight some less-explored features of the dual-use dilemma in the life sciences—features I believe deserve significant exploration. Though we may continue to worry about the potential for state-sponsored biological weapons programs, a primary concern in the dual-use dilemma is that of small groups or individuals committing acts of bioterrorism. My examples are chosen to identify issues in regulating these small-scale operations. This does not mean that there is nothing to be learned from the nuclear sciences about state-level regulations, or that my analysis doesn't cover these issues at all. Rather, my focus is meant to target the trajectory I see the (rapidly) evolving life sciences following, in an effort to identify issues that will arise or are already coming to light regarding dual-use research.

Materials control

Materials control is an important part of preventing the misuse of science and technology. If access to the constituent materials of biological or nuclear weapons is restricted, their use can be restricted presumably to good uses by good and responsible actors. As the subject of the Fink Report that led to their conclusion that the nuclear sciences cannot inform our thought on the dual-use dilemma, materials control is a good first point of inquiry for this chapter. The differences between fissile materials and biological pathogens listed in Table 16.1 are a paradigmatic way of contrasting nuclear science and biology. The differences between the materials used in both fields present a picture in which nuclear science is far removed from biology.

There are definitely some pathogens that possess the same type of threat as nuclear weapons. A smallpox attack could potentially cause up to 167 000 deaths—greater than the number who died when the United States dropped a nuclear weapon on Hiroshima.[8] An anthrax attack on an unprepared population could cause more than half a million deaths, but only if the emergency response

---

8   See Samuel, G., Selgelid, M. J. and Kerridge, I. 2010, 'Back to the future: controlling synthetic life sciences trade in DNA sequences', *Bulletin of the Atomic Scientists*, vol. 66, p. 10.

to the act was mismanaged.[9] The botulinum toxin—a biological toxin and the most potent poison in the world—released into the milk supply of the United States could kill up to 200 000 people, the majority children or the elderly.[10] Most recently, there are concerns that genetically modified avian influenza could kill millions.[11]

Those listed above, however, are some of only a few pathogens capable of the devastation similar to that visited on Hiroshima. Moreover, comparisons to Hiroshima are dated and misleading: the W76 warhead, a mainstay of the US nuclear arsenal, has a yield of 100 kilotons,[12] roughly five times that of the Hiroshima bomb. The W76 bomb is, further, small in yield compared with others in the US and Russian arsenals. Nuclear weapons range from 'tactical' warheads of a fraction of a kiloton, to megaton (where one megaton is equivalent to 1000 kilotons) scales.[13]

Biological attacks, further, suffer from a greater range of contingencies than nuclear attacks, and are prone to more variable yields than nuclear weapons. Bad weather—or, in the case of spores that are sensitive to sunlight, good weather—can seriously mitigate the effect of a bioweapons attack, to say nothing of other factors. Damage caused by biological weapons can be further mitigated by strong public health policies, smart post-attack vaccinations, and so on.[14] This is contrasted with the indiscriminate and destructive nature of nuclear weapons, which are more or less impervious to weather conditions or most attempts to defend against them.

This is not to say that biological pathogens should be treated less seriously than fissile materials. A key difference listed in the Fink Report is that the destructive potential of nuclear weapons is balanced to an extent by the difficulty of procuring fissile materials. Bioweapons, on the other hand, are likely to be more attainable and more augmented for causing death in years to come. For the most part, however, the distinction between the (broad) class of things called

9   Wein, L. M., Craft, D. L. and Kaplan, E. H. 2003, 'Emergency response to an anthrax attack', *Proceedings of the National Academies of Science*, vol. 100, no. 7, pp. 4346–51. It is worth, in the interest of disclosure, noting that 660 000 people are predicted to die in the modelled attack in the case where only individuals symptomatic with the disease are given antibiotic treatment. Wein et al. are quite clear in their study that there are a number of factors that, taken together, could lower this death count to approximately 10 000.
10   Wein, L. M. and Liu, Y. 2005, 'Analyzing a bioterror attack on the food supply: the case of botulinum toxin in milk', *Proceedings of the National Academies of Science*, vol. 102, no. 28, pp. 9984–9.
11   Enserink, M. 2011, 'Grudgingly, virologists agree to redact details in sensitive flu papers', *Science Insider*, 20 December 2011, <http://news.sciencemag.org/scienceinsider/2011/12/grudgingly-virologists-agree-to.html> (viewed 1 February 2012).
12   Where 1 kiloton = 1000 tonnes of TNT equivalent energy of explosion.
13   Norris, R. S. and Kristensen, H. N. 2007, *Estimates of the U.S. Nuclear Weapons Stockpile* [2007 and 2012], <http://www.fas.org/programs/ssp/nukes/publications1/USStockpile2007-2012.pdf> (viewed 24 February 2010).
14   See, for example, Wein et al., 2003, op. cit.; Chen, L.-C. et. al. 2004, 'Aligning simulation models of smallpox outbreaks', *Intelligence and Security Informatics*, vol. 3073, pp. 1–16; Wein et al., 2005, op. cit.

'biological pathogens' and the (narrow) class of things called 'fissile materials' is inaccurate; we are working with two classes of materials for which the comparisons we can make are not strong ones.

Radionuclides

A better comparison, I contend, is between biological pathogens and radionuclides described above. Though radionuclides more generally are not dual-use materials of the same degree as fissile materials—which form a small but significant set of radionuclides—radionuclides generally pose a series of dual-use issues, some of which I have described above.

Radionuclides are significant for two reasons. First, much as we think there are dual-use issues with life-sciences materials, technology and research that are worthy of attention even though they are not dangerous on the scale of, say, weaponised smallpox, there are dual-use issues in the nuclear sciences that do not involve nuclear weapons. Second, the way fissile materials were and are regulated is, as the Fink Report argued, somewhat unique. It is useful to examine the regulation of other areas of the nuclear sciences, which may have experienced issues with regulation more in line with what we would expect from the life sciences.

Though nowhere near as powerful as a thermonuclear detonation, or the expert use of weaponised smallpox, dirty bombs would cause damage comparable with less spectacular—and paradigmatic—examples of bioterrorism. First, the 'Amerithrax' attacks that occurred following the 11 September 2001 attacks on the World Trade Centre killed five and infected a further 17 people. Second, the attempted attack by the Aum Shinrikyo cult through the dispersion of anthrax failed,[15] but could have easily inflicted hundreds of casualties had it worked; the Sarin gas attack in 1995 on the Tokyo subway by the same cult killed 12 and injured thousands.[16] Finally, the lacing of salad bars with salmonella by the Bagwasrinesh cult in Oregon did not kill anyone, but hospitalised 45 and infected 750.[17] These cases are of serious concern, but the number of casualties (or hospitalisations, in those cases that lacked fatalities) is much smaller than from a thermonuclear explosion.

Sufficient exposure to radioisotopes can also lead to radiation poisoning. The most high-profile recent case of this is the death of KGB defector Alexander Litvinenko from a lethal dose of polonium-210. Polonium-210 is found in

---

15   See, for example, Miller, J., Broad, W. and Engelberg, S. 2001, *Germs: Biological Weapons and America's Secret War*, Simon & Schuster, New York.
16   Tu, A. T. 1999, 'Overview of the sarin terrorist attacks in Japan', in A. T. Tu and W. Gaffield (eds), *Natural and Selected Synthetic Toxins: Biological Implications*, ACS Symposium Series, 20 December, pp. 304–17.
17   Miller et al., op. cit.

antistatic brushes used in photography, and in the United States is sold in quantities large enough to be used to administer a fatal dose to someone. Large quantities of powdered radionuclides, even those that emit alpha radiation (which is typically considered harmless as it cannot penetrate the skin), distributed in a water supply could cause significant harm.

The malevolent use of radionuclides in dirty bombs or as poisons makes many industrial, medical and commercial technologies dual use. Determined terrorist groups presently could accrue large amounts of radionuclides through looting old or abandoned radiotherapy devices.[18] In 1987, for example, two men stole a capsule containing caesium-137 from an abandoned hospital site in Goiânia, Brazil.[19] The amount of caesium they stole would cause a significant number of casualties and economic damage if used in a dirty bomb. It is believed that a terrorist organisation would be able to easily procure quantities of radionuclides similar to those in Goiâna, if not larger amounts, through the purchase of radioactive refuse or radionuclides from former Soviet states that once functioned as disposal sites for highly radioactive nuclear waste.[20] UN inspectors have raised concerns that there are still highly problematic quantities of unguarded, unaccounted for radiological material in the world.[21]

One might object that the issues involved with securing biological pathogens are not the same as those with radionuclides. After all, influenza if left absent a host can die out relatively quickly, whereas caesium-137 will continue to exist and decay with a half-life of approximately 30 years until, via beta and gamma decay, it has turned into the stable barium-137. I don't deny that there are differences in the practicalities of securing different types of dual-use material—but this is both within and between fields. What is more interesting are the challenges presented by attempts to subvert regulations, especially when we consider materials that—befitting their classification as dual use—are used in public, commercial or industrial applications.

Acquiring materials piecemeal

Even if existing large sources of radioactive material were secured there is a possibility that someone could acquire significant nuclear materials to, for example, create a dirty bomb by acquiring dangerous radionuclides *piecemeal*. By this, I mean that dangerous radioactive compounds could be acquired in

---

18 Zimmerman and Loeb, op. cit.; International Atomic Energy Agency (IAEA) 2002, *Inadequate Control of World's Radioactive Sources*, <http://www.iaea.org/newscenter/pressreleases/2002/prn0209.shtml> (viewed 24 February 2011).
19 International Atomic Energy Agency (IAEA) 1988, *The Radiological Accident in Goiânia*, <http://www-pub.iaea.org/mtcd/publications/pdf/pub815_web.pdf> (viewed 24 February 2011).
20 Zimmerman and Loeb, op. cit.
21 IAEA, 2002, op. cit.

quantities small enough or clandestine enough to avoid the oversight of a body like the Nuclear Regulatory Commission (NRC). To illustrate this, consider the case of David Hahn, who at seventeen years of age, attracted the attention of the NRC and the Environmental Protection Agency for acquiring dangerous amounts of radium, americium and thorium, when he irradiated his mother's potting shed in an attempt to create a nuclear breeder reactor.[22] Though radium and americium are both problematic, as they could be used in a dirty bomb, David's acquisition of significant quantities of thorium is especially worrisome, as thorium-232 can be converted into uranium-233, a fissile material.

Hahn failed in his attempts to create a reactor (breeders require liquid sodium as a heat-transfer mechanism and are notoriously difficult to create even in ideal, well-financed conditions). What he did succeed in was creating a highly active neutron source. Hahn's neutron source was made by encasing americium that he had extracted from smoke detectors (and, later, radium extracted from the paint used in antique clocks) in lead with a pinhole in it, and using a strip of beryllium to convert the alpha radiation from the americium/radium into neutrons. Hahn targeted his neutron source at a grid of thorium-232 bricks he had purified from the ash of the mantles of camping lanterns. His purification method was relatively sophisticated considering his resources; he heated bars of lithium sourced from lithium-ion batteries surrounded by the thorium ash, causing the thorium to reduce and lithium to oxidise, leaving lithium oxide ash and pure thorium.[23]

Hahn's story is important because it shows how little training and money are needed to extract radioisotopes from commonplace technologies. Smoke detectors, camping lanterns and antique clocks do not typically fall into the purview of 'dual-use technologies', and yet they were sufficient for a seventeen-year-old to acquire thorium in quantity and purity that he (had he bought the thorium) would have violated materials control regulations of the NRC; his experiment caused enough of a hazard to warrant a Superfund hazardous materials clean-up.

The case of David Hahn raises a problem of acquisition of materials from existing technologies and natural sources. Piecemeal acquisition of dual-use materials in the life sciences has already received some attention.[24] There are common and naturally occurring viruses that could conceivably function as a natural template for a modified, virulent pathogenic organism. For instance, a potential terrorist *could* attempt to synthesise genetically modified influenza—a presently

---

22   See Silverberg, R. 2004, *The Radioactive Boy Scout: The Frightening True Story of a Whiz Kid and his Homemade Nuclear Reactor*, Random House, New York.

23   Ibid.

24   Samuel et al., op. cit. See also National Academies of Science 2010, *Sequence Based Classification of Select Agents: A Brighter Line*, The National Academies Press, Washington, DC.

technically difficult proposition—or they could wait until flu season to gain living native samples and modify them. Anthrax, to use another example, is still a common disease amongst cattle in developing countries.

The advent of 'do-it-yourself' (DIY) or 'garage' biology, where individuals practise biology at home or in private hobby groups, further undermines the efficacy of materials control. As the cost of sequencing and synthesis continues to decline, and the methods of DIY biologists in procuring samples either individually or through communal open labs ('biological commons') or creating their own technologies become increasingly sophisticated,[25] materials control will become progressively more costly, or even completely ineffective. Current regulatory approaches should take this into account when attempting to design overlapping regulatory measures—the so-called 'web of prevention'[26]—to prevent the misuse of biotechnology.

Moreover, if present expectations for biology are fulfilled, biological materials stand to become a ubiquitous part of modern lifestyles not in the way of nuclear materials, but in the way of microchips.[27] Many of these biological materials will not pose a significant dual-use threat, but there is a good chance that many of them will. Understanding how biological materials will appear in the everyday world and how motivated individuals might use them for malevolent or just manifestly reckless purposes is an important step in understanding dual use beyond the paradigm of state-level weapons conventions, deterrence and a field-wide security enterprise—a paradigm that is already recognised as untenable or insufficient for the dual-use threat in the life sciences.

Restricting access to technology

Another regulatory option is to restrict access to dual-use technologies. Biological technologies such as modern DNA sequencers have serious dual-use potential. DNA sequencers are able to sequence code in aid of doing beneficial research or creating non-pathogenic varieties of viruses for study or medicine; however, they may also find use in developing pathogens for use in harming others—for example, in manufacturing novel or existing pathogens to use in biological weapons.

---

25   See Carlson, R. 2010, *Biology is Technology: The Promise, Peril, and New Business of Engineering Life*, Harvard University Press, Cambridge, Mass.

26   Feakes, D., Rappert, B. and McLeish, C. 2007, 'Introduction: a web of prevention?' in B. Rappert and C. McLeish (eds), *A Web of Prevention: Biological Weapons, Life Sciences and the Future Governance of Research*, Earthscan, London, pp. 1–14.

27   See Carlson, op. cit.

Restricting access to technologies, much like materials control, is no less sure a strategy than materials control against talented, curious or motivated individuals. Famously, in 1976, Dr Abdul Qadeer Khan stole the plans for a centrifuge from his workplace at Urenco in the Netherlands. This theft, and Khan's later work, would ultimately mark the genesis of the Pakistani atomic weapons project.[28] In today's communicative environment, there are many more opportunities for determined individuals like Khan to enable the proliferation of technologies. The ubiquity of portable electronic storage devices and the potential of online so-called 'cloud' storage today make information containment very difficult.

In weighing the benefits created by restricting technologies against the costs of such policies, we must also account for the possibility that determined individuals, as in the case of David Hahn, will simply adopt indigenous or DIY methods to achieve their goals. This is already an emerging phenomenon in biology; 'garage biologists' have created DIY and low-cost solutions to expensive technologies in biology, such as Lava Amp, a low-cost, USB-powered PCR thermocycler.[29] For those devices that can't be made at a low cost, second-hand laboratory equipment is now available online. The sophistication and availability of low-cost or DIY biological laboratory tools are only likely to increase in the near future.

The creativity and drive to create lower-cost, more accessible tools in biology are surely things to be valued, for their own sake and for the gains we could have with a larger and more accessible research culture. This rapid lowering of access barriers to technologies, however, comes at a cost to security. That is, as biological technologies become ubiquitous, indigenous approaches to dual-use technologies will emerge as much from small enterprises as from large projects. While in the nuclear sciences, it required a state to manufacture indigenous technologies where restrictions on dual-use technology were already in place, individuals may conceivably be able to manufacture their own technologies.

These gains through low-cost, accessible technologies are surely not the only things we value. These gains must still be weighed against other things we value, such as our collective wellbeing, how that collective wellbeing is distributed (that is, equality), if those who are harmed as a result had the capacity to understand the risks of the proliferation of these technologies and consent to their proliferation (that is, fairness), and how tightly we think these matters should be regulated in the public interest (that is, liberty). Running roughshod over these complex ethical issues reduces us to a simple appeal to utility, and

---

28   See Corera, G. and Myers, J. J. 2006, *Shopping for Bombs: Nuclear Proliferation, Global Insecurity, and the Rise and Fall of A. Q. Khan's Nuclear Network*, Oxford University Press, London.

29   'Lava Amp', <http://www.lava-amp.com/> (viewed 27 February 2011).

a narrow sense of utility at that. Most of us, reflecting on the nuclear sciences, would admit that the nuclear sciences, even if they have brought gains to our lives, have brought gains unequally and at times unfairly.

Restriction of information

At times, it may be that the knowledge produced by a particular piece of research is dual use. The polio virus case, for example, functioned as a 'proof of principle' that viruses could be synthesised de novo. This is independent of which virus—or strain of a particular virus—is to be synthesised. It means that vaccines could be constructed rapidly, but also that virulent pathogens could be made. Even then, this could be good, enabling more individuals to study a virus with an eye to creating medicines against it; however, the knowledge could be also used to harm others. Jeronimo Cello and Eckard Wimmer, two of the researchers who worked on the polio virus synthesis, noted that one of the important outcomes of the research was confirming that viruses, in fact, behave in a chemical fashion and can be synthesised like chemicals.[30] It is this knowledge, we might think, which is dangerous as much as any physical materials used in the research.

One option is to censor information. We could restrict what projects are undertaken, or we could restrict—either temporarily or permanently, depending on the seriousness of the dual-use concern—the publication of research. For example, a 2005 study in the *Proceedings of the National Academies of Science* modelling a terrorist attack in which botulinum toxin was released in the US milk supply was embargoed for approximately a month as the journal, the National Academy of Sciences and government officials debated the security concerns in releasing such a report.[31] In 2011, the National Science Advisory Board for Biosecurity (NSABB) recommended the censorship of two studies in which highly pathogenic H5N1 avian influenza virus was modified to be transmissible between ferrets (close immunological analogues to humans), though this recommendation was later reversed.[32]

Regulating the production of such knowledge by restricting its creation or its dissemination is typically met with suspicion or outrage. Denying scientists

---

30   Selgelid, M. J. and Weir, L. 2010, 'Reflections on the synthetic production of poliovirus', *Bulletin of the Atomic Scientists*, vol. 66, no. 3, pp. 1–9.

31   Alberts, B. 2005, 'Modeling attacks on the food supply', *Proceedings of the National Academy of Sciences*, vol. 102, no. 29, pp. 9737–8. See also Wein, L. M. 2008, 'Homeland security: from mathematical models to policy implementation: the 2008 Philip McCord Morse lecture', *Operations Research*, vol. 57, no. 4, pp. 801–11.

32   Evans, N. G. 2013, 'Great expectations—ethics, avian flu and the value of progress', *Journal of Medical Ethics*, vol. 39, pp. 209–13.

their right to freedom of speech and their right to freedom of intellectual inquiry is seen as depriving them of basic rights. Moreover, there is concern that censorship or restriction of knowledge will not only deprive scientists of their rights, but also prevent beneficial research from occurring. Until the avian flu recommendation, scientific self-censorship was considered the norm for regulation of dual-use information.[33]

Though an important consideration, it would be a mistake to take self-governance on its own as a feasible regulatory strategy; no field demonstrates this more aptly than the nuclear sciences. For example, in 1939, Enrico Fermi, working with Szilard, showed that a nuclear chain reaction was experimentally possible. Szilard pressed Fermi at Columbia University and Frederic Joliot in Paris to keep their results secret. Fermi agreed hesitantly to comply on the condition that others followed suit,[34] but Joliot's team rejected Szilard's proposal, Joliot stating that if he withheld publication an implication would be that 'Hitler ... succeeded in destroying another precious liberty'.[35]

Following publication of this information in May 1939, a September meeting of the German military began an investigation into the plausibility of a chain reaction leading to weapons[36] and banned exports of uranium from German-controlled Czechoslovakia, effectively beginning the arms race for the bomb.[37]

This is a prime example of why relying on self-imposed censorship of results—the kind mentioned in contemporary reports on the dual-use dilemma[38]—is problematic. When regulatory measures are self-imposed without a way of binding members of a community to a decision, single individuals can undermine the greater projects of a community without good justification. Most of us would, I think, say that Joliot was unjustified in his assessment of the situation—surely if Hitler had succeeded in creating atomic weapons, more than just Joliot's liberty to publish would have been at risk.

Admitting that scientific self-censorship is not enough, there is genuine concern about permitting government control over scientific information. Allowing governments to control scientific information can be both ineffective and prone to abuse. For example, in 1979, the Department of Energy and the US Government attempted to censor the *Progressive* magazine and suppress information relating to the construction of the hydrogen bomb. This censorship was attempted under the auspices of the 'born secret doctrine', a policy of

---

33   National Academies of Science, 2004, op. cit., p. 101.
34   Rhodes, R. 1995, *The Making of the Atomic Bomb*, Simon & Schuster, New York, p. 295.
35   Goldsmith, M. 1976, *Frédéric Joliot-Curie: A Biography*, Lawrence & Wishart, London, p. 74.
36   Powers, T. 2000, *Heisenberg's War: The Secret History of the Atomic Bomb*, Penguin Books, London, p. 65.
37   Weart, S. 1976, 'Scientists with a secret', *Physics Today*, vol. 29, no. 2, p. 28.
38   For example, National Academies of Science, 2004, op. cit. See recommendation three of the Fink Report. See also recommendations three and four of National Academies of Science, 2006, op. cit.

secrecy in the *Atomic Energy Act of 1954*, restricting: 1) the design, manufacture or utilisation of atomic weapons; 2) the production of special nuclear material (uranium-233 or 235 or plutonium in natural or enriched states); and 3) the use of special nuclear material in the production of energy.[39]

The article of concern, written by Howard Morland, was based on information in the public domain.[40] Morland had not, in writing his article, used any information that was *secret*. Much was still considered 'born classified', but had existed in the public domain long enough that it was more or less commonly accessible knowledge. The information of concern was the description of radiation implosion, the technical process that allowed the 'super'—better known as the hydrogen bomb—to operate.[41]

Morland's stated intentions in releasing the article were to promote discussion about the arms race and the extensive secrecy of the nuclear complex.[42] In this sense, the knowledge was used in a manner that many would consider to be good. Nuclear secrecy, while justified in certain areas, has been criticised for subverting discourse on arms reductions and control of nuclear weapons.[43]

The US Government's reasons for attempting to censor Morland's article were that public knowledge was not complete. Potential builders needed more than just what was publicly available—knowledge that Morland's article, they claimed, provided. The court, however, rejected this claim, stating that any deduction that Morland had made regarding the technical mechanisms of the hydrogen bomb were clearly independently deducible: four countries already had independently created hydrogen bombs.

Lessons for biology

The lesson of nuclear science is that restrictions on information, technology or materials are not, as is commonly described, tensions between freedom and control. Rather, the tension is all about control—namely, whose control should take priority. The above cases highlight the problems that arise when individual scientists or governments have unilateral control over the dissemination of information. Despite the increasing proliferation of information and communications technologies, we should resist the idea that individuals having

39   US Congress, *The Atomic Energy Act of 1954*, ss. 1–15.

40   Knoll, E. 1979, 'The "secret" revealed', *The Progressive*, vol. 43, pp. 1–2; Morland, H. 2005, 'Born secret', *Cardozo Law Review*, vol. 26, p. 1401.

41   Knoll, op. cit.; Morland, op. cit.

42   See 'Symposium on the progressive case', *Cardozo Law Review*, vol. 26, no. 4 (Special Issue 2005).

43   See, for example, Rotblat, J. (ed.) 1982, *Scientists, the Arms Race, and Disarmament: A Pugwash/ UNESCO Symposium*, Taylor & Francis, London; Masco, J. 2002, 'Lie detectors: on secrets and hypersecurity in Los Alamos', *Public Culture*, vol. 14, no. 3, p. 441.

unilateral control over the dissemination of scientific information is always better than allowing government control over some scientific information; and even if it is, that individual control is the best strategy all things considered.

Rather, we should aim to moderate between these two positions. What the balance between these two extremes should be is a live issue. Moreover, it is an *ethical* issue, not just one of efficiency or technical capacity to control information (though it surely is that as well). Yet just as we would deny the desirability of subjecting science to centralised, complete government control, we can deny the desirability of individuals making unilateral decisions about scientific information.

Further, we can broaden our view to include the views and concerns of citizens. Much of the debate around dual use stems from dual use being—by definition— beneficial to humanity. Consulting humanity on what constitutes 'beneficial' would presumably be a step in the right direction. Morland's contention was that lacking in information due to the secrecy of the nuclear complex, the public was not able to engage with decision-making about the role of the nuclear sciences in public life. This, then, leads us to discussions about education.

Education

Education is advocated as a worthwhile strategy to deal with the dual-use dilemma.[44] Typically, the proposed venues for ethics education are institutions of higher learning, where professional or apprentice practitioners can be trained in the ethical implications of their discipline. The aim of this education, then, is to engage bench scientists or students of science in the ethical issues associated with their research.

There are, however, other types of education that are crucial to mitigating the harms of the dual-use dilemma, as demonstrated by the nuclear sciences. These measures are associated with the education of the public in both ethics and scientific literacy. Recall that Morland's contention in the *Progressive* case was that public engagement and reasonable knowledge of the technical as well as the political aspects of the regulatory debates and structures in the nuclear sciences were necessary to functioning scientific policy.

Moreover, ordinary public citizens may at times be the only witnesses to problematic acts that are not the purview of conventional regulations. Without appropriate education, ordinary public citizens are not well placed to know what

---

44 See, for example, Rappert, op. cit.; National Science Advisory Board for Biosecurity (NSABB), *Strategic Plan for Outreach and Education on Dual Use Research Issues*, National Science Advisory Board for Biosecurity, Bethesda, Md.

to look for or, sometimes, even what they are looking at. One of the observations that came from the David Hahn case is that, had they been suitably attentive, David's parents, teachers and classmates should have known something was amiss. Particularly at the point at which David began showing up with burns caused by exposure radiation, the public who interacted with David, we might think, should have become concerned enough to notify someone.[45]

Literature on dual use in the life sciences has discussed the merits—albeit to a lesser degree than the education of lab scientists—of public outreach.[46] Such outreach typically aims to create a culture of acceptance of novel technologies. The Biotechnology Industry Organization (BIO), for example, notes that 31 per cent of Americans surveyed by the Robert Wood Johnson Foundation are at least somewhat concerned about becoming ill from a bioterror attack.[47] The aim of public outreach is often to calm fears about biotechnology, often by showing how the emerging science will contribute to security.[48]

Another reason to pursue extensive public education is that as the life sciences become ever more ubiquitous in the technological and social landscapes, the public's comprehension may become an important part of security. In the life sciences, this is particularly important as 'garage' or 'DIY' biology comes into its own. DIY biologists are biologists who pursue biological research from their homes, usually as a hobby or as part of an entrepreneurial activity. While DIY biologists can create communities that have strong commitments to safety and security, such as the organisation DIYBio,[49] this is not a necessary—much less mandatory—part of being a DIY biologist. It remains to be seen how such a community can deal with members who would actively seek to cause harm, or perform reckless or dangerous experiments.

It may be that in some cases individuals engaged in DIY biology who pursue reckless or malevolent acts are only subject to observation by their peers, friends and family. In these cases, oversight is only plausible if those around an individual engaged in reckless or malevolent activity are suitably knowledgeable to recognise that something is amiss. If the public is not educated, activity done outside official laboratories may be overlooked with disastrous consequences.

Moreover, lack of public education may lead to powerful stakeholders dominating regulatory discourse in ways that are not conducive to the promotion of the good

45   See Silverberg, op. cit., ch. 8 and 'Epilogue'.
46   For example, Presidential Commission for the Study of Bioethical Issues 2010, *New Directions: The Ethics of Synthetic Biology and Emerging Technologies*, Washington, DC.
47   Biotechnology Industry Organization (BIO) 2010, *Healing, Fueling, Feeding: How Biotechnology is Enriching Your Life*, Biotechnology Industry Organization, Washington, DC, pp. 23–24, 29 n. 63.
48   Carlson, op. cit.; Carlson, R. 2003, 'The pace and proliferation of biological technologies', *Biosecurity and Bioterrorism: Biodefense Strategy, Practice, and Science*, vol. 1, no. 3, pp. 203–14.
49   'DIYBio', <http://diybio.org> (viewed 27 February 2011).

of society as a whole. A longstanding criticism of the nuclear sciences is that regulatory debates were dominated by militant voices.[50] In a similar fashion, much has been made recently of the potential for biology to create a new form of 'bio-economy'.[51] Without input from a range of actors, we might see money come to dominate the debate in the life sciences, much as the military came to dominate the debate in the nuclear sciences.

Conclusion: From ethical comparison to ethical practice

In this chapter, I have presented a number of cases that inform our thinking on issues created by, and issues that arise in dealing with, the dual-use dilemma. This, as I noted at the beginning of this chapter, is not a comprehensive account of the history of the nuclear sciences that might apply to the biological sciences. It is a snapshot to demonstrate the issues that can emerge from a comparison between the two, and lessons learnt about the success or failure of various attempts at regulation.

The above cases, I propose, provide four points that ought to guide our ethical practice

1. regulation of objects or information provides a barrier to access, but cannot be future-proof

2. the genius of the determined should not be underestimated—motivated and gifted individuals and groups will work around technical or political hurdles

3. restriction of information should seek to moderate between the risks of government and of self-imposed controls on information

4. education must focus on public education as well as the education of scientists.

These recommendations are not particularly controversial, but nor should they be. The history of regulating the malevolent use of science and technology is one that is as old as science and technology. In addition to any suggestions that directly bear on the types of regulation we should adopt, nuclear science serves as a reminder of the pitfalls regulation can succumb to.

A common theme in these points is that—as the nuclear sciences learned the hard way—traditional regulatory arguments that pit absence of formal

---

50 Rotblat, op. cit.; Chan, A. H. 1993, 'Team B: the trillion-dollar experiment', *Bulletin of the Atomic Scientists*, vol. 49, no. 3, pp. 24–7, <http://www.channelingreality.com/NWO_WTO/Team_B_The_trillion-dollar_experiment.pdf> (27 February 2011).

51 Carlson, 2010, op. cit.

regulation against absolute government control are deeply problematic. Nuclear science, I believe, tells us that all levels of stakeholder need to be engaged, most importantly the public. Public citizens are the end users and beneficiaries of biotechnology, and are likely to be victims of the accidents and misdeeds that motivate biosecurity debates. As biology becomes increasingly corporate and ubiquitous amongst private practitioners—the aforementioned 'garage' biologists—the public not only has the most to gain or lose from biology, but will also become witness to the misuse or abuse of biology and biotechnology. It is therefore imperative that regulation and biosecurity take into account the role the public will play.

This public education is surely to the benefit of everyone. For scientists, it presents an ever more educated public who values science and the benefits it brings. For a government, it presents a well-educated public who is better able to contribute to public life, and better able to understand the consequences of the misuse of biology, and react accordingly (for example, by supporting appropriate policy).

A better-educated public also provides the grounding for what I will refer to as 'open security'. As the pace of technological change increases, maintaining security will increasingly rely on surveillance, rather than attempting to build unwieldy regulations. In a world where the average personal computer can run complex simulations of nuclear weapons or epidemiological analysis; where desktop biological synthesis devices are not merely speculative but being discussed as real technologies;[52] where individuals are endlessly and perpetually connected, debates about security must honestly weigh the trade-offs and costs of restrictions to materials and information. As I have discussed often in this chapter, restrictions are not future-proof, and are only valuable when their benefits and costs are weighed against other important social goods. Increasingly, the cost of even small amounts of justified control[53] is prohibitive. Life will overtake security if security does not become adaptive.

Finally, it is important to take a broader view of my analysis. First, I have argued that a number of the 'usual suspects' in regulation are problematic, but not necessarily regulatory methods particular to nuclear science. Current discourse in the life sciences, from the Fink Report on, still treats materials control, technology control, information control, education and so on as primary methods of controlling the dual-use threat. But these methods were not foolproof in the nuclear sciences; today they are almost invitations to practitioners in the life

---

52   Carlson, R. 2007, 'Laying the foundations for a bio-economy', *Systems and Synthetic Biology*, vol. 1, no. 3, pp. 109–17, <http://www.ncbi.nlm.nih.gov/pubmed/19003445>.
53   As opposed to unjustified amounts of control, which we might think is easier to affect, but costs a population much more.

sciences to innovate around regulation. The nuclear sciences, in their failure to regulate the spread of dual-use technologies, should serve as a point of inquiry to figure out what doesn't work, and why.

Second, we should think beyond the nuclear sciences and begin drawing lessons of regulation from other technological cultures. If the genetic age stands to replace the information age, we should look at lessons learnt in the past 10 years of online culture to determine how regulation succeeded and failed to address the malevolent use of online technologies. The more information we can draw on, the more adaptive regulatory discourse can be.

Nuclear science is a broad and complex field with a long history. How much we can learn from it is as much a matter of how much care we take to investigate this history and apply lessons learnt to our current thinking and practice. As the debate over dual use in the life sciences is one in which we wish to prevent catastrophe, it seems prudent to take any source of guidance we can. Of those experiences in regulating science, none should be more important than the hard-earned lessons of nuclear science.

17. Considering Contextuality in Dual-Use Discussions: Is There a Problem?

Louise Bezuidenhout

Introduction

As dual-use ethics continues to grow as a topic of discussion, a number of features are increasingly becoming identifiable in the discourse. While many of these have been well discussed in a number of other volumes,[1] this chapter focuses on a little-examined characteristic: how issues relating to contextuality in life-science research are currently addressed in dual-use ethics.

The issue of contextuality in dual-use ethics is an interesting topic for consideration because it may be simultaneously argued that there is too much focus as well as too little. Those suggesting that dual-use ethics has been predominantly context driven will point to the central role that the 'web of prevention' rhetoric and policy development have had in the evolving discussions on responsibility. And they would not be wrong; indeed, much of the discourse in dual-use ethics has evolved out of control and regulatory discussions and continues to be strongly influenced by them.

The alternative—that current approaches to dual-use ethics are largely de-contextualised—is more difficult to defend (and ultimately much less popular). As this chapter will elaborate, however, examining the contextual oversights in dual-use ethics raises some extremely important considerations. In particular, these considerations shed light on how and why scientists in developing countries remain marginalised from dual-use ethics discourse and are unlikely to gain more prominence if current approaches are continued.

In order to elaborate on this position, the chapter will start by briefly examining dual-use ethics in light of the latter position. It will then go on to highlight how assumptions made through this position may impact on scientists in non-Western countries. These issues are supported by fieldwork observations from

---

1 Such as National Research Council 2011, *Challenges and Opportunities for Education about Dual-Use Issues in the Life Sciences*, The National Academies Press, Washington, DC.

dual-use ethics research conducted in Africa. The chapter will conclude by highlighting some areas that need to be further examined if dual-use ethics is to become a globally debated topic in the scientific community.

Contextuality in dual-use ethics

Dual-use ethics is an expanding field of study following the recent widespread endorsement of increasing capacity in ethics education for the life sciences— particularly from the West. Since 2001 international bodies such as the UN Policy Working Group on the United Nations and Terrorism, national organisations like the British Medical Association, and international agencies including the International Committee of the Red Cross, have made calls for the enhanced education of scientists, administrators, physicians and others about the potential for destructive application of the biomedical and medical sciences.[2]

Importantly, many of these calls have emphasised the need for promoting ethical decision-making, which serves to indicate a responsibility for scientists beyond legal and regulatory compliance.[3] In *Globalization, Biosecurity, and the Future of the Life Sciences*,[4] the report commissioned by the National Research Council of the United States, it was argued that it was prudent to establish a 'decentralized, globally distributed, network of informed and concerned scientists who have the capacity to recognize when knowledge or technology is being used inappropriately or with the intent to cause harm'.[5] This implies that the familiarisation of scientists with the legal requirements of bio-risk management is the starting point for ethics education, rather than an end in itself, and that the development of a sense of individual responsibility is desirable within the science community.[6] Indeed, as the National Science Advisory Board

---

2   Rappert, B. and Davidson, E. M. 2008, 'Improving oversight: development of an educational module on dual-use research in the west', in *Uganda National Academy of Sciences Promoting Biosafety and Biosecurity within the Life Sciences: An International Workshop in East Africa*, Uganda National Academy of Sciences, Kampala, p. 127.

3   The extent of individual responsibility of scientists within dual-use dilemmas is a subject of considerable debate, but will not be addressed here due to space limitations. It is well reasoned in Miller, S. and Selgelid, M. J. 2007, 'Ethical and philosophical considerations of the dual-use dilemma in the biological sciences', *Science and Engineering Ethics*, vol. 13, pp. 523–80.

4   Committee on Advances in Technology and the Prevention of their Application to Next Generation Biowarfare Threats 2006, *Globalization, Biosecurity, and the Future of the Life Sciences*, National Research Council, Washington, DC.

5   Ibid.

6   Uganda National Academy of Sciences 2008, *Promoting Biosafety and Biosecurity within the Life Sciences: An International Workshop in East Africa*, Uganda National Academy of Sciences, Kampala, p. 6. This is in line with Article IV of the Biological and Toxin Weapons Convention.

for Biosecurity (NSABB) stated: 'an enhanced culture of awareness is essential to an effective system of oversight and is a critical step in scientists taking responsibility for the dual-use potential of their work.'[7]

Current approaches to dual-use ethics discourse and pedagogy

Current approaches to dual-use ethics have therefore focused predominantly on identifying and understanding the responsibilities that scientists have towards ameliorating the dual-use potential of their research. These discussions of responsibility have been heavily influenced by the development of the 'web of prevention' model that promotes multiple stakeholders in biosecurity. This approach dates to the early 2000s when the International Committee of the Red Cross (ICRC) launched an initiative on 'Biotechnology, Weapons and Humanity', calling for the reaffirmation of norms against biological weapons and better controls on potentially dangerous biotechnology.[8] This web of prevention emphasised the crucial need for the involvement of security, health and judicial communities in addressing the dual-use issue.[9]

Crucially, the web of prevention concept has built on existing biosafety and biosecurity initiatives to include security, law-enforcement and life-science organisations, and the coordination of international oversight. In one formulation it was suggested that any 'web of prevention' include (and/or improve) initiatives such as

- export controls
- disease detection and prevention
- effective threat intelligence
- biosafety and biosecurity initiatives
- international and national prohibitions
- oversight of research
- education and codes of conduct.[10]

Dual-use ethics has played an important role in shaping understanding of the notions of both distributed responsibility and partial responsibility that the

7   National Science Advisory Board for Biosecurity (NSABB) 2007, *Report of the NSABB Working Group on Oversight Framework Development*, National Science Advisory Board for Biosecurity, Bethesda, Md.
8   Feakes, D., Rappert, B. and McLeish, C. 2007, 'Introduction: a web of prevention?' in B. Rappert and C. McLeish (eds), *A Web of Prevention: Biological Weapons, Life Sciences and the Future Governance of Research*, Earthscan, London, pp. 1–14.
9   Ibid.; and International Committee of the Red Cross (ICRC) 2003, *Biotechnology, Weapons and Humanity*, International Committee of the Red Cross, Geneva.
10   <http://www.brad.ac.uk/bioethics/EducationalModuleResource/EnglishLanguageVersionofEMR/> (viewed 25 August 2013), see lecture 2.

web of prevention model engenders for scientists. When considering the notion of partial responsibility, it is important to note that dual-use ethics has widely endorsed the idea that although '(t)he misapplication of peacefully intended research may cause moral distress among scientists ... it is difficult to argue that researchers should (solely) be held morally accountable for harm caused by unforeseen acts of misuse'.[11] Indeed, scientists are usually suggested only to have a limited amount of responsibility regarding the reuse of their data, and bioterrorist activities are thought of as 'beyond the responsibility of most life scientists either to prevent or to respond to'.[12]

This notion of partial responsibility is intimately connected to the idea of distributed responsibility. As suggested by Ehni,[13] 'only a mixed authority which is constituted by the scientific community together with governmental bodies, but with the participation of scientists meeting their responsibilities so far as possible, can solve the problem'. Indeed, the web of prevention model engages a wide range of stakeholders who bear some responsibility towards addressing and controlling the dual-use potential of the life sciences, including security, health and judicial communities. The web of prevention model has been very influential in structuring discussions regarding dual-use controls and most commonly includes a number of different areas of interventions including public health initiatives, security surveillance, biosafety and biosecurity controls, and the education of scientists and the development of codes of conduct.[14]

The web of prevention model has thus been influential in promoting the idea that scientists bear only partial responsibility for dual-use issues, and that they cannot be expected to address the dual-use potential of their research alone. While there has been general agreement on this, there remains considerable discussion on how this idea of a partial responsibility may be understood.

In recent years, there have been a number of attempts to determine 'lists of (conditional) duties' for scientists that will clearly elucidate the expectations that they have towards dual-use concerns. These, as promoted by Kuhlau et al.,[15] may be summarised as follows

- the duty to prevent bioterrorism
- the duty to engage in response activities
- the duty to consider the negative implications of their work

---

11 Kuhlau, F., Eriksson, S., Evers, K. and Hoglund, A. T. 2008, 'Taking due care: moral obligations in dual use research', *Bioethics*, vol. 22, no. 9, pp. 477–87 at p. 483.
12 Ibid., p. 477.
13 Ehni, H.-J. 2008, 'Dual use and the ethical responsibility of scientists', *Archivum Immunologiae Et Therapiae Experimentalis*, vol. 56, pp. 147–52 at p. 151.
14 As discussed in Rappert and McLeish, op. cit.
15 Kuhlau et al., op. cit., pp. 483–6.

- the duty not to publish or share sensitive information
- the duty to oversee or limit access to dangerous materials
- the duty to report activities of concern.[16]

These duties (and similar ones, such as Ehni's)[17] are predominantly presented as deontological duties and thus come with two important characteristics. First, these duties are usually presented as globally applicable (or at least, any discussion to the contrary is absent) and may be applied in any laboratory context around the world. Second, by virtue of being deontological, these duties are presented as moral obligations for scientists. Thus, there is the (explicit or implicit) understanding that the failure to fulfil them has ethical import.

This duty approach has made a considerable impression on current dual-use ethics, and formulations similar to the one presented above are often used in ethics pedagogy to inform discussion on responsibility and expectations.[18] This presents an interesting contrast in current dual-use educational initiatives, and for ethics discourse in general. Although there is considerable discussion about the interplay between the (contextual) web of prevention and the ethical responsibilities of scientists, and although there is extensive debate on contextually suitable styles of pedagogy,[19] dual-use responsibilities continue to be presented to scientists as a list of 'globally applicable' duties with little discussion about how they are applied in a contextual fashion. The next section considers this idiosyncrasy in further detail.

'Web of prevention' aside ...

Of course these duties are an excellent means of presenting dual-use responsibility discourse to scientists, and should not be viewed otherwise; however, that is not to say that this approach is not without its problems. In order to understand what these problems are, it is important to go back to these duties and examine them properly in light of the broader web of prevention. In particular, it becomes crucial to ask: what expectations about the research environment do these duties make?

---

16 Ehni proposed similar duties (Ehni, op. cit., p. 150): 'not to carry out a certain type of research; systematically to anticipate dual-use applications in order to warn of dangers generated by them; to inform public authorities about such dangers; not to disseminate results publicly, but keep dangerous scientific knowledge secret.'
17 Ibid.
18 National Research Council, 2011, op. cit.
19 There is a considerable amount of discussion about how dual-use ethics should be taught and who should be teaching the scientists. For an extensive discussion, see Rappert, B. (ed.) 2010, *Education and Ethics in the Life Sciences*, ANU E Press, Canberra.

The duties for scientists, as proposed by Kuhlau and her colleagues above, were subjected to a number of different conditions[20]

- it must be within their professional responsibility
- it must be within their professional capacity and ability
- it must be reasonably foreseeable
- it must be proportionally greater than the benefits
- it must be not more easily achieved by other means.

A brief survey of these conditions, it must be noted, does not make provision for any deficiencies in research environments, lack of support or problems with carrying out the duties. Rather, these conditions seem to delineate student scientists from principle investigators, and technicians from researchers. Thus, these conditions provide little in the way of support for scientists working under research conditions that may be markedly different from the 'Western norm'.[21]

It therefore becomes important to question whether 'dual-use responsibility duties' such as those proposed above make implicit assumptions about research environments and the implementation of webs of prevention. If the duties discussed above are therefore re-examined in light of this, a number of key considerations are noted. First, the duty to prevent bioterrorism, while a laudable goal, may be seen to be largely dependent on the existence of a web of prevention in action. Without the combined efforts of the security, health and judicial stakeholders, it is difficult to conceptualise how this duty may be carried out. Without the integrated involvement of governmental and international bodies,[22] it is difficult to conceptualise how such a duty would be acted upon.

Second, the duty to engage in response activities raises important questions about the responsibility of scientists in the absence of coordinated activities. Is it their responsibility to lobby for the establishment of response activities, or have they fulfilled their obligations solely due to the absence? Furthermore, if scientists are obligated to engage in response activities, are they similarly obliged to be involved in those not created in their own milieu? Are scientists, for example, in developing countries morally obliged to actively participate in any Western response activity, or is that in fact a form of ethical imperialism?

Third, although the duty to consider the negative implications of their work may at the outset be seen as self-explanatory, it is vital to consider that risk and benefit are interpreted quite differently around the world. Therefore, it is

---

20   Kuhlau et al., op. cit., pp. 481–2.
21   As will be discussed in the following sections, these differences could be in regulatory controls, funding, extra-laboratory service provision, governmental involvement and support, and access to the international life-science community.
22   As proposed by Miller and Selgelid, op. cit.

highly likely that scientists in non-Western countries—particularly countries experiencing food insecurity, considerable healthcare challenges and (non-bioterrorism) security issues—may view the negative implications of their work in a totally different light to their Western colleagues.[23] Thus, are scientists to consider what they perceive to be the negative (and positive) implications of their work, or simply the negative implications according to a Western perspective?

The third, fourth and fifth duties (not to publish or share sensitive information, to oversee or limit access to dangerous materials and to report activities of concern) all depend on the provision of national structures that will allow scientists to report concerns, control their research and manage their security. In the absence of such structures, it becomes crucial to question where scientists' responsibilities lie. Furthermore, will issues such as the fear of losing international funding or collaboration mean that, in such circumstances, scientists will not seek international alternatives?

Even the briefest of critiques of these duties in light of differing research contexts highlights the importance of a more deeply contextualised dual-use ethics discourse. As not all research environments are equal in their social, physical and extra-laboratory support provisions, it becomes important to ask: what happens when scientists cannot fulfil these duties through no fault of their own?

Properly considering research environments

Such considerations, of course, present difficulties to any notion of 'global duties' for dual-use responsibility amongst scientists. In turn, it may be suggested that this presents a crossroads to dual-use ethics discourse. Either the notion of 'global duties' must be abandoned in favour of more contextually sensitive suggestions or the 'global duties' must be thoroughly excavated to eliminate the implicit expectations that they contain about research environments.

Such debates are, of course, extremely complicated and beyond the scope of this chapter. Nonetheless, although not offering answers, the rest of the chapter will concern itself with highlighting why such issues need to be taken extremely seriously—particularly in the realm of dual-use ethics pedagogy. If one considers the issues raised above, a number of questions immediately spring to mind. These may include whether it is unfair to expect scientists to act as whistleblowers when the likelihood of losing their job is near 100 per cent; whether it is feasible to expect scientists to report their concerns when there

---

23 Bezuidenhout, L. (forthcoming), 'Moving life science ethics debates beyond national borders: some empirical observations'.

are no structures in place for them to do so; and whether, in the face of extreme public health crises in many countries, the risk of losing funding outweighs any threat of terrorism and thus perceptions of risk.

These concerns all require serious consideration. They suggest not only that achieving a 'common culture of awareness and a shared sense of responsibility'[24] amongst the global scientific community may be more complicated than initially envisioned, but also that current methods of raising dual-use awareness may alienate—rather than incorporate—scientists from non-Western research environments. If scientists are presented with such duties during dual-use ethics education without accompanying discussion on the strengths and limitations of implementing them contextually, it is just possible that such initiatives may do more harm than good.

Such hesitations relate to another chapter in this volume, by Judi Sture, which examines the concept of ethical erosion within communities of learners. It is possible that the presentation of 'idealised' duties or 'unattainable' standards of behaviour may significantly detract from attempts to engage scientists in discourse about dual-use responsibilities. Indeed, studies with scientists in a number of African laboratories strongly suggest that the wholesale importing of Western ethical approaches to teach dual use to these scientists was limited in success.[25]

These issues are further complicated by the lack of capacity in most developing countries to invest in home-grown ethics initiatives—at least for the moment. Thus, as it stands, within developing countries ethics education often remains largely in the hands of foreign funding agencies or interest groups. It is thus plausible to reiterate the question: is a lack of understanding of the structure of research within developing countries hampering efforts to build capacity within ethics?

Researching in developing countries

One of the reasons that these issues of contextuality are so poorly represented in dual-use ethics discourse is because of its historical legacy. Dual use has really only become a topic of concern in the life sciences since the terrorism events of 2001. In the subsequent years of the 'war on terror', the majority of dual-use discussion, quite naturally, occurred in the United Kingdom, the United States and a small number of other developed Western countries.

---

24   National Science Advisory Board for Biosecurity (NSABB) 2006, *Globalization, Biosecurity and the Future of the Life Sciences*, The National Academies Press, Washington, DC.
25   Bezuidenhout, op. cit.

Within these countries, the physical research environments of laboratories and the regulations governing biosafety and biosecurity have high degrees of similarity. Indeed, widely endorsed prerequisites for biosafety and biosecurity provisions[26] have further strengthened the harmony and standardisation between these laboratories. Thus, in the case of most dual-use discussions, the presence of a minimum level of biosafety and biosecurity regulation (that is implemented effectively) is not discussed *because it is already in place*.

Furthermore, when one considers the process of daily life-science research in light of any web of prevention, it becomes evident that a number of additional assumptions are also made about research environments *because of key similarities between laboratories in the United States, the United Kingdom and the European Union*. These include issues such as

- the existence of core funding for research facilities
- stable, adequate and reliable water, electricity and transport services
- stable, adequate and reliable postal, telecommunications and internet provision
- skilled support staff, such as technicians and those who maintain or repair equipment
- efficient and informed customs controls
- national regulations governing and guiding biosafety and biosecurity measures
- capacity and protection for the reporting of misconduct.

Unfortunately, outside a Western context, such provisions should not be automatically assumed. Indeed, many laboratories around the world represent extremely different working conditions—ones that challenge these assumptions. Many laboratories in developing countries, as will be discussed below, struggle daily with problems that range from a lack of core funding for facilities to unreliable electricity supplies.

Despite these different working conditions, it is vital to note that the research in these laboratories should not necessarily be considered unethical, unsafe or insecure. Rather, scientists have often found innovative ways to work around the limitations of their environments[27] and produce high-quality research;

---

26 Laboratory biosafety includes areas such as recruiting and retaining qualified individuals, training, laboratory work practices (for example, disinfection, waste handling, material control and accountability), personal protective equipment, medical surveillance, maintenance, access controls, self-assessments, documentation, corrective actions, reporting requirements, and incident response plans. Attempts to standardise such procedures have been spearheaded by the World Health Organisation (WHO), which has published guidelines such as World Health Organisation (WHO) 2004, *Laboratory Biosafety Manual*, World Health Organisation, Geneva.

27 There are a number of innovative low-cost alternatives to many biosafety and biosecurity requirements while still ensuring that the laboratories comply with international standards.

however, what does need to be considered is whether the current Western-centric approach to dual-use responsibility as presented in current ethics education may potentially alienate scientists by presenting them with duties they cannot fulfil.

Research environments in developing countries

Literature on laboratory life in developing countries is comparatively scarce. Indeed, the majority of discussion of laboratory environments, provisions for research and national support structures often comes from research and development (R&D) focused policy reports aimed at capacity building and investment. These reports tend to focus more on the shortcomings of these research environments rather than innovative alternatives; however, they provide some important considerations about research environments in developing countries.

The issues identified in these reports include[28]

- small, undifferentiated institutions
- lack of funding and lack of effective mechanisms for utilisation of funds
- high teaching burden
- lack of experienced mentors due to brain-drain
- history of poor investment in higher education and research
- lack of buy-in by institutions for new initiatives (such as centres of excellence)
- need for networking and networks
- corruption, mismanagement and institutional rigidity
- lack of governmental support, funding and control
- lack of vetted information about possible collaborators and institutions
- need for strong administrative and managerial skills
- inadequate resources and allocation thereof.

Furthermore, very few developing countries currently contribute even 1 per cent of their gross domestic product (GDP) to science R&D. Because of this lack of governmental involvement, many of these laboratories lack dedicated core funding, meaning that all day-to-day research expenses must somehow be

---

28 As informed by the fieldwork; Kiringia, J. M., Wambebe, C. and Baba-Moussa, A. 2005, 'Status of national research bioethics committees in the WHO African region', *BMC Medical Ethics*, vol. 6, no. 10; Fine, J. C. 2007, 'Investing in STI in sub-Saharan Africa: lessons from collaborative initiatives in research and higher education', Global Forum: Building Science, Technology and Innovation Capacity for Sustainable Growth and Poverty Reduction, Washington, DC; and Council on Health Research for Development (COHRED) 2010, *Fact Sheet: NEPAD-COHRED Strengthening Pharmaceutical Innovation in Africa*, COHRED, Johannesburg.

covered by grants for dedicated projects. This has far-reaching implications, as often salaries for researchers and technicians are not guaranteed by the research institutions.

Taken together, such characteristics raise some important considerations for the duties presented above. In particular, the disjunction between the 'ideal and real' research environments as well as the distance between the 'desired and actual' behavioural outcomes may have significant impacts on attempts to engage developing-country scientists in dual-use discussion.

Between 2011 and 2012, I conducted a large number of interviews with scientists in a range of African life-science laboratories.[29] This research, which is also discussed in a number of publications,[30] aimed to examine whether developing-country scientists engaged easily with the dual-use debate as it is currently presented in educational modules. This study thus, incorporating the issues raised above, questioned whether the current approaches to dual-use responsibility were too 'Westernised' for scientists in these countries and needed re-examining.

The majority of the participants in this study had problems with the concept of dual use as it was presented in current educational modules. In particular, the strong focus on bioterrorism and the perceived lack of sensitivity towards the problems they experienced in their daily research were often used as justifications for this negative reaction. The two following sections detail a couple of the most prevalent responses: first, that dual-use was not a problem for African scientists, and second, that even if it was a problem, there was nothing that could be done about it.

Why is dual use my problem?

Many of the scientists interviewed approached the concept of dual use as a problem far removed from their personal research. While many of the scientists expressed a strong academic interest in the problems associated with dual use, they nonetheless maintained that it was not a problem for African scientific research. In many interviews participants made comments similar to: 'I think that in Africa we just don't deal with such questions. I think it's more in the domain of the Western world, America, UK, where the threat of bioterrorism is a very real threat and so I think this issue is poignant there.'

Some of the reactions, on the other hand, were more extreme and at least 20 per cent of interviewees expressed hostility regarding the concept and its presence

---

29 At least 40 interviews in four different sites (three countries) with participants ranging from postgraduate students to heads of departments and institute directors.
30 See Bezuidenhout, op. cit.; and Bezuidenhout, L. (forthcoming), 'Ethics in the minutiae: examining the impact of daily laboratory processes on ethical behaviour and ethics education'.

in international debates. In many cases, as below, scientists expressed frustration at what they saw as a skewed Western perspective of the state of science, emphasising the other serious issues that they felt should take precedence. One participant was emphatic in their opinion, stating:

> I thought it was totally irrelevant and paranoid on the part of the Western world for this threat that often doesn't materialise and it's just huge amounts of money that go into fighting this phantom threat where I feel like we have more important things to do here as we're in the middle of such a huge HIV and TB epidemic and we just want to get on with doing the research. It was not an issue that I'd ever considered before and quite frankly I don't feel it's very relevant.

Many similar responses mentioned that dual use was 'just not a topic for discussion' because of the considerable healthcare and food-security challenges within these countries that the scientists were attempting to address through their research. Such perceptions of risk and benefit obviously differ from the expectations of current dual-use discourse, and may significantly challenge how the 'duty to consider the negative implications of their work'[31] was interpreted by these African scientists. Furthermore, one must question how the 'duty to prevent bioterrorism' might be interpreted if, as with the participant quoted above, scientists perceived the topic to be 'totally irrelevant and paranoid'.

Such considerations have far-reaching consequences for the success of ethics education and the perpetuation of dual-use awareness. Lack of personal buy-in to the concept of dual use due to differing perceptions of risk and benefit has severe consequences for ethics pedagogy, as students will struggle to make the connection between the information received during their instruction and their daily behaviour. This undermines the concept of a culture of responsible awareness and questions the effectiveness of concepts such as codes of conduct for the life sciences.

Nonetheless, during this fieldwork a number of the participants manipulated the dual-use concept to fit the concerns within their own environment. Thus, still taking dual use to refer to 'the potential for beneficial scientific research to be misused for nefarious purposes by a third party',[32] the scientists connected the harm caused by a third party to fearmongering within irresponsible scientific journalism instead of bioterrorism. By contextualising the concept within their own research environment, the participants were then able (and willing) to critically re-examine the concept and its utility within their daily research.[33]

---

31 One of the duties proposed by Kuhlau et al., op. cit.
32 As proposed by Miller and Selgelid, op. cit.
33 This is extensively examined in Bezuidenhout, 'Ethics in the minutiae', op. cit.

Even if dual use is a problem, what can I do about it?

Another characteristic of many of the participants interviewed was that they removed themselves from responsibility for the dual-use problem by emphasising their lack of agency for engaging with any sort of dual-use control. This was often related to a number of characteristics within their environments. In many cases, participants suggested that '[b]eing alone you can't change the system [and] you just become a problem to the institution. It has to come from above. If you come somewhere personally and tell them about standards you are causing problems.' The participants commonly associated these feelings with issues within their research environments.

Those at all the field sites mentioned the difficulties of getting reagents and samples in and out of their countries. They mentioned that poorly trained border officials, unrealistic foreign export and import requirements and poor transport infrastructure significantly complicated their daily research. Statements such as '[i]t already takes four to six weeks to get a delivery through, so any extra restrictions will make it even worse' regularly appeared in the interviews. In such cases, it must be asked whether expecting scientists to 'raise awareness of dual-use concerns'[34] and draw attention to 'the negative implications of their work'[35] are remotely feasible.

At one of the field sites I was made aware of the fact that the government did not provide any core funding for the facility, and that the entirety of their running costs and research budget came from foreign grants. This, fieldwork participants suggested, raised two important problems for the dual-use control duties. First, as the possibility of losing their funding had implications far beyond stopping a project, many participants stated that they would be unwilling to 'report activities of concern'[36] due to the possibility that 'misunderstandings might shut us down entirely'. Second, the participants highlighted that any expectations of improving security or 'limiting the access to dangerous materials'[37] were unrealistic as general facility maintenance and improvement were 'not included in project-specific grants'. Thus, the participants viewed themselves as in a difficult position in which not raising dual-use concerns was indeed the 'lesser of two evils'.

In all of the facilities participants also regularly asked me what 'response activities' they had a duty to engage in, and to whom they could 'report activities of concern'. In both cases it was patently clear that a lack of government involvement and low levels of institutional buy-in for dual-use control meant

---

34 One of the duties proposed by Kuhlau et al., op. cit.
35 One of the duties proposed by ibid.
36 One of the duties proposed by ibid.
37 One of the duties proposed by ibid.

that there were often no answers to be had. It was my impression that presenting these duties without a proper, contextually considered understanding of how it may be implemented often turned the participants off the dual-use discussion in its entirety, as it was once again perceived as 'not a problem for Africa'.

If, as is the case in dual-use ethics, the duties continue to be presented as moral obligations, it is also easy to see how these scientists are placed in ethically untenable positions that compromise their ethical development. Thus, structural issues within the research environment, if unaddressed, have the potential to undermine ethical training. Continually facing deficits in the ethical conflicts between expected duties and the characteristics of the research environment may cause frustration and resignation amongst the scientists and lower the potential for them to get involved in ethical discussions. It must be asked how to reflect these environmental issues within ethics training.

Starting to re-contextualise dual-use ethics

This chapter thus presents a contrasting perspective of dual-use ethics discourse and pedagogy to prevailing norms. Based on the theoretical analysis and the fieldwork discussed here and elsewhere it strongly suggests that the current dual-use ethics discourse is largely de-contextualised due to the absence of discussion on variations in research environments. This has significant implications for discussions on responsibility and duty rhetoric and consequently for the building of a 'common culture of awareness and a shared sense of responsibility within the global community of life scientists'.[38]

The majority of responsibility and duty rhetoric within dual-use discussions is strongly influenced by the web of prevention model of dual-use oversight. Despite a strongly deontological, global approach to these discussions, the chapter highlighted that these duties have a number of implicit assumptions about the laboratory (and extra-laboratory) environments in which life-science research takes place. Unsurprisingly, due to the historical legacy of the dual-use discussion, these assumptions are based predominantly on a Western understanding of a 'minimum level of research environment'. Such an environment includes key contributions from the government, the research institution and the general surrounding society.

As this chapter points out, however, such an environment only reflects a portion of laboratories around the world, and in many different countries high-quality research is occurring under significantly different conditions. In such

---

38   National Science Advisory Board for Biosecurity 2006, Globalization, Biosecurity and the Future of the Life Sciences, Washington D. C., The National Academies Press. See pp 5.

cases, the implicit assumptions about research environments inherent within ethics discourse may serve to alienate scientists from engaging within the dual-use debate. Paradoxically, as the life sciences become increasingly global, these previously marginalised communities of scientists are precisely the ones that urgently require representation.

So, it remains to be asked, what can be done? As mentioned above, it may be that dual-use ethics—particularly responsibility ethics—needs to be critically re-evaluated. Either the notion of 'global duties' for dual-use control must be abandoned in favour of more contextually sensitive suggestions or the 'global duties' must be thoroughly excavated to eliminate the implicit expectations that they contain about research environments. Of course, this is a complicated discussion and will require considerable attention from the dual-use ethics community.

Much else, however, can be done. Within much dual-use discussion there is a tendency to talk about research environments as largely homogenous bodies. Although, of course, much of the discourse remains centred in (and on) Western countries, developing a sense of awareness of the heterogeneity of research environments (even within these countries) cannot help but strengthen current discourse. Furthermore, cultivating an awareness of the challenges faced by developing-country researchers will no doubt facilitate their inclusion in international discourse.

Within ethics pedagogy, any gestures towards contextual sensitivity will no doubt be much appreciated. In particular, educationalists should consider how discussions about *implementing duties into daily life* may be fostered by pedagogical initiatives. With regards to developing-country scientists, however, future educational initiatives should also consider whether developing 'more context specific case studies'[39] can really fix current problems, or whether considerable and sustained attention needs to be paid to the problems identified above.

It has often been quoted that dual use cannot be addressed within national borders or in institutions, but requires a regional—indeed an international—response due to the nature of the threat and the characteristics of modern life sciences. In order to adequately realise this need, a global body of dual use-aware scientists is vital. It is therefore becoming increasingly important that the previously marginalised groups of scientists from developing countries become key players in the development of a truly international dual-use discussion. In order to do so, the first step is to recognise the differences inherent in others.

---

39   National Research Council, op. cit.

Part IV: Ethical Futures

18. Exploring the Role of Life Scientists in Combating the Misuse of Incapacitating Chemical and Toxin Agents

Michael Crowley

Introduction

Over the past two decades there has been a revolution in the life sciences with extremely rapid advances in genomics, synthetic biology, biotechnology, neuroscience and the understanding of human behaviour. The speed of progress is staggering. For example, in 1999 a special meeting of the National Academies of Sciences and the Society of Neuroscience noted that '[t]he past decade had delivered more advances than all previous years of neuroscience research combined'.[1] Many of these developments have great potential to benefit humankind—in, for example, the production of more effective, safer medicines.[2] Concern has been raised, however, by a growing number of those in the scientific and medical communities regarding the 'dual-use' nature of certain advances with the consequent danger of the new technologies being misused for the development of a new range of chemical or biological weapons. Meselson has stated that '[d]uring the century ahead, as our ability to modify fundamental life processes continues its rapid advance, we will be able not only to devise additional ways to destroy life, but also be able to manipulate it including the processes of cognition, development and inheritance'.[3]

And he added: 'A world in which these capabilities are widely employed for hostile purposes would be a world in which the very nature of conflict had radically changed. Therein could lie unprecedented opportunities for violence, coercion, repression or subjugation.'[4]

---

1 Society of Neuroscience 1999, Neuroscience 2000: A New Era of Discovery, Symposium organised by the Society of Neuroscience, Washington, DC, 12–13 April.
2 Andreasen, N. 2004, *Brave New Brain: Conquering Mental Illness in the Era of the Genome*, Oxford University Press, USA.
3 Meselson, M. 2000, 'Averting the hostile exploitation of biotechnology', *The CBW Conventions Bulletin*, no. 48, pp. 16–19.
4 Ibid., pp. 16–19.

Table 18.1 (Bio)chemical Threat Spectrum Chart

Classical chemical weapons	Industrial pharmaceutical chemicals	Bio-regulators peptides	Toxins	Genetically modified biological weapons	Traditional biological weapons
Cyanide Phosgene Mustard Nerve agents	Fentanyl Carfentanil Remifentanil Etorphine Dexmedetomidine Midazolam	Substance P Neurokinin A	Staphylococcal enterotoxin B (SEB)	Modified/tailored bacteria and viruses	Bacteria Viruses Rickettsia Anthrax Plague Tularemia

← Poison →

← Chemical Weapons Convention →

← Biological and Toxin Weapons Convention →

← Infect →

Source: Adapted from: Pearson, G. 2002, 'Relevant scientific and technological developments for the first CWC Review Conference: the BTWC Review Conference experience', *CWC Review Conference Paper*, no. 1, University of Bradford, UK, p. 5.

As the ongoing revolution in the life sciences has proceeded, the boundary between chemistry and biology, and consequently the distinction between certain chemical and biological weapons, has become increasingly blurred. Rather than thinking of chemical and biological weapons threats as distinct, certain analysts including Aas,[5] Dando,[6] Davison[7] and Pearson[8] believe it is more useful to conceptualise them as lying along a continuous biochemical threat spectrum. This chapter will focus upon research and development of those mid-spectrum agents (pharmaceutical chemicals, bio-regulators and toxins featured in Table 18.1) that some consider as having potential utility as incapacitating weapons (incapacitants). The chapter will employ a holistic arms control (HAC) approach to examine the potential dangers and proposed utility of such agents and explore the obligations and opportunities for the life-science community to ensure that such agents are not utilised for hostile purposes.

Incapacitants: A primer

Although certain states and multilateral organisations such as the North Atlantic Treaty Organisation (NATO)[9] have sought to characterise incapacitants, there is currently no internationally accepted definition for these chemical agents. Indeed certain leading international experts believe that such a technical definition is not possible.[10] Whilst recognising the contested nature of this discourse, as a provisional working description, they can be considered as substances whose chemical action on specific biochemical processes and physiological systems, especially those affecting the higher regulatory activity of the central nervous

---

5   Aas, P. 2003, 'The threat of mid-spectrum chemical warfare agents', *Prehospital and Disaster Medicine*, vol. 18, no. 4, pp. 306–12.

6   Dando, M. 2007, 'Scientific outlook for the development of incapacitants', in A. Pearson, M. Chevrier and M. Wheelis (eds), *Incapacitating Biochemical Weapons*, Lexington Books, Lanham, Md, p. 125.

7   Davison, N. 2009, *Non-Lethal Weapons*, Palgrave Macmillan, Basingstoke, UK, pp. 106–7.

8   Pearson, op. cit.

9   NATO defines an incapacitant as a 'chemical agent which produces temporary disabling conditions which (unlike those caused by riot control agents) can be physical or mental and persist for hours or days after exposure to the agent has ceased. Medical treatment, while not usually required, facilitates a more rapid recovery.' North Atlantic Treaty Organisation (NATO) 2000, *Glossary of Terms and Definitions (AAP-6 (V), Modified version 02)*, 7 August 2000.

10   A report of an expert meeting organised by Spiez Laboratory concluded that 'because there is no clear-cut line between (non-lethal) ICA [incapacitating chemical agents] and more lethal chemical war-fare agents, a scientifically meaningful definition cannot easily be made. One can describe several toxicological effects that could be used to "incapacitate", but in principle there is no way to draw a line between ICAs and lethal agents.' See Mogl, S. (ed.) 2011, *Technical Workshop on Incapacitating Chemical Agents*, Spiez, Switzerland, 8–9 September; Spiez Laboratory 2012, op cit., p. 10; The Royal Society 2012, *Brain Waves Module 3: Neuroscience, Conflict and Security*, Science Policy Centre, The Royal Society, London, pp. 44–5; Organisation for the Prohibition of Chemical Weapons (OPCW) 2013, *Report of the Scientific Advisory Board on Developments in Science and Technology for the Third Special Session of the Conference of the States Parties to the Chemical Weapons Convention*, Third Review Conference RC-3/DG.1, 8–19 April 2013, 29 October 2013, p. 4.

system, produce a disabling condition (for example, can cause incapacitation or disorientation, incoherence, hallucination, sedation or loss of consciousness) or, at higher concentrations, death.[11]

There is a wide variety of agents that could potentially be utilised as incapacitants including anaesthetic agents, skeletal muscle relaxants, opioid analgesics, anxiolytics, antipsychotics, antidepressants and sedative-hypnotic agents,[12] many of which are currently legitimately utilised by the medical or veterinary professions.[13] Table 18.2 summarises the results of a literature study and analysis of biomedical research into a range of pharmaceutical agents, published in 2000, by the Applied Research Laboratory and the College of Medicine at Pennsylvania State University, to identify the range of drug classes that had potential utility as incapacitants. The study clearly illustrates and indeed actively explores the potential dual-use applications of drugs initially developed for medical purposes, noting that:

> It is well known that for every one new compound successfully proceeding from the discovery phase through all phases of clinical trials and on to market, perhaps hundreds, if not thousands, of compounds are discarded or shelved by the pharmaceutical industry [for example, as a result of their side effects] … However, in the variety of situations in which non-lethal techniques are used there may be less need to be concerned with side-effects; indeed, perhaps a calmative may be designed that incorporates a less than desirable side-effect … as part of the drug profile.[14]

Furthermore, the study recommends explicit collaboration in this area, stating that 'it may be appropriate to develop a working relationship with the pharmaceutical industry to better incorporate their knowledge and expertise in developing a non-lethal calmative technique'.[15] The ethical implications of such relationships are, however, not explored.

---

11   Adapted from Pearson et al., op. cit., p. xii. Incapacitants have also been called advanced riot-control agents, biochemical agents, biotechnical agents, calmatives, incapacitating biochemical weapons and immobilising agents.
12   See, for example: Lakoski, J., Bosseau Murray, W. and Kenny, J. 2000, *The Advantages and Limitations of Calmatives for Use As A Non-Lethal Technique*, College of Medicine Applied Research Laboratory, Pennsylvania State University.
13   See Aas, op. cit., p. 309.
14   Lakoski et al., op. cit., p. 48.
15   Ibid., p. 48.

Figure 18.2 Indicative Drug Classes and Agents with Potential Utility as Incapacitants

Drug class	Selected compounds	Site of action
Benzodiazepines	Diazepam Midazolam Etizolam Flumazenil (antagonist)	GABA receptors
Alpha$_2$ adrenergic receptor agonists	Clonidine Dexmedetomidine Fluparoxan (antagonist)	Alpha$_2$ adrenergic receptors
Dopamine D3 receptor agonists	Pramipexole CI-1007 PD 128907	D3 receptors
Selective serotonin reuptake inhibitors	Fluoxetine Sertraline Paroxetine WO-09500194	5-HT transporter
Serotonin 5-HT$_{1A}$ receptor agonists	Buspirone Lesopitron Alnespirone MCK-242 Oleamide WAY-100, 635	5-HT$_{1A}$ receptor
Opioid receptors and mu agonists	Morphine Carfentanil Naloxone (antagonist)	Mu opioid reception
Neurolept anaesthetics	Propofol Droperidol and fentanyl combination Phencyclidines	GABA receptors DA, NE and GABA receptors Opioid receptors
Corticotrophin-releasing factor receptor antagonists	CP 154,526 (antagonist) NBI 27914 (antagonist) CRF-BP (binding protein)	CRF receptor
Cholecystokinin B receptor antagonists	CCK-4 CI-988 (antagonist) CI-1015 (antagonist)	CCKB receptor

Source: Adapted from Lakoski, J., Bosseau Murray, W. and Kenny, J. 2000, *The Advantages and Limitations of Calmatives for Use As A Non-Lethal Technique*, College of Medicine Applied Research Laboratory, Pennsylvania State University, pp. 15–16.

There is a long history, dating from the late 1940s, of certain state programs attempting to develop incapacitant weapons employing a range of pharmaceutical chemicals or toxins.[16] Analysis of open-source information from the mid 1990s onwards indicates that a number of states including China,[17] the Czech Republic,[18] Russia[19] and the United States[20] appear to have conducted research relating to incapacitants and/or possible means of delivery at some stage during this period. It is, however, difficult to establish the current situation, and certain states that have previously shown an interest in developing such agents—such as the United States—have recently declared that no such activities currently take place.[21]

According to the International Committee of the Red Cross (ICRC): 'There is clearly an ongoing attraction to "incapacitating chemical agents" but it is not easy to determine the extent to which this has moved along the spectrum from academia and industrial circles into the law enforcement, security and military apparatuses of states.'[22]

---

16 See, for example: Crowley, M. 2009, *Dangerous Ambiguities: Regulation of Riot Control Agents and Incapacitants under the Chemical Weapons Convention*, University of Bradford, UK; Dando, M. and Furmanski, M. 2006, 'Midspectrum incapacitant programs', in M. Wheelis, L. Rózsa and M. Dando (eds), *Deadly Cultures: Biological Weapons Since 1945*, Harvard University Press, Cambridge, Mass.; Davison, op. cit.; Furmanski, M. 2007, 'Historical military interest in low-lethality biochemical agents', in Pearson et al., op. cit.; Pearson, A. 2006, 'Incapacitating biochemical weapons: science, technology, and policy for the 21st century', *Nonproliferation Review*, vol. 13, no. 2.

17 Crowley, op. cit., p. 82; Guo Ji-Wei and Xue-sen Yang 2005, 'Ultramicro, nonlethal and reversible: looking ahead to military biotechnology', *Military Review*, July–August, as cited in Pearson, A. 2007, 'Late and post-Cold War research and development of incapacitating biochemical weapons', in Pearson et al., op. cit.

18 Hess, L., Schreiberová, J., Málek, J. and Fusek, J. 2007, 'Drug-induced loss of aggressiveness in the macaque rhesus', *Proceedings of 4th European Symposium on Non-Lethal Weapons, 21st–23rd May 2007, Ettlingen, Germany*, European Working Group on Non-Lethal Weapons, Pfinztal: Fraunhofer ICT, V15; Hess, L., Schreiberova, J. and Fusek, J. 2005, 'Pharmacological non-lethal weapons', *Proceedings of the 3rd European Symposium on Non-Lethal Weapons, 10th–12th May 2005, Ettlingen, Germany*, European Working Group on Non-Lethal Weapons, Pfinztal: Fraunhofer ICT, V23; Davison, N. and Lewer, N. 2006, *Bradford Non-Lethal Weapons Research Project (BNLWRP)—Research Report No. 8*, University of Bradford, UK, p. 50.

19 Klochikin, V., Pirumov, V., Putilov, A. and Selivanov, V. 2003, 'The complex forecast of perspectives of NLW for European application', *Proceedings of the 2nd European Symposium on Non-Lethal Weapons, 13th–14th May 2003, Ettlingen, Germany*, European Working Group on Non-Lethal Weapons, Pfinztal: Fraunhofer ICT; Klochinkhin, V., Lushnikov, A., Zagaynov, V., Putilov, A., Selivanov, V. and Zatekvakhin, M. 2005, 'Principles of modelling of the scenario of calmative application in a building with deterred hostages', *Proceedings of the 3rd European Symposium on Non-Lethal Weapons, 10th–12th May 2005, Ettlingen, Germany*, European Working Group on Non-Lethal Weapons, Pfinztal: Fraunhofer ICT.

20 Crowley, op. cit., pp. 76–8; Davison, op. cit., pp. 105–42; Furmanski, op. cit.; Pearson, 2007, op. cit.; Furmanski and Dando, op. cit.

21 Organisation for the Prevention of Chemical Weapons (OPCW), Executive Council 2013, Statement by Ambassador Robert P. Mikulak, United States Delegation to the OPCW at the Seventy Second Session of the Executive Council, Seventy-Second Session EC-72/NAT.8, 6–7 May 2013, 6 May 2013. In his statement, ambassador Mikulak declared '[i]n this context, I also wish very clearly and directly to reconfirm that the United States is not developing, producing, stockpiling, or using incapacitating chemical agents'.

22 International Committee of the Red Cross (ICRC) 2010, *Expert Meeting: Incapacitating Chemical Agents, Implications for International Law*, Montreux, Switzerland, 24–26 March, p. 3.

Box 18.1 Contemporary Czech Republic research into incapacitating chemical agents

The Czech Republic has had a longstanding research program into incapacitants dating from at least 2000, part of which was funded by the military.[a] In May 2005, at the Third European Symposium on Non-Lethal Weapons, Czech researchers delivered a paper[b] describing their investigations over several years, administering rhesus monkeys with various pharmacological cocktails in order to determine which combinations and doses resulted in 'fully reversible immobilization'. The paper also described how '[f]ully reversible analgesic sedation was ... tested in man', utilising the triple combination of dexmedetomidine, midazolam and fentanyl given to patients undergoing surgery, and a second combination of dexmedetomidine, midazolam and ketamine, which was tested on 10 nurses.[c]

In a 2007 follow-up paper, Czech researchers described how they 'decided to test new combinations [of drugs] for suppression or complete abolition of aggressive behaviour' in macaque monkeys.[d] The researchers claim that 'the results can be used to pacify aggressive people during medical treatment (mental disease), terrorist attacks and during production of new pharmacological nonlethal weapons'.[e] In July 2010, Czech researchers published a paper describing their studies inducing immobilisation in orang-utans and chimpanzees utilising a naphthylmedetomidine-ketamine-hyaluronidase combination.[f] Although the results of this research were presented in terms of facilitating the relocation and painless medical examination of the animals, such research may also potentially be applicable to incapacitant development. Czech researchers have also investigated a number of alternative means of agent delivery including via inhalation administration, which was initially tested on rats and then on human 'volunteers',[g] who were reported to have been children in hospital.[h] Researchers have also explored conjunctival, nasal, transbucal, sublingual and transdermal administration.[i]

[a] According to Davison and Lewer, research to develop sedative and anaesthetic agent combinations for use as weapons had been funded by the Czech Army under Project No: MO 03021100007. See Davison and Lewer, op. cit., p. 50.

[b] Hess et al., 2005, op. cit.

[c] Ibid., pp. 8–9.

[d] Hess et al., 2007, op. cit., p. 6.

[e] Ibid., p. 7.

[f] Hess, L., Votava, M., Schreiberová, J., Málek, J. and Horáček, M. 2010, 'Experience with a naphthylmedetomidine-ketamine-hyaluronidase combination in inducing immobilization in anthropoid apes', *Journal of Medical Primatology*, vol. 39, no. 3 (June), pp. 151–9.

[g] Hess et al., 2005, op. cit., pp. 11–12.

[h] Davison, op. cit., p. 128.

[i] Hess et al., 2005, op. cit., pp. 10–14.

Potential dangers and proposed utility

Proponents of incapacitants have promoted their development and use in certain law-enforcement scenarios (such as hostage-taking situations) where there is a need to rapidly and completely incapacitate a single or a group of individuals without causing death or permanent disability. Incapacitants have also been raised as a possible tool in a variety of military operations, especially in situations where combatants and noncombatants are mixed.[23]

23   See, for example: Fenton, G. 2007, 'Current and prospective military and law enforcement use of chemical agents for incapacitation', in Pearson et al., op. cit., pp. 103–23; Whitbred, G. 2006, 'Offensive use of chemical

A broad range of observers, however—including scientific and medical professionals, arms-control organisations, international legal experts, human rights monitors and humanitarian organisations, as well as a number of states—is highly sceptical about the development and utility of incapacitants, highlighting the fact that such weapons are not inherently nonlethal, even if they were to be used with a nonlethal intent. In their 2003 study conducted under the auspices of the Federation of American Scientists, Klotz et al. developed a predictive model illustrating 'why seemingly non-lethal incapacitating agents may be quite lethal in actual use'.[24] In their conclusion, they stated: 'We have shown, at least within the approximations of our simple (but generous) two receptor equilibrium model, that even with a therapeutic index of 1,000 (above any known anaesthetic or sedative agent), a chemical agent used as an incapacitating weapon can be expected to cause about 10% fatalities.'[25]

Furthermore, as Pearson has noted, even such predictive modelling will potentially underestimate fatalities when an incapacitant is used in real-life situations where there is uncontrollable variability 'both in terms of exposure (uneven concentration and exposure time) and within the target population (age, size, gender, health status and individual susceptibility)'.[26] As a result of such considerations, the British Medical Association believes:

> The agent whereby people could be incapacitated without risk of death in a tactical situation does not exist and is unlikely to in the foreseeable future. In such a situation, it is and will continue to be almost impossible to deliver the right agent to the right people in the right dose without exposing the wrong people, or delivering the wrong dose.[27]

Similarly, Klotz et al. concluded that 'genuinely non-lethal chemical weapons are beyond the reach of current science'.[28]

technologies by US special operations forces in the global war on terrorism', *Maxwell Paper*, no. 37, Air University Press, Maxwell Air Force Base, Ala. It should be noted that other authors have questioned the utility of incapacitants in certain proposed scenarios such as premeditated hostage situations, due to the availability of countermeasures. See Wheelis, M. 2007, 'Non-consensual manipulation of human physiology using biochemicals', in Pearson et al., op. cit., p. 6.

24 Klotz, L., Furmanski, M. and Wheelis, M. 2003, 'Beware the siren's song: why "non-lethal" incapacitating chemical agents are lethal', *Federation of American Scientists Paper*, <http://www.fas.org/bwc/papers/sirens_song.pdf> (viewed 8 February 2012).

25 Ibid., p. 7.

26 Pearson, 2007, op. cit., p. 70.

27 British Medical Association (BMA) 2007, *The Use of Drugs As Weapons: The Concerns and Responsibilities of Healthcare Professionals*, BMA, London, p. 1.

28 Klotz et al., op. cit., p. 1.

Box 18.2 Use of Chemical Incapacitant by the Russian Federation

Concerns about incapacitants were heightened following the use of a presumed derivative of fentanyl by Russian security forces to free more than 800 hostages held by heavily armed Chechen separatists in the Dubrovka Theatre in Moscow, in October 2002.[a] According to reports, 30 minutes after an incapacitant was pumped into the theatre, the building was stormed by Russian Spetsnaz special forces who killed all of the Chechen hostage takers, including those left unconscious from the incapacitant, in apparent contravention of international humanitarian law.[b] Although the hostages were released, more than 120 died as a result of the direct effects of the agent used or of airway constriction due to their incapacitation. An undetermined, but large, additional number of hostages suffered long-term damage, or died prematurely in the years after the siege.[c]

Treatment of the hostages who had been poisoned was delayed and compromised by the refusal of the Russian authorities to state publicly what type of incapacitant had been used in the theatre for four days after the siege had ended.[d] On 30 October 2002, the Russian health minister, Yuri Shevchenko, identified the incapacitating agent as 'a mixture of derivative substances of the fast action opiate Fentanyl'.[e] Shevchenko further stated that 'I officially declare: chemical substances which might have fallen under the jurisdiction of the international convention on banning chemical weapons were not used during the special operation'.[f] The minister refused, however, to be more precise about the chemicals used even on 11 December 2002 when faced with a parliamentary question. He said it was a 'state secret'.[g] In 2012, results of trace analysis undertaken in the United Kingdom of extracts of clothing and urine from survivors indicated that the aerosol comprised a mixture of two anaesthetics, carfentanil and remifentanil.[h] At the time of writing, the Russian authorities have still not publicly stated exactly what chemical or chemicals were used.

[a] For descriptions of the incident, see, for example: Koplow, op. cit.; Pearson et al., op. cit.; British Broadcasting Corporation (BBC) 2004, *Horizon: The Moscow Theatre Siege*, BBC 2, 15 January, <http://www.bbc.co.uk/science/horizon/2004/moscowtheatretrans.shtml> (viewed 30 July 2009); Amnesty International 2003, *Rough Justice: The Law and Human Rights in the Russian Federation*, October, Amnesty International, London, AI Index EUR 46/054/2003.

[b] Specifically, the prohibition against attacking those recognised as *hors de combat*. See: Henckaerts, J. 2005, 'Study on customary international humanitarian law: a contribution to the understanding and respect for the rule of law in armed conflict', *International Review of the Red Cross*, vol. 87, no. 857, p. 203.

[c] Wheelis, M. 2010, 'Human impact of incapacitating chemical agents', in ICRC, op. cit.; Levin, D. and Selivanov, V. 2009, 'Medical and biological issues of NLW development and application', *Proceedings of the 5th European Symposium on Non-Lethal Weapons, 11th–13th May 2009, Ettlingen*, Germany, European Working Group on Non-Lethal Weapons, Pfinztal: Fraunhofer ICT, V23, p. 7.

[d] See, for example: Human Rights Watch 2002, 'Independent commission of inquiry must investigate raid on Moscow theater: inadequate protection for consequences of gas violates obligation to protect life', Press release, 30 October 2002, Human Rights Watch.

[e] 'Russian experts discuss use of fentanyl in hostage crisis', ITAR-TASS, [from Moscow in English], 2112 hrs GMT, 30 October 2002, FBIS-SOV-2002-1030, as cited by Perry Robinson, October 2007, op. cit.

[f] Alison, S. 2002, 'Russian confirms siege gas based on opiate fentanyl', [from Moscow for *Reuters*], 1257 hrs ET, 30 October 2002, as cited in Perry Robinson, 2007, op. cit.

[g] Amnesty International, op. cit., p. 53.

[h] Riches, J., Read, R., Black, R., Cooper, N. and Timperley, C. 2012, 'Analysis of clothing and urine from Moscow theatre siege casualties reveals carfentanil and remifentanil use', *Journal of Analytical Toxicology*, vol. 36, pp. 647–56.

In addition, Nixdorff and Melling have surveyed the potential long-term physiological consequences of exposure to incapacitants. Although insufficient research has been undertaken to produce conclusive results, they believe that:

> Numerous human and animal studies have shown that exposure to incapacitating biochemical agents may induce heterogeneous cognitive and physiological impairments and [may] lead to long term health effects. This is even more pronounced when exposures to incapacitating agents are combined with other factors such as stress or activation of the immune system.[29]

Even if all technical barriers to the development of a truly 'nonlethal' incapacitant were overcome, there are a number of serious risks and damaging consequences that could follow from the development of such weapons. These include the following.

- Creeping legitimisation: Perry Robinson believes that attempts by certain states, particularly the United States, to legitimise the development and use of incapacitants threaten to erode the norm against the weaponisation of toxicity.[30] He believes that this 'creeping legitimisation' presents the greatest danger to the existing prohibitions on chemical and biological weapons and to the re-emergence of chemical and biological warfare.[31]

- Proliferation and legitimisation by states: Pearson has warned that efforts to develop incapacitating weapons 'may well gather steam as more nations become intrigued by them and, observing the efforts of Russia and the United States, become convinced not only that effective and acceptably "non-lethal" incapacitating agents can be found, but that their use will be legitimized'.[32]

- Proliferation to, and misuse by, non-state actors: Analysts have highlighted the potential utility of incapacitants to a range of non-state actors including criminals, terrorists, paramilitary organisations and armed factions in failing or failed states many of whom would not feel as constrained as states by international law and concerns about lethality.[33]

- Use as a lethal force multiplier: There are concerns that incapacitants will be used by both military and law-enforcement agencies, not as an alternative

---

29 Nixdorff, K. and Melling, J. 2007, 'Potential long-term physiological consequences of exposure to incapacitating biochemicals', in Pearson et al., op. cit., p. 165.

30 Perry Robinson, J. 2007, 'Categories of challenge now facing the Chemical Weapons Convention', *52nd Pugwash CBW Workshop: 10 Years of the OPCW: Taking Stock and Looking Forward*, Noordwijk, the Netherlands, 17–18 March, p. 20; Perry Robinson, J. October 2007, *Non-lethal Warfare and the Chemical Weapons Convention, Further Harvard Sussex Program submission to the OPCW Open-Ended Working Group on Preparations for the Second CWC Review Conference*, Harvard Sussex Program.

31 Perry Robinson, 2007, op. cit., p. 19.

32 Pearson, 2006, op. cit., p. 172.

33 See, for example: ibid., p. 169; Wheelis, M. and Dando, M. 2005, 'Neurobiology: a case study of the imminent militarization of biology', *International Review of the Red Cross*, vol. 87, no. 859, p. 564.

to lethal force, but as a means to make lethal force more deadly. During the October 2002 Moscow theatre siege, those Chechen hostage takers who were rendered unconscious by the incapacitant were then reportedly shot where they lay by Russian forces rather than being arrested.[34]

- Facilitation of torture and other human rights violations: As well as potentially being utilised for torture and ill treatment of individuals, incapacitants could also facilitate repression of groups by, for example, allowing the capture, en masse, of large numbers of people participating in peaceful demonstrations.

- Militarisation of biology: Analysts[35] have warned that the continuing utilisation of the life sciences in the development of incapacitants could potentially open the way to more malign objectives, such as the widespread repression of entire populations.

- Camouflage for lethal chemical weapons programs: States could exploit the limited transparency mechanisms required under the Chemical Weapons Convention (CWC), for incapacitants and other toxic chemicals designated for use in law enforcement, to hide illicit activities.[36]

- Confusion between lethal and 'nonlethal' chemical weapons: A state using a 'nonlethal' incapacitant during an armed conflict may be perceived by another party as having used a lethal chemical weapon and thus initiate an escalating cycle of retaliation leading to actual use of lethal chemical agents.[37]

Advances in science and technology

In the light of the previous research that has been conducted by certain states into incapacitants, the potential application of the current advances in the life sciences, particularly genomics, biotechnology, synthetic biology and neuroscience,[38] to incapacitating weapons development is a cause for concern. Trapp has highlighted the potential implications of the misuse of such research:

> The explosion of knowledge in neuroscience, bioregulators, receptor research, systems biology and related disciplines is likely to lead to the discovery, amongst others, of new physiologically-active compounds that can selectively interfere with certain regulatory functions in the brain or other organs, and presumably even modulate human behavior

---

34 Koplow, D. 2006, 'The Russians and the Chechens in Moscow in 2002', in *Non-Lethal Weapons: The Law and Policy of Revolutionary Technologies for the Military and Law Enforcement*, Cambridge University Press, Cambridge, pp. 100–13.

35 BMA, op. cit., p. 1; Perry Robinson, October 2007, op. cit., p. 32; Wheelis and Dando, op. cit., pp. 553–71.

36 Perry Robinson, 2007, op. cit., p. 31.

37 Pearson, 2006, op. cit., p. 170.

38 See, for example: The Royal Society 2011, *Brain Waves, Module 1: Neuroscience, Society and Policy*, Science Policy Centre, The Royal Society, London; The Royal Society, 2012, op. cit.; Andreasen, op. cit.; Society of Neuroscience, op. cit.

in a predictable manner. Some of these new compounds (or selective delivery methods) may well have a profile that could make them attractive as novel candidate chemical warfare agents.[39]

Wheelis and Dando had previously surveyed developments and future trends in neurobiology and concluded that there were indications that military interest was already directed towards the next generation of substances affecting the brain and central nervous system:

> In addition to drugs causing calming or unconsciousness, compounds on the horizon with potential as military agents include noradrenaline antagonists such as propranolol to cause selective memory loss, cholecystokinin B agonists to cause panic attacks, and substance P agonists to induce depression. The question thus is not so much when these capabilities will arise—because arise they certainly will—but what purposes will those with such capabilities pursue.[40]

Indeed in 2005, Boken, a toxicologist with the Croatian Ministry of Defence, published a paper warning of the potential use of a range of bio-regulators in warfare and terrorist activities:

> Recent years have seen a rapid advance in the discovery of new bioregulators, especially of the incapacitating ones, in the understanding of their mode of action and synthetic routes for manufacture. Some of these compounds may be many hundreds of times more potent than the traditional chemical warfare agents. Some very important characteristics of new bioregulators that would offer significant military advantages are novel sites of toxic action; rapid and specific effects; penetration of protective filters and equipment, and militarily effective physical incapacitation.[41]

Advances in discovery or synthetic production of potential incapacitating agents have occurred in parallel with developments in particle engineering and nanotechnology that could allow the delivery of biologically active chemicals to specific target organs or receptors. The implications of this were highlighted in the 2008 report by the National Research Council on *Emerging Cognitive Neuroscience and Related Technologies*,[42] which warned that nanotechnologies could be used to overcome the blood–brain barrier and thereby 'enable

39 Trapp, R. 2010, '"Incapacitating chemical agents": some thoughts on possible strategies and recommendations', in ICRC, op. cit., p. 65.
40 Wheelis and Dando, op. cit., p. 10.
41 Boken, S. 2005, 'The toxicology of bioregulators as potential agents of bioterrorism', *Arh Hig Tokiskol*, vol. 56, pp. 205–11.
42 National Research Council 2008, *Emerging Cognitive Neuroscience and Related Technologies*, The National Academies Press, Washington, DC.

unparalleled access to the brain. Nanotechnologies can also exploit existing transport mechanisms to transmit substances into the brain in analogy with the Trojan horse.'[43] The report also highlighted the potential threats resulting from developments in nanotechnologies or gas-phase techniques that allow dispersal of highly potent chemicals over wide areas. It noted that at present 'pharmacological agents are not used as weapons of mass effect, because their large-scale deployment is impractical' as it is 'currently impossible to get an effective dose to a combatant'. The report states, however, that 'technologies that could be available in the next 20 years would allow dispersal of agents in delivery vehicles that would be analogous to a pharmacological cluster bomb or a land mine'.[44] Despite the interest in the development of incapacitants by certain states, and the ongoing advances in relevant science and technology with dual-use application, the international governmental and scientific communities currently seem unwilling or unable to establish and implement effective regulatory controls in this area.

The following sections of this chapter seek to apply a holistic arms-control approach to this issue in order to explore potential routes for effective regulation of such agents.

Applying a holistic arms-control approach to the regulation of incapacitants

For many years the governmental and non-governmental arms-control communities have sought to develop strategies to combat the proliferation of chemical and biological weapons to state and non-state actors. Recognising that reliance upon a single disarmament or arms-control agreement alone would not guarantee success, scholars have explored a number of concepts seeking to broaden the range of possible regulatory mechanisms. Utilising and building upon such work, particularly the concepts of preventative arms control[45] and webs of prevention (or protection),[46] the author has sought to develop a 'holistic arms control' (HAC) framework for regulation. Although the proposed HAC analytical framework concentrates upon existing arms-control

---

43 Ibid., p. 135.
44 Ibid., p. 137.
45 See, for example: Altmann, J. 2006, 'Preventive arms control: concept and design', in J. Altmann, *Military Nanotechnology: Potential Applications and Preventive Arms Control*, Routledge, London.
46 See, for example: Pearson, G. 1998, 'The vital importance of the web of prevention', in *Proceedings of the Sixth International Symposium on Protection against Chemical and Biological Warfare Agents*, Stockholm, 11–15 May; Pearson, G. 2001, 'Why biological weapons present the greatest danger', *Seventh International Symposium on Protection against Chemical and Biological Warfare Agents*, Stockholm; McLeish, C. and Rappert, B. (eds) 2007, *A Web of Prevention: Biological Weapons, Life Sciences, and the Governance of Research*, Earthscan, London.

and disarmament measures,[47] it attempts to widen the range of applicable mechanisms for regulation, and also the nature of the actors involved in such regulatory measures.

Consequently, HAC can be thought of as a framework for analysis to aid the development of a comprehensive, layered and flexible approach to arms control that

- is developed for and unique to the specific type of weapon or technology under consideration rather than for a broad grouping of weapons

- is potentially applicable to all stages of a weapon's existence (that is, research, development, mass production, stockpiling, deployment, use, transfer and destruction)

- is responsive to developments in science and technology (and will be able to regulate weapons that have not yet been invented)

- seeks to identify the types of permissible and non-permissible weapons, acknowledges where ambiguity lies, and seeks to develop mechanisms for resolving such ambiguities

- seeks to clearly identify existing constraints upon the permitted use of weapons—that is, legitimate targets (for example, whether this would include armed combatants, terrorist organisations, criminals, civilians), legitimate types of operations (for example, whether this would include law enforcement, military operations other than war, armed conflict) and how such operations should be conducted (rules of engagement); acknowledges where existing ambiguity lies and highlights potential mechanisms for resolving such ambiguities

- is responsive to developments in the nature of the use/misuse of weapons in practice

- is not necessarily limited to a single existing arms-control or disarmament treaty, but actively explores and seeks to incorporate states' existing responsibilities under the full range of relevant international law and applicable agreements

- is responsive to developments in international law, particularly those limiting types and use of weapons

- incorporates measures to facilitate effective national implementation as well as verification, enforcement and transparency mechanisms

---

47 This chapter does not, therefore, explore a range of parallel processes that can potentially play important roles in preventing or ameliorating the effects of incapacitant weapons attack, including broadband defence measures, strengthening public health surveillance and response, and so on. For further discussion of these issues, see, for example, Pearson, 1998, op. cit.; Pearson, 2001, op. cit.; McLeish and Rappert, op. cit.

- recognises that states are the prime actors in existing regulatory regimes, and allows for and encourages participation by the full range of relevant stakeholders.

Applying this approach to the case of incapacitants, a HAC regime can be envisioned comprising

- state-led activities
- adherence to comprehensive legal prohibitions against chemical and biological weapons (CBW) enshrined in the Geneva Protocol, the Biological and Toxin Weapons Convention (BTWC) and the Chemical Weapons Convention (CWC)
- adherence to international humanitarian law (notably, the four Geneva conventions and two additional protocols) and international human rights law (including the Convention against Torture, the International Covenant on Civil and Political Rights and the Universal Declaration on Human Rights)
- adherence to other relevant international law and agreements including the Rome Statute of the International Criminal Court, the Single Convention on Narcotic Drugs and the UN Convention on Psychotropic Substances
- effective monitoring, verification, investigation and enforcement of the above obligations
- application of stringent export controls and interdiction measures.
- engagement by civil society
- conducting societal monitoring and verification
- developing a 'culture of responsibility' amongst the scientific and medical communities built upon strong normative and ethical standards
- developing and advocating mechanisms to strengthen the regime.

Whilst certain authors have previously examined the application of a range of state-led mechanisms for the regulation of incapacitants,[48] the importance of the life-science community's engagement in this issue has been underexplored. In the following sections, the author will, therefore, briefly examine the potential application of the two most pertinent arms-control regimes—the Chemical Weapons Convention and the Biological and Toxin Weapons Convention— before concentrating upon the potential roles that civil society, and particularly

---

48   See, for example: Crowley, op. cit.; Casey-Maslen, S. 2010, *Non-Kinetic-Energy Weapons Termed 'Non-Lethal': A Preliminary Assessment under International Humanitarian Law and International Human Rights Law*, Geneva Academy of International Humanitarian Law and Human Rights, Geneva; ICRC, op. cit.; International Committee of the Red Cross (ICRC) 2012, *Toxic Chemicals As Weapons for Law Enforcement: A Threat to Life and International Law?* Synthesis paper, September, International Committee of the Red Cross, Geneva; International Committee of the Red Cross (ICRC) 2013, *'Incapacitating Chemical Agents': Law Enforcement, Human Rights Law and Policy Perspectives*, Expert Meeting, Montreux, Switzerland, 24–26 April 2012, ICRC, Geneva, 13 January 2013.

the scientific and medical communities, can play in preventing the misuse of biomedical research for hostile purposes, most notably, the development of incapacitating weapons.

Biological and Toxin Weapons Convention

Article I of the BTWC declares that:

> Each State Party to the Convention undertakes never in any circumstances to develop, produce, stockpile or otherwise acquire or retain:
>
> 1. Microbial or other biological agents, or toxins, whatever their origin or method of production, of types and in quantities that have no justification for prophylactic, protective or other peaceful purposes.
>
> 2. Weapons, equipment or means of delivery designed to use such agents or toxins for hostile purposes or in armed conflict.[49]

Article I, together with the extended understandings agreed at successive BTWC review conferences,[50] makes it clear that the convention is comprehensive in its scope and that all naturally or artificially created or altered microbial and other biological agents and toxins, as well as their components, regardless of their origin and method of production are covered. Because some possible candidate incapacitants, such as bio-regulators including neurotransmitters, could be considered biological agents or toxins, a range of such incapacitants would be covered by the BTWC.

Although the BTWC does appear to cover certain incapacitants, there are ambiguities regarding the nature and scope of such coverage. For example, although the use of incapacitants of a biological origin in armed conflict or for 'hostile purposes' would be banned, the delineation between prohibited 'hostile purposes' and permitted 'peaceful purposes' has not been fully established under the convention. Consequently, it is unclear how the use of incapacitants of biological origin for 'military operations other than war' (MOOTW), would

---

49 *Convention on the Prohibition of the Development, Production and Stockpiling of Bacteriological (Biological) and Toxin Weapons and their Destruction*, 1972, 1015 UNTS 163, art. 1.

50 See, for example, *Seventh BWC Review Conference Final Document* (2012): '1. ... Conference declares that the Convention is comprehensive in its scope and that all naturally or artificially created or altered microbial and other biological agents and toxins, as well as their components, regardless of their origin and method of production and whether they affect humans, animals or plants, of types and in quantities that have no justification for prophylactic, protective or other peaceful purposes, are unequivocally covered by Article I. 2. The Conference reaffirms that Article I applies to all scientific and technological developments in the life sciences and in other fields of science relevant to the Convention ... 3. The Conference reaffirms that the use by the States Parties, in any way and under any circumstances, of microbial or other biological agents or toxins, that is not consistent with prophylactic, protective or other peaceful purposes, is effectively a violation of Article I.' United Nations 2012, *Final Document of the Seventh Review Conference*, Geneva, 5–22 December 2011, BWC/CONF.VII/7, 13 January 2012, art. I, paras 1–3.

be regulated by the BTWC. To date, there have been no determinations of these issues by the BTWC states parties.[51] Further important limitations on the value of the BTWC (and its control regime) as a tool to regulate incapacitants arise from its current lack of effective verification and compliance mechanisms, and also the absence of an international organisation that could coordinate such activities and facilitate implementation by states parties.[52]

Chemical Weapons Convention

Although certain states in their background scientific papers to the seventh BTWC Review Conference highlighted the potential dangers of the misuse of biologically active agents such as bio-regulators and peptides that could be used as incapacitants, currently discussions on the regulation of such weapons have largely concentrated on the Chemical Weapons Convention (CWC). The CWC prohibits the development, production, stockpiling, transfer and use of chemical weapons.[53] In addition, it requires that all existing stocks of chemical weapons[54] and chemical weapons production facilities be destroyed.[55] The treaty is of unlimited duration and is designed to be far more comprehensive in scope and application than any prior international agreement on chemical weapons. It is overseen by its own treaty body, the Organisation for the Prevention of Chemical Weapons (OPCW), including a technical secretariat of more than 500 inspectors, scientists, legal experts and ancillary staff headed by the director-general, which carries out the daily work of monitoring, verifying and facilitating implementation of the convention.

Although the convention prohibits chemical weapons, it allows for the controlled peaceful use of toxic chemicals. Article II.2 of the convention defines a 'toxic chemical' as 'any chemical, regardless of its origin or method of production, which, through chemical action on life processes, can cause death, temporary

---

51   Analysis was undertaken of all relevant documents pertaining to this issue publicly available up to 10 September 2013.

52   Although there is no equivalent of an OPCW for the BTWC, the Sixth BTWC Review Conference decided to create and fund a (three-person) Implementation Support Unit (ISU) within the UN Office for Disarmament Affairs (UNODA) of the UN Office at Geneva. The ISU was launched in August 2007 and its mandate was renewed and extended by the Seventh BTWC Review Conference to run until 2016. The ISU provides administrative support to, and prepares documentation for, meetings agreed by the BTWC Review Conference. The ISU also facilitates communication among states parties, international organisations, and scientific and academic institutions, as well as NGOs. It also acts as a focal point for submission of information by and to states parties, and will support, as appropriate, the implementation by the states parties of the decisions and recommendations of the Sixth and Seventh BTWC Review Conferences. The ISU, however, has no authority to undertake verification or compliance activities.

53   *Convention on the Prohibition of the Development, Production, Stockpiling and Use of Chemical Weapons and on their Destruction (Chemical Weapons Convention)*, 1993, art. I.1.

54   Ibid., art. I.3.

55   Ibid., art. I.4.

incapacitation or permanent harm to humans or animals'.[56] The convention therefore covers a wide range of chemicals within its scope of regulation including certain chemical agents that could be used as incapacitants.

To determine whether the use of a toxic chemical such as an incapacitant would be in conformity with the CWC, the intention or purpose for its use needs to be determined. Under Article II.1 of the convention, chemical weapons are defined as 'toxic chemicals or their precursors, except where intended for purposes not prohibited by the Convention, as long as the types and quantities are consistent with such purposes'.[57]

Such 'purposes not prohibited' are defined under Article II.9 and include:

> (c) Military purposes not connected with the use of chemical weapons and not dependent on the use of the toxic properties of chemicals as a method of warfare;

- Law enforcement including domestic riot control purposes.[58]

It is therefore clear that the use of toxic chemicals such as incapacitants for purposes not provided for in Article II.9 (for example, as a method of warfare) would be prohibited, as would development, production, acquisition, stockpiling, retention or transfer of these chemicals for such purposes (under Article I of the CWC).

There are, however, a number of ambiguities in the CWC and limitations in its current implementation that could seriously restrict its ability to effectively regulate incapacitants. Although the use of toxic chemicals is permitted for law enforcement 'as long as the types and quantities are consistent with such purposes',[59] there is no definition of 'law enforcement' in the convention. Furthermore, no OPCW policymaking organ has made any interpretative statements elaborating the scope or nature of permitted law-enforcement activities or regarding which toxic chemicals (if any)—save riot-control agents—could be used for such purposes. Consequently, the extent to which incapacitants could be used (if at all) for activities such as counterterrorist operations is contested.[60] Unfortunately, and despite attempts made by a number of CWC states parties to raise this issue at the 2013 Third Review Conference,[61] there

---

56 Ibid., art. II.2.
57 Ibid., art. II.1.
58 Ibid., art. II.9.
59 Ibid., art. II.1.
60 For divergent argumentation on this issue, see: Fidler, D. 2007, 'Incapacitating chemical and biochemical weapons and law enforcement under the Chemical Weapons Convention', in Pearson et al., op. cit.; von Wagner, A. 2007, 'Toxic chemicals for law enforcement including domestic riot control purposes under the Chemical Weapons Convention', in Pearson et al., op. cit.
61 See, in particular, the statements by Germany, Switzerland, the United Kingdom and the United States made on 8 and 9 April 2013 in the General Debate of the Third Special Session of the Conference of the States Parties to Review the Operation of the Chemical Weapons Convention (Third Review Conference), The Hague, 8–19 April 2013, all available from the OPCW web site: <www.opcw.org>.

has been a collective failure by the CWC states parties and policymaking organs to effectively address the regulation of incapacitants under the convention. It is therefore left to individual states parties to interpret the scope and nature of their obligations with regard to the regulation of such agents.

Box 18.3 The ICRC and Incapacitants: A call to action

A range of civil society organisations and researchers has urged the international governmental community—particularly the OPCW and the CWC states parties—to take action to address the dangers of the potential proliferation and inappropriate use of incapacitants.[a] The ICRC, however, has noted that '[a]lthough there have been exhortations by some States and by some in academic circles to address these issues, there has been little or no movement to date in the relevant multilateral fora'.[b] Consequently in March 2010, the ICRC convened the first of two expert meetings to explore the implications of these issues for international law. The meeting brought together a group of 33 government and independent experts who were joined by ICRC staff members.[c] For details of the second ICRC meeting, see: ICRC, 2013, op. cit. In the subsequent report of the meeting, the ICRC urged '[s]tates to give greater attention to the implications for international law of *"incapacitant chemical agents"'*. The organisation also noted that '[t]here is currently an opportunity to address *preventatively* the challenges and risks identified'[d] (emphasis added). In addition, the report concluded that '[t]here is a clear need to tackle the issues raised by *"incapacitating chemical agents"* in appropriate fora engaging a broad range of experts including policy makers, law-enforcement professionals, security personnel, military personnel, *health professionals, scientists* and lawyers with IHL [international humanitarian law], human rights and disarmament expertise'[e] (emphases added).

---

[a] See, for example: Crowley, op. cit., esp. pp. 57–101, 117–19; Pearson et al., 2007, op. cit.; Perry Robinson, 2007 and October 2007, op. cit.

[b] ICRC, 2010, op. cit., p. 7.

[c] Ibid., p. 3.

[d] ICRC, 2010, op. cit., p. 75.

[e] Ibid., p. 75.

Engagement by the medical and scientific communities

Given the evident ambiguities and limitations of the existing state-centric chemical and biological weapons control regimes, compounded by their inadequate implementation by certain states parties, a great responsibility lies upon the life-science community as a whole and individual researchers in particular to ensure that their activities do not contribute to the development of a new generation of chemical and biological weapons, including incapacitants. Furthermore, it can be argued that there is an obligation upon life scientists to move beyond such 'personal regulation' and to take a far more proactive role in monitoring and regulating incapacitating agent research and combating the potential misuse of such agents. This section will explore potential avenues for greater engagement and action through the application of societal verification as a complement to the existing official verification mechanisms, the development of a culture of responsibility amongst the scientific and medical communities, and finally explore the possible roles that scientists, academics and other civil-society actors can play in informing and influencing the actions of states in this area.

Societal monitoring and verification

Although there is no agreed formal definition of societal monitoring and verification, Diseroth describes societal verification as

> connot[ing] the involvement of civil society in monitoring national compliance with, and overall implementation of, international treaties or agreements. One important element is citizens' reporting of violations or attempted violations of agreements by their own government or others in their own country ... A more recent development is civil society monitoring of global compliance with international agreements. In contrast to official verification organisations employing professional experts, societal verification may involve the whole of society or groups within it.[62]

The establishment of a global societal monitoring and verification network—involving large numbers of civil society actors resident in all states party to relevant chemical or biological treaties who are able to monitor their state's implementation of treaty obligations—appears to be unlikely in the near to medium terms. As Rotblat acknowledges, '[e]ven if governments were persuaded to pass laws to make reporting legitimate', which itself would be a revolutionary development and counter to existing practice in arms control and disarmament policy and the practice of many states, 'this goes so much against traditional loyalties that it would require a considerable educational effort to induce people to act on it voluntary'.[63] Consequently, he believes that implementation of societal verification 'requires a change in certain attitudes of the general public, which may take time'.[64]

A more limited form of societal verification, however, can be envisaged, comprising a smaller number of activist researchers who have access to the relevant technical expertise and can, at the very least, undertake open-source monitoring and analysis, and potentially conduct field missions. Due to resource, personnel, political and security constraints, such groups are likely to be limited

---

62   Deiseroth, D. 2000, 'Societal verification: wave of the future?' in VERTIC, *2000 Verification Yearbook*, p. 265.

63   Rotblat, J. 1993, 'Societal verification', in J. Rotblat, J. Steinberger and B. Udgaonkar, *A Nuclear-Weapon-Free World: Desirable? Feasible?* Westview Press, Boulder, Colo., p. 108.

64   Ibid., p. 105.

in terms of the countries from which they can operate and consequently the quantity and quality of information they are able to receive, particularly from inaccessible regions and closed or semi-closed authoritarian countries.

Open-source monitoring and analysis

A small number of academic and non-governmental organisations[65] have undertaken monitoring of open-source data, often utilising a range of national oversight and transparency mechanisms, to obtain information relating to the research, development and utilisation of incapacitants.[66] Such open-source monitoring and analysis are time-consuming, resource intensive and the information obtained is often limited as a result of national security restrictions, commercial confidentiality considerations and limited access to research published in certain countries. In addition, there is much inaccurate or biased reporting disseminated by both proponents and opponents of such weapons.

Despite the methodological difficulties and the limitations in the information obtained, such work is vital to the formation of an informed public discourse on the existing threats and potential dangers of the proliferation and misuse of these weapons. In addition it can also help in the development of timely and realistic publicly available threat assessments relating to R&D, deployment or utilisation of such weapons in specific countries. Furthermore, information derived from civil society research can be sent to relevant intergovernmental organisations, most notably the OPCW.

Field missions and witness testimony

Independent scientists, health professionals and non-governmental organisations (NGOs) can sometimes collect their own information, first hand, from onsite investigations or may be able to utilise information (for example, witness testimony) and analyse materials (for example, weapons shells, clothing fragments, soil samples) obtained from other civil society actors operating in the field (for example, journalists, national NGOs). There are several potential constraints upon such investigations including access, logistics and translation; safety considerations for researchers and witnesses; and difficulties ensuring chain of custody, as well as establishing the representativeness of the information

---

65   See, for example, the publications of: Biological Weapons Prevention Programme, Bradford Non-Lethal Weapons Research Project; Centre for Arms Control and Non-Proliferation; Federation of American Scientists; Harvard Sussex Project; International Network of Engineers and Scientists for Global Responsibility; Pugwash Conferences on Science and World Affairs; and the Sunshine Project.

66   For discussion, see: Crowley, M. 2010, 'Monitoring and opposing the misuse of incapacitants—exploring the potential roles for independent scientists', in J. Finney and V. Slaus (eds), *Assessing the Threat of Weapons of Mass Destruction*, Nato Science for Peace and Security Series E, Human and Societal Dynamics—Volume 61, IOS Press, Amsterdam, pp. 114–32.

obtained. Despite such constraints, material collected during field missions can provide information that could not be obtained by any other means—for example, allowing identification of toxic chemical agents utilised during a military or law-enforcement operation.

Box 18.4 Independent Analysis of a New Riot-Control Agent Used in the West Bank

In July 2005, the Israeli Army reportedly employed a new riot-control agent against Palestinian and Israeli civilian protesters that resulted in severe skin injuries. The Israeli Army refused to identify the agent; however, scientists based in the United Kingdom obtained one of the munitions utilised, and following physical and chemical analyses, were able to identify the contents as capsaicin with an inert carrier and a dispersal agent.[a] The results were found to correspond with the commercially available 'Pepperball Tactical Powder'. The paper noted that '[s]kin injuries of the severity described had not previously been reported with this agent, and would be difficult to manage for clinicians who were unaware of the nature of the agent'.[b] As well as alerting clinicians to the nature and effects of chemical agents they may face in the future, such research can also help to identify possible international transfers of chemical agents and devices.

[a] Hay, A., Giacaman, R., Sansur, R. and Rose, S. 2006, 'Skin injuries caused by new riot control agent used against civilians on the West Bank', *Medicine,Conflict and Survival*, vol. 22, no. 4 (October–December).

[b] Ibid.

Building a culture of responsibility within the life-science and biomedical communities

In its 2004 public statement 'Preventing hostile use of the life sciences', the ICRC declared: 'If measures to prevent the hostile use of advances in the life sciences are to work, a culture of responsibility is necessary among individual life scientists. This applies whether these scientists are working in industry, academia, health, defense or in related fields such as engineering and information technology.'[67]

According to the ICRC, such a culture of responsibility is also needed 'within the institutions that employ scientists and fund research in the life sciences'.[68] Similar calls to the scientific and medical communities have also been made by the states parties to the BTWC and the CWC at review conferences.[69]

---

67   International Committee of the Red Cross (ICRC) 2004, *Preventing Hostile Use of the Life Sciences: From Ethics and Law to Best Practice*, 11 November 2004, International Committee of the Red Cross, Geneva, <http://projects.exeter.ac.uk/codesofconduct/ Chronology/Principles_Actionpoints_11Nov04.pdf> (viewed 14 January 2012).

68   Ibid.

69   For example, see: Third Review Conference of the Parties to the Convention on the Prohibition of the Development, Production and Stockpiling of Bacteriological (Biological) and Toxin Weapons and on their Destruction, *Final Document*, Part II, BWC/CONF.III/23, 9–2 September 1991, p. 3; *Report of the Second Special Session of the Conference of the States Parties to Review the Operation of the Chemical Weapons Convention (Second Review Conference)*, 7–18 April 2008, RC-2/4, 18 April 2008.

The remaining sections of this chapter will explore the current range of initiatives being undertaken by those in the scientific and medical communities to nurture a culture of responsibility, beginning with the growing recognition of the dual-use dilemma and the consequent requirement for effective oversight of research. This will then be followed by a discussion of the potential utility of oaths, codes and pledges and the parallel processes of education and awareness-raising in building the appropriate norms of behaviour for the scientific and biomedical communities. The practical application of such principles by individual scientists through such practices as whistleblowing will be explored as well as the duty of individual scientists to inform the policies and practices of governments in this area.

Research oversight and the 'dual-use dilemma'

To date, much of the discourse amongst the life-science community concerning how best to combat the proliferation and misuse of chemical and biological weapons has concentrated on tackling the dual-use dilemma and has been framed in terms of regulating the actions of individual life scientists conducting 'academic' research projects and publishing 'academic' articles. Highly influential in this discourse have been the 2003 Fink Report (*Biotechnology Research in An Age of Terrorism*)[70] and the 2006 Lemon Report (*Globalisation, Biosecurity and the Future of the Life Sciences*), both produced under the auspices of the National Research Council of the US National Academies. Both reports highlighted the importance of taking a comprehensive approach to analysing 'dual-use' research of potential concern. The Lemon Report recommended the adoption of 'a broadened awareness of threats beyond the classical "select agents" and other pathogenic organisms and toxins, so as to include, for example, approaches for disrupting host homeostatic and defense systems and for creating synthetic organisms'.[71] The broad threat spectrums enunciated by both reports, particularly that of the Lemon Committee, appear to capture incapacitants within their scope.

As a result of the concerns and recommendations outlined in the Fink and Lemon reports and the work of others,[72] a range of oversight structures and processes

---

70   National Research Council 2004, *Biotechnology Research in an Age of Terrorism*, [Fink Report], The National Academies Press, Washington, DC, p. 114.
71   National Research Council 2006, *Globalisation, Biosecurity and the Future of the Life Sciences*, [Lemon Report], The National Academies Press, Washington, DC, p. 216.
72   A range of analysts has subsequently highlighted the need to address further potential actors of concern beyond individual life-science researchers. Garfinkel et al., for example, defined three major points for potential policy intervention—namely: 'commercial firms that sell synthetic DNA (oligonucleotides, genes or genomes) to users; owners of laboratory "bench-top" DNA synthesizers, with which users can produce their own DNA; the users (consumers) of synthetic DNA themselves and the institutions that support and oversee their work.' See: Garfinkel, M., Endy, D., Epstein, G. and Friedman, R. 2007, *Synthetic Genomics: Options for Governance*, J. Craig Venter Institute, Rockville, Md, and San Diego, Calif.

was established by governments, scientific bodies, academic institutions, funders and publishers to review potential dual-use research, assessing the risks and benefits of such research to determine whether they need to be modified or withdrawn.[73]

Analysing the application of oversight measures in practice, however, Rappert believes that 'such procedures rarely conclude that manuscripts, grant applications or experiment proposals should not be undertaken or restricted'. In 2009, van Aken and Hunger analysed the application of biosecurity policies agreed by a group of 32 influential science journals under which manuscripts could be modified or rejected where 'the potential harm of publication outweighs the potential societal benefits'.[74] Despite such policies having been established in 2003, van Aken and Hunger found that no manuscript has ever been rejected on security grounds.[75] Rappert believes the 'same could be said' of those funders who have established submission-oversight systems.[76] Furthermore, Rappert believes that 'even more notable with these review processes is the infrequency with which they have identified items "of concern" in the first place'.[77] Whilst information relating to the research controls of government departments (especially defence-related ones) is not readily available, Rappert believes that in relation to universities and other publicly funded agencies, 'it seems justifiable to conclude that—barring dramatic changes—oversight processes will identify little research as posing security concerns and will stop next to nothing'.[78]

Others have criticised the voluntary nature of the existing controls on life-science dual-use research. For example, commenting upon the release of a draft of the National Science Advisory Board for Biosecurity (NSABB) 'Proposed Framework for the Oversight of Dual Use Life Sciences Research: Strategies for Minimizing the Potential Misuse of Research Information',[79] the Sunshine Project director, Hammond, stated that the 'NSABB is divorced from reality if its members believe that another set of voluntary NIH [National Institutes of Health] guidelines is sufficient, and would be remotely effective, at preventing

73 For a discussion of such initiatives, see: Rappert, B. 2008, 'The benefits, risks, and threats of biotechnology', *Science & Public Policy*, vol. 35, no. 1, pp. 37–44.
74 Journal Editors and Authors Group 2003, 'Uncensored exchange of scientific results', *Proceedings of the National Academy of Sciences*, vol. 100, no. 4, p. 1464.
75 van Aken, J. and Hunger, I. 2009, 'Biosecurity policies at international life-science journals', *Biosecurity and Bioterrorism*, vol. 7, no. 1, pp. 61–72.
76 Rappert, B. 2010, 'Introduction: education as...', in B. Rappert (ed.), *Education and Ethics in the Life Sciences: Strengthening the Prohibition of Biological Weapons*, ANU E Press, Canberra, pp. 8–9. Rappert cites the UK Biotechnology and Biological Sciences Research Council, the UK Medical Research Council, the Wellcome Trust, the Centers for Disease Control, and the Southeast Center of Regional Excellence for Emerging Infectious Diseases and Biodefence.
77 Ibid., p. 9.
78 Ibid., p. 9.
79 National Science Advisory Board for Biosecurity (NSABB) 2007, *Proposed Framework for the Oversight of Dual Use Life Sciences Research: Strategies for Minimizing the Potential Misuse of Research Information*, National Science Advisory Board for Biosecurity, Bethesda, Md.

dual-use disasters'.[80] Research conducted by the Sunshine Project from 2004 to 2007 indicated that many US organisations obliged to follow NIH guidelines did not do so.[81] A 2007 Sunshine Project survey discovered that 18 of the top-20 US biotechnology companies did not comply with existing voluntary NIH biotechnology guidelines.[82] Instead of a voluntary approach, Hammond stated that '[e]ffective federal management of dual-use risks requires making safety and security oversight truly mandatory and subject to the sobering light of public scrutiny'.[83]

As well as concerns about the implementation of such voluntary oversight systems in practice, a further concern relates to the limited range of issues being considered by such bodies. The discourse and much of the current activity appear to be concentrated upon preventing the diffusion of dual-use knowledge, skills and materials to various non-state actors with malign intent—principally, terrorist organisations. Insufficient attention has been given to utilising existing dual-use monitoring mechanisms or adopting additional processes to specifically combat the misuse of dual-use expertise in state programs, even though national CBW research and development programs arguably pose a greater danger to the CWC and BTWC than non-state actors.

Oaths, codes and pledges for the life-science community

One approach to building a culture of responsibility has been through the development of a range of non-binding ethical codes, codes of conduct and oaths or pledges. An early advocate of such activities was Joseph Rotblat, who declared in his 1995 Nobel acceptance speech that:

> The time has come to formulate guidelines for the ethical conduct of scientists, perhaps in the form of a voluntary Hippocratic Oath … At a time when science plays such a powerful role in the life of society, when the destiny of the whole of mankind may hinge on the results of scientific research, it is incumbent on all scientists to be fully conscious of that role, and conduct themselves accordingly. I appeal to my fellow scientists to remember their responsibility to humanity.[84]

The development of ethical codes of conduct became one of the priority areas during the BTWC inter-sessional process. Subsequently, initiatives supporting

---

80 Sunshine Project 2007, 'Earth calling NSABB: voluntary compliance won't work', Press release, 17 April 2007.
81 See, for example: Sunshine Project 2004, Mandate for Failure: The State of IBCs in an Age of Bioweapons Research, <http://www.sunshine-project.org/biodefense/ibcreport.html> (viewed 11 January 2012).
82 Supporting correspondence available at <http://www.sunshine-project.org/publications/pr/support/deregistries.pdf> (viewed 11 January 2011).
83 Sunshine Project, 2007, op. cit.
84 Rotblat, J. 1995, Remember Your Humanity, Nobel Lecture, Oslo, 10 December 1995, <http://nobelprize.org/nobel_prizes/peace/laureates/1995/rotblat-lecture.html> (viewed 11 January 2012).

such codes have been undertaken by a wide range of scientific associations and organisations including the American Society of Microbiology,[85] the National Academy of Sciences, The Royal Society,[86] the International Centre for Genetic Engineering and Biotechnology, the International Union of Biochemistry and Molecular Biology and the International Council for the Life Sciences. These activities have been complemented and stimulated by the ICRC as well as the work of individual scientists[87] and academics.[88]

Such codes may well help to sensitise life scientists to the dangers of dual-use research, and reinforce the importance of, and promulgate, ethical 'red lines' where the legal prohibitions or normative taboos are already clearly defined and widely accepted. The effectiveness of such an approach, to date, has, however, been questioned by a range of scholars including Corneliussen, Dando, Perry Robinson and Rappert.[89] One important limitation of the majority of such initiatives is that the resulting codes are aspirational and non-binding in nature with no clearly identified penalties elaborated for those individuals who breach the prohibitions, or mechanisms established to monitor and enforce such prohibitions.

Recognising the limitations of self-governance initiatives to effect change in this area, some have called for codes of conduct to become binding, with those breaching such codes facing sanction from their peers (or the state). For example, Rotblat, in a letter to a Pugwash Workshop on Science, Ethics and Society in 2004, stated his belief that:

> [S]ome believe that the search for knowledge overrides all other considerations and that scientists should be entitled to ignore the ethical elements of their work … The harm to society that has resulted from such attitudes has brought science into disrepute, and action is needed to restore the proper image of science. The introduction of a 'hippocratic' oath is our example of such action, but it should perhaps be given more than a symbolic value. *Perhaps the time has come for a*

---

85   American Society of Microbiology (ASM) 2005, *Code of Ethics (Revised and approved by ASM Council in 2005)*, American Society of Microbiology, Washington, DC.
86   The Royal Society 2005, *The Roles of Codes of Conduct in Preventing the Misuse of Scientific Research*, The Royal Society, London.
87   See Atlas, R. and Somerville, M. 2007, 'Life sciences or death science: tipping the balance towards life with ethics, codes and laws', in McLeish and Rappert, op. cit.
88   See, for example, Revil, J. and Dando, M. 2006, 'A Hippocratic oath for life scientists', *EMBO Reports*, vol. 7.
89   Rappert, B. 2010, *Biological Weapons and Codes of Conduct*, <http://projects.exeter.ac.uk/codesofconduct/Chronology/index.htm> (viewed 22 December 2011).

binding code of conduct, where only those who abide by the code should be entitled to be practicing scientists, something which applies now to medical practice [emphasis added].[90]

Examination of the literature, however, reveals that no such binding codes of conduct prohibiting research, development or utilisation of incapacitating agents have been established by any national or international scientific organisation to date.

Furthermore, Rappert has noted that 'if codes are to go beyond reiterating platitudes about the abhorrence of using modern biology toward malign ends, then they are likely to confront major issues of controversy. For instance, codes could comment on the acceptability of disputed attempts to develop "non-lethal" incapacitating agents.'[91] Areas of dispute or controversy, such as incapacitants, are the ones, however, where such codes remain silent or at best provide ambiguous guidance. Similarly, as with dual-use research oversight, whilst numerous codes condemn and seek to prevent the involvement of scientists in the development of biological and chemical weapons by non-state actors, it is questionable whether enough energy has been devoted to targeting the involvement of life scientists in state-run weapons programs.

Corneliussen has asked

> why voluntary self-governance regimes—and codes of conduct in particular—are being given so much attention in policy discussions about preventing the misuse of biological research when they appear to have significant shortcomings in practice. Indeed, why have individual scientists become the target of the policy discussions when it is generally accepted within the disarmament community that the greatest risk of misuse is at the level of national biological weapons programmes ... Preventing these state-level programmes in the future should therefore be a primary concern, rather than implementing codes of conduct for life scientists.[92]

Corneliussen further contends that:

> [T]he current sole focus on codes, and the extensive investment of resources that accompanies it, might well serve to detract from other more

---

90   Rotblat, J. 2004, 'Letter to workshop', in *Report of 2nd Pugwash Workshop: Science, Ethics and Society*, Ajaccio, Corsica, 10–12 September 2004, <http://www.pugwash.org/reports/ees/corsica2004/corsica2004. htm#attach2> (viewed 22 December 2011).
91   Rappert, B. 2005, 'Biological weapons and the life sciences: the potential for professional codes', *Disarmament Forum. Volume 1: Science Technology and the CBW Regimes*, p. 56.
92   Corneliussen, F. 2006, 'Adequate regulation, a stop-gap measure, or part of a package? Debates on codes of conduct for scientists could be diverting attention away from more serious questions', *EMBO Reports*, vol. 7 (Special Issue), p. 53.

crucial regulatory measures that target not only individual scientists but also state programmes. Without this plurality of regulatory measures in place, codes of conduct are doomed to fail.[93]

More recently there have been some indications of an increasing awareness amongst certain life-science communities of the dangers of cooption into state programs and initiatives undertaken to address these dangers (see Box 18.5).

Box 18.5 A Pledge for Neuroscientists

In January 2010, Dr Curtis Bell, senior scientist emeritus at the Oregon Health and Science University, circulated a pledge intended to foster opposition amongst neuroscientists to the application of neuroscience to 'torture and other forms of coercive interrogation or manipulation that violate human rights and personhood'.[a] According to Bell, 'such applications could include drugs that cause excessive pain, anxiety, or trust, and manipulations such as brain stimulation or inactivation'.[b] Furthermore, signatories would oppose the application of neuroscience to 'aggressive war ... illegal under international law'. The pledge states that a 'government which engages in aggressive wars should not be provided with tools to engage more effectively in such wars. Neuroscience can and does provide such tools. Examples include ... drugs which damage the effectiveness of soldiers on the other side.'[c]

Under the pledge, neuroscientists commit to making themselves aware of the potential misuse of neuroscience for violations of 'basic human rights or international law such as torture and aggressive war' and commit to refusing to 'knowingly participate in the application of Neuroscience to violations of basic human rights or international law'.[d] Bell acknowledges that '[s]igning this pledge will not stop aggressive wars or human rights violations, or even the use of neuroscience for these purposes'; however, he believes that 'by signing, neuroscientists will help make such applications less acceptable'.[e]

---

[a] Bell, C. 2010a, *Pledge by Neuroscientists to Refuse to Participate in the Application of Neuroscience to Violations of Basic Human Rights or International Law*, <http://spreadsheets.google.com/viewform?formkey=dEF4RFhhSWZwNktCakYtbTdkd1cxckE6MA> (viewed 4 December 2011); Bell, C. 2010b, 'Letter to International Society for Neuroethology', <http://neuroethology.org/cgi-bin/dada/mail.cgi/archive/emaillist/20100106211821/> (viewed 4 December 2011); Bell, C. n.d., 'Responsibilities of neuroscientists concerning aggressive war and torture', Poster display, <http://files.me.com/curtiscbell/pu0ycj> (viewed 4 December 2011).

[b] Bell, 2010a, op. cit.

[c] Ibid.

[d] Ibid.

[e] Bell, C. 2010c, 'Neurons for peace: take the pledge, brain scientists', *New Scientist*, no. 2746 (8 February).

Despite the energy and resources expended upon the development and promotion of codes, however, Rappert's stark conclusion is that 'efforts to devise meaningful codes have largely floundered'. Rappert believes that '[i]n no small part, this has been due to the lack of prior awareness and attention by researchers as well as science organisations to the destructive applications of the life sciences. Before codes can help teach, education is needed.'[94]

---

93   Ibid., p. 54.
94   Rappert, 2010, op. cit., p. 14.

Education and awareness-raising

In 2005, following a series of interactive seminar discussions held with life scientists at 15 UK universities, Dando and Rappert concluded there was little evidence that participants 'regarded bioterrorism or bioweapons as a substantial threat … considered that developments in the life sciences research contributed to biothreats … were aware of the current debates and concerns about dual-use research; or … were familiar with the BWC'.[95] In 2006, Rappert et al. presented a report to the Sixth BTWC Review Conference, which stated that 'despite the recent international attention given to the problem of the potential misuse of the life sciences', the initial UK findings were essentially replicated in later seminars in the United Kingdom and in Finland, Germany, the Netherlands, South Africa and the United States.[96] Subsequent experience by Dando and Rappert 'of carrying out seminars in 16 different countries with a few thousand life scientists in over 110 different departments has consolidated these findings'.[97]

In a 2009 *Nature* article, Dando highlighted the 'lack of engagement with this issue among life scientists', which he considered 'alarming',[98] specifically with regard to the consequent dangers of the misuse of scientific and technological advances for the development of incapacitant weapons. Dando and colleagues have identified the current lack of adequate biosecurity teaching as an important factor contributing to such low levels of awareness amongst life scientists. Following their review of the effectiveness of existing education and awareness-raising initiatives, Rappert et al. conclude:

> In such circumstances it is quite unrealistic to expect that simply, for example, adding a lecture to a standard course in the life sciences will make a great deal of difference … in depth implementation of the BWC within States Parties requires a significant effort on education and outreach for such implementation to be effective. To achieve this, a simple declaration as at previous Review Conferences about the

---

95   Dando, M. and Rappert, B. 2005, *Codes of Conduct for the Life Sciences: Some Insights from UK Academia*, Bradford Briefing Paper No. 16 (2nd series), University of Bradford, UK, p. 23.

96   Rappert, B., Chevrier, M. and Dando, M. 2006, *In-Depth Implementation of the BTWC: Education and Outreach*, Report presented at Sixth BTWC Review Conference, University of Bradford, UK, para. 89.

97   Whitby, S. and Dando, M. 2010, 'Biosecurity awareness-raising and education for life scientists: what should be done now?' in B. Rappert (ed.), *Education and Ethics in the Life Sciences: Strengthening the Prohibition of Biological Weapons*, ANU E Press, Canberra. For a detailed discussion, see: Rappert, B. 2007, *Biotechnology, Security and the Search for Limits: An Inquiry into Research and Methods*, Palgrave, Basingstoke, UK; Rappert, B. 2009, *Experimental Secrets: International Security, Codes, and the Future of Research*, University Press of America, Lanham, Md.

98   Dando, M. 2009, 'Biologists napping while work militarized', *Nature*, vol. 460, no. 7258, p. 951. Dando's concerns are echoed in an accompanying *Nature* editorial entitled: 'A question of control: scientists must address the ethics of using neuroactive compounds to quash domestic crises.'

importance of education will be insufficient and States Parties will need to take concerted action to ensure increased educational provision and outreach.[99]

Similarly, Whitby and Dando believe that 'correcting this deficiency in education- and awareness-levels of life scientists will be a massive task'.[100] It is one that will require action by a broad range of constituencies involved in life-science education including, governments, bodies responsible for the administration of standards in higher education, funders of life-science education, civil society groups and NGOs involved in the production of educational material, and teachers and trainers.[101]

Whilst a number of albeit relatively small and isolated initiatives are now being undertaken to educate and raise awareness amongst life scientists regarding dual-use dilemmas and the potential dangers of research being misused for chemical or biological weapons development,[102] there do not appear to be any sustained activities designed to foster greater awareness and knowledge of such issues beyond the life-science community. Rappert has noted that '[s]cant efforts made prior to 2001 (and even since) by scientists to popularise how their work might aid the production of bioweapons indicate the historical pattern of not seeking to foster wider debate and awareness'.[103]

It is, however, worth considering whether and how those scientists, academics and educators concerned about incapacitants and who are currently conducting CBW education and awareness-raising activities can also engage with key civil society actors in areas such as human rights, humanitarian law and medical ethics who at present have limited or no knowledge of the dangers posed by the potential harnessing of advances in chemistry and the life sciences to hostile purposes. The education and engagement of such expert communities would enrich and inform the existing discourse concerning incapacitants and broaden the range of actors seeking effective restrictions of such weapons. In considering such issues it may be worth exploring the roles of civil society awareness-raising initiatives and public education in helping to build successful multidisciplinary coalitions dedicated to addressing complex issues such as the prohibition of antipersonnel landmines and cluster munitions, addressing climate change, and promoting the establishment of the International Criminal Court.

---

99   Rappert et al., 2006, op. cit., paras 89, 90.
100   Whitby and Dando, op. cit., p. 182.
101   For further discussion, see: Dando, M. 2009, 'Dual-use education for life scientists', *Disarmament Forum*, vol. 1, pp. 41–4.
102   Project on Building a Sustainable Capacity in Dual-Use Bioethics, available from Bradford University web site, <http://www.brad.ac.uk/bioethics/About/> (viewed 4 December 2011).
103   Rappert, 2010, op. cit., p. 16.

Developing and applying ethical standards for health professionals

In addition to life scientists, the participation of physicians and other health professionals in previous state-run incapacitant weaponisation programs has been documented in a range of countries including the United Kingdom, the United States and South Africa.[104] More recently, a number of health professionals in certain countries have voiced support for medical involvement in the research and development of incapacitant weapons. Anaesthesiologists who were engaged in the Czech Republic's incapacitant weapons development program have highlighted counterterrorism considerations: 'many agents used in everyday practice in anaesthesiology can be employed as pharmacological non-lethal weapons. An anaesthetist familiar with the pharmacokinetics and pharmacodynamics of these agents is thus familiar with this use. As a result, he or she can play a role in combating terrorism.'[105]

The necessity of medical participation for the viability of incapacitant research has been highlighted by Gross, who believes that chemical incapacitants are among a limited range of 'nonlethal' weapons that are in effect '"medicalized" in that they rely on advances in neuroscience, physiology, and pharmacology and on the active participation of physicians and other medical workers'.[106]

In comparison with the life-science community, amongst certain sectors, at least, of the medical professional community, ethical discourse regarding the involvement in development and use of 'nonlethal' chemical weapons appears to be more advanced. There are a number of ethical codes and declarations that are potentially applicable and may well constrain the involvement of health professionals in this area. First, there is a range of declarations and regulations adopted by the World Medical Association (WMA) that guide health professionals in situations of conflict and unrest[107] and specifically prohibit their involvement

---

104   See, for example: Maclean, A. 2006, *Historical Survey of the Porton Down Volunteer Programme*, Ministry of Defence, London, <http://www.mod.uk/NR/rdonlyres/7211B28A-F5CB-4803-AAAC-2D34F3DBD961/0/part_iv.pdf> (viewed 17 August 2011), pp. 109–42; Ketchum, J. 2006, *Chemical Warfare: Secrets Almost Forgotten*, James S. Ketchum (self-published); Gould, C. and Folb, P. 2002, *Project Coast: Apartheid's Chemical and Biological Warfare Programme*, United Nations Institute for Disarmament Research, Geneva.
105   Hess et al., 2005, op. cit.
106   Gross, M. 2010, 'Medicalized weapons and modern war', *Hastings Center Report*, vol. 40, no. 1, pp. 34–5.
107   World Medical Association (WMA) 1956, *Regulations in Times of Armed Conflict*, Adopted by the 10th World Medical Assembly, Havana, Cuba, October 1956, last amended by the 35th World Medical Assembly, Tokyo, Japan, 2004, and editorially revised at the 173rd Council Session, Divonne-les-Bains, France, May 2006, para. 2.

in torture, ill treatment and other forms of human rights abuse.[108] The WMA has also sought to develop ethical guidelines prohibiting medical involvement in the development of chemical or biological weapons.[109]

In addition, guidance has also been developed by certain national medical associations and other medical bodies[110] on the ethical considerations surrounding health professionals' involvement in specific weapons research or weapons development more generally. A number of such bodies have also established a range of mechanisms and structures to implement ethical standards including ethics boards that have the authority to suspend or disbar physicians from practising medicine in cases of extreme misconduct.

There are thus ethical frameworks and mechanisms in place that describe and regulate the duty of health professionals to abide by and promote aspects of human rights and international humanitarian law, and specifically prohibit engagement in acts such as torture or the development of biological and toxin weapons. Whilst these and other ethical standards—particularly those concerned with medical research involving human subjects[111]—can in theory be applied to the development and utilisation of incapacitants, this does not appear to have occurred in a consistent manner to date. There are no widely accepted guidelines specifically determining the permissibility or non-permissibility of physician involvement in the development, testing or utilisation of so-called 'nonlethal' weapons in general and incapacitants in particular. Indeed the issue appears, at present, to be both underexplored and contentious, with a spectrum of opinions held by health professionals and medical ethicists.

Amongst national medical associations, it is the British Medical Association (BMA) that has taken the lead in the development of ethical guidance for the health community on the issue of 'nonlethal' chemical weapons, and in

---

108 World Medical Association (WMA) 1975, *Declaration of Tokyo*, Adopted by the 29th World Medical Assembly, Tokyo, Japan, October 1975, and editorially revised at the 170th Council Session, Divonne-les-Bains, France, May 2005, and the 173rd Council Session, Divonne-les-Bains, France, May 2006.

109 World Medical Association (WMA) 1990, *Declaration on Chemical and Biological Weapons*, Adopted by the 42nd World Medical Assembly, Rancho Mirage, Calif., October 1990, and rescinded at the WMA General Assembly, Santiago, 2005; World Medical Association (WMA) n.d., *Washington Declaration on Biological Weapons*, paras 18 and 19.

110 For example, the US Army textbook of military ethics urges medical professionals 'to stay in the business of healing and not hurting, which includes not participating in or contributing to weapons research and development'. Frisina, M. E. 2003, 'Medical ethics in military biomedical research', in T. E. Beam and L. R. Sparacino (eds), *Military Medical Ethics*, vol. 2, Office of the Surgeon General/Borden Institute, Washington, DC, pp. 533–61, as cited in Gross, op. cit., p. 37.

111 See, for example: World Medical Association (WMA) 1964, *Declaration of Helsinki*, Adopted by the 18th WMA General Assembly, Helsinki, , June 1964, and last amended at the 59th WMA General Assembly, Seoul, October 2008, <http://www.wma.net/en/30publications/10policies/b3/index.html> (viewed 22 July 2011); *Nuremberg Code*, 1949, as detailed in *Trials of War Criminals before the Nuremberg Military Tribunals under Control Council Law*, vol. 2, no. 10, pp. 181–2, US Government Printing Office, Washington, DC, 1949, available from: Office of Human Subjects Research, US National Institutes of Health, <http://ohsr.od.nih.gov/guidelines/nuremberg.html> (viewed 22 July 2011).

particular incapacitants. In its 2007 publication *Drugs As Weapons*, which explored the implications of incapacitant research, development and use,[112] the BMA declared that:

> [D]octors should not knowingly use their skills and knowledge for weapons' development for the same reasons that these ethical considerations oppose doctors' involvement in torture and the development of more effective methods of execution. In other words, the duty to avoid doing harm rises above, for instance, a duty to contribute to national security.[113]

The report specifically recommends that national organisations that represent healthcare professionals should

> [w]ork to promote the norms prohibiting the use of poisons, and therefore the BWC and the CWC. They should further promote understanding that the use of drugs as weapons would violate such norms ... Advocate against the use of drugs as weapons and not be involved in the training of military or law enforcement personnel in the administration of drugs as weapons.[114]

Despite the activities of the BMA, the issue does not appear to have been specifically addressed formally by the WMA during its general assembly or in any other public WMA policy body.[115] Given the importance of medical participation to research, development, testing and utilisation of incapacitant weapons, the development of clear guidance in this area is needed.

Harvard Sussex draft convention

The Harvard Sussex Project (HSP) has proposed an alternative approach to addressing individual responsibility and culpability in the development or misuse of chemical and biological weapons, through the development of criminal sanctions. Meselson and Robinson have argued that '[a]ny development, production, acquisition, or use of biological and chemical weapons is the result of decisions and actions of individual persons, whether they are government officials, commercial suppliers, weapons experts, or terrorists'. They contend, however, that the BTWC and CWC 'are directed primarily to the actions of states, and address the matter of individual responsibility to only a limited degree'.[116]

---

112   BMA, op. cit.
113   Ibid., p. 20.
114   Ibid., p. 24.
115   A review was undertaken of all relevant publicly available World Medical Association documentation from January 1990 to December 2011.
116   Meselson, M. and Perry Robinson, J. 2004, 'A draft convention to prohibit biological and chemical weapons under international criminal law', *Fletcher Forum of World Affairs*, vol. 28, no. 1 (Winter), p. 57.

In order to address this failing, HSP has developed a Draft Convention to Prohibit Biological and Chemical Weapons under International Criminal Law. If enacted, the HSP draft convention would make it a crime under international law for any person knowingly: to develop, produce, acquire, retain, transfer or use biological or chemical weapons; to order, direct or knowingly render substantial assistance to those activities; or to threaten to use biological or chemical weapons.[117]

Under the HSP draft convention, each state party would be required to 'establish jurisdiction with respect to such crimes according to established principles of judicial law', and where the state has jurisdiction and is satisfied that the facts warrant such action, 'to submit those cases to competent authorities for the purpose of extradition or prosecution'. Furthermore, with respect to the actual use of CBWs, each state party would be required to 'establish jurisdiction over all persons found on its territory regardless of their nationality or place of the offence'.[118]

The HSP draft convention is a civil society initiative that will require considerable state support for it to be adopted by the international governmental community. One potential route previously explored by the HSP is 'for a group of states to submit the proposed convention or a similar draft in the form of a resolution for consideration by the UN General Assembly [UNGA], seeking its referral to the UNGA Sixth (legal) Committee for negotiation of an agreed text'.[119] Robinson has stated that '[o]nce we are satisfied that the political environment is favourable, our plan is to convene an international conference that will bring together policy makers, jurists and exponents of the Draft Convention'.[120]

Nonparticipation/whistleblowing

Any serious attempt by state or non-state actors to develop new or indeed existing chemical or biological weapons, be they considered 'nonlethal' or otherwise, would require the involvement of an array of scientists, engineers, technicians and other ancillary workers. Whilst such staff are essential to the development, production, stockpiling, transfer and use of such weapons, they are also potentially capable of 'blowing the whistle' on such research and development programs through public denunciations, leaking information to

---

117 For the full text of the present draft of the convention, see <http://www.sussex.ac.uk/Units/spru/hsp/Harvard-Sussex-Program-draft-convention-Text.html> (viewed 4 June 2010). Also see accompanying *Draft Legal Commentary*, <http://www.sussex.ac.uk/Units/spru/hsp/Draft%20Convention%20supporting%20docs/HSP%20papers/Legal%20Commentary.pdf> (viewed 4 June 2010).
118 *Preamble to HSP Convention*, <http://www.sussex.ac.uk/Units/spru/hsp/Harvard-Sussex-Program-draft-convention-Text.html> (accessed 4 June 2010).
119 Perry Robinson, J. 2011, 'Criminalization of biological and chemical armament', *CBW Conventions Bulletin*, (Special Issue), February 2011, p. 4.
120 Ibid., p. 5.

journalists or by reporting concerns about potential or realised breaches of national regulations or violations of international treaties directly to the relevant national or international regulatory bodies.

In his Nobel acceptance speech, Rotblat stated that:

> The purpose of some government or industrial research is sometimes concealed, and misleading information is presented to the public. It should be the duty of scientists to expose such malfeasance. 'Whistle-blowing' should become part of the scientist's ethos. This may bring reprisals; a price to be paid for one's convictions. The price may be very heavy.[121]

Deiseroth highlights the particular vulnerability of whistleblowers:

> Compared with normal citizens, employees are in a special situation because they owe their employer a certain loyalty and, by law, are normally not allowed to disclose internal or confidential information. Whistle-blowers, therefore, need protection if they make a disclosure in good faith and on the basis of reliable evidence.[122]

Similarly, Falter notes that 'it is neither realistic nor legitimate to put the full burden of whistle-blowing and potential retaliation on individual scientists and their moral sensibilities'.[123] Indeed, whilst it is the duty of individual scientists to make known their concerns about the misuse of scientific research for activities that breach ethical standards or international law, it is the responsibility of the scientific community as a whole to ensure that such whistleblowers are fully protected. This was recognised by the ICRC in their 2004 statement, which declared that '[t]hose working in life sciences who voice concern and take responsible action require and deserve political and professional support and protection', and the corresponding action point, which was to 'ensure that adequate mechanisms exist for voicing such concerns without fear of retribution'.[124]

Although a number of states such as South Africa,[125] the United Kingdom[126] and the United States[127] have legislation relating to whistleblowing activities

---

121   Rotblat, 1995, op. cit.
122   Deiseroth, op. cit., p. 266.
123   Falter, A. 2010, 'Including civil society into confidence building: protecting whistleblowers and societal verification', in Finney and Slaus, op. cit., p. 289.
124   ICRC, 2004, op. cit.
125   Government of South Africa 2000, *Protected Disclosures Act*, <http://www.workinfo.com/free/Sub_for_legres/data/Disclosure/protected.htm> (viewed 23 January 2012).
126   Government of the United Kingdom 1998, *Public Interest Disclosure Act*, available on the UK National Archives web site: <http://www.legislation.gov.uk/ukpga/1998/23/contents> (viewed 23 January 2012).
127   Government of the United States of America 1989, *US Federal Whistleblower Protection Act*, 5 USC, s. 1201.

on their statute books, the effectiveness of such legislation and its enforcement are variable.[128] Furthermore, Martin, who has long experience of working with, and seeking to protect, whistleblowers in many different spheres, believes that 'the track record of whistle-blower protection measures—whistle-blower laws, hot-lines, ombudsmen and the like—is abysmal. In many cases, these formal processes give only an illusion of protection. Codes of ethics seem similarly impotent in the face of the problems.'[129]

It is important that independent scientists, health professionals and professional bodies—in cooperation with human rights, civil liberties and whistleblowing organisations—promote the establishment of truly effective mechanisms under international and domestic law that provide legal protection against discrimination and criminal prosecution for whistleblowers. Furthermore, given the failings of the current systems of whistleblower protection, the life-science and biomedical communities have a duty to support those individuals who refuse to participate in what they consider immoral research and development projects, and those who blow the whistle on such activities. A number of scientific associations and professional bodies have mechanisms for promoting ethical standards amongst their members, which can also be utilised to support colleagues facing reprisals for acting ethically.[130]

Conclusion

In its public statement 'Preventing hostile use of the life sciences, from ethics and law to best practice', the ICRC has noted that '[a]dvances in the life sciences hold great promise for humanity', but that 'there is also great risk if these same advances are put to hostile use'.[131] Whether the life sciences will be employed for the further development and proliferation of incapacitant weapons with the consequent dangers of the misuse of such weapons for internal repression or offensive military operations will depend to a significant degree on the actions of individual scientists and biomedical researchers and on the role of the life-science community as a whole. At a minimum this must entail stringent personal and community reflection and regulation of the trajectories

---

128 See, for example: Calland, R. and Dehn, G. (eds) 2004, *Whistleblowing around the World*, Open Democracy Advice Centre, Cape Town, & Public Concern at Work, London; Deiseroth, op. cit.; Devine, T. 1997, *The Whistleblower's Survival Guide*, Fund for Constitutional Government, Washington, DC; Devine, T. 2004, 'Whistleblowing in the United States: the gap between vision and lessons learned', in Calland and Dehn, op. cit.; Martin, B. 1999a, 'Suppression of dissent in science', *Research in Social Problems and Public Policy*, vol. 7, pp. 105–35; Martin, B. 1999b, *The Whistleblower's Handbook*, Jon Carpenter, Charlbury, UK.
129 Martin, B. 2007, 'Whistle-blowing: risks and skills', in McLeish and Rappert, op. cit., p. 6.
130 For example, the AAAS Science and Human Rights Program (SHRP), which works with scientists to 'advance science and serve society' through human rights, has successfully campaigned on behalf of a number of scientific whistleblowers. For further information, see: <http://shr.aaas.org/>.
131 ICRC, 2004, op. cit.

and possible applications of relevant research to ensure it cannot be used for hostile purposes. In addition, it can be argued that now is the time for the life-science community to take a far more active and engaged stance on this issue and join with the arms control, disarmament, human rights and humanitarian law communities to explore the full range of voluntary, normative and legally binding mechanisms that can be applied to protect humanity from the potential of abuse of the relevant technologies they are developing, whilst preserving the beneficial applications.

19. WHO Project on Responsible Life-Sciences Research for Global Health Security: Why Ethics and How to Strengthen It?

Emmanuelle Tuerlings and Andreas Reis[1]

Introduction

There has been much discussion over the past years about global health security and how to strengthen it. One area that has raised much activity revolves around the risks posed by accidents and the potential deliberate misuse of life-sciences research.[2] Different actors have proposed a variety of measures to manage such potential risks.[3] Yet little information is available about the needs and capacities of countries, laboratories and research institutions in this area.

The World Health Organisation (WHO) has developed a self-assessment questionnaire for laboratory managers and researchers to assess their needs and

---

1   The author is a staff member of the World Health Organisation. The author alone is responsible for the views expressed in this chapter and he does not necessarily represent the decisions or policies of the World Health Organisation.

2   Research accidents are understood as research activities that may unexpectedly pose some risks via 'accidental' discoveries. World Health Organisation (WHO) 2010, *Responsible Life Sciences Research for Global Health Security. A Guidance Document*, WHO/HSE/GAR/BDP/2010.2, World Health Organisation, Geneva.

3   See, for instance: National Research Council 2004, *Biotechnology Research in An Age of Terrorism*, Committee on Research Standards and Practices to Prevent the Destructive Application of Biotechnology, The National Academies Press, Washington, DC; National Science Advisory Board for Biosecurity (NSABB) 2007,*Proposed Framework for the Oversight of Dual Use Life Sciences Research: Strategies for Minimizing the Potential Misuse of Research Information*, National Science Advisory Board for Biosecurity, Bethesda, Md; Miller, S. and Selgelid, M. 2008, *Ethical and Philosophical Consideration of the Dual-Use Dilemma in the Biological Sciences*, Springer, Dordrecht; Report prepared by the Centre for Applied Philosophy and Public Ethics at The Australian National University for the Australian Department of Prime Minister and Cabinet, National Security Science and Technology Unit, November 2006; Steinbruner, J. et al. 2007, *Controlling Dangerous Pathogens. A Prototype Protective Oversight System*, March, Center for International and Security Studies at Maryland (CISSM), University of Maryland, College Park; InterAcademy Panel on International Issues (IAP) 2005, *IAP Statement on Biosecurity*, November; 'Statement on the scientific publication and security', *Science*, vol. 299 (2003), p. 1149; United Nations Meeting of the States Parties to the Convention on the Prohibition of the Development, Production and Stockpiling of Bacteriological (Biological) and Toxin Weapons and on their Destruction; *Managing Risks of Misuse Associated with Grant Funding Activities. A Joint Biotechnology and Biological Sciences Research Council (BBSRC), Medical Research Council (MRC) and Wellcome Trust Policy Statement*, September 2005; Wellcome Trust 2005, *Guidelines on Good Research Practice*, November; Dando, M. and Rappert, B. 2005, 'Codes of conduct for the life sciences: some insights from UK academia', *Briefing Paper No. 16 (Second series)*, May, University of Bradford, UK.

capacities in regards to responsible life-sciences research. This self-assessment questionnaire is part of a larger project on 'Responsible life-sciences research for global health security', which published several documents on this issue.[4] The responsible life-sciences research framework and its associated self-assessment questionnaire are based on three pillars supporting public health: research excellence, ethics, and biosafety and laboratory biosecurity. The bio-risk framework for responsible life-science research is an integrated approach, where each pillar is equally important and should complement and reinforce the others. This chapter will focus on the ethics pillar because it is often regarded as an area that is complex, underfunded and an obstacle to undertaking research. This chapter starts by giving an overview of the bio-risk management framework for responsible life-sciences research. It then underlines the role of ethics in this area and it ends by discussing how the self-assessment questionnaire can provide useful feedback to countries and institutions to identify needs and capacities in ethics, and ways to strengthen them.

The bio-risk management framework for responsible life-sciences research

The WHO project on responsible life-sciences research for global health security aims at promoting excellent, safe, secure and responsible life-sciences research through an integrated approach that recommends investing capacities in three pillars supporting public health: research excellence; ethics; biosafety and laboratory biosecurity.[5]

- Pillar 1: Research excellence. This concerns fostering quality in life-science activities, which is the basis for developing new treatments and therapeutics, strengthening health research systems, and promoting public health surveillance and response activities. These elements are essential to protecting and improving the health and wellbeing of all people.

---

4   World Health Organisation (WHO) 2005, *Life Sciences Research: Opportunities and Risks for Public Health. Mapping the Issues*, WHO/CDS/CSR/LYO/2005/20, World Health Organisation, Geneva, <http://www.who.int/csr/resources/publications/deliberate/WHO_CDS_CSR_LYO_2005_20/en/index.html> (viewed 7 February 2012); World Health Organisation 2007, *Scientific Working Group on Life Science Research and Global Health Security. Report of the First Meeting*, WHO/CDS/EPR/2007.4, World Health Organisation, Geneva, <http://www.who.int/csr/resources/publications/deliberate/WHO_CDS_EPR_2007_4/en/index.html> (viewed 7 February 2012); World Health Organisation (WHO) 2008, *Research Policy and Management of Risks in Life Sciences Research for Global Health Security. Report of the Meeting. Bangkok, Thailand, 10–12 December 2007* WHO/HSE/EPR/2008.4, World Health Organisation, Geneva, <http://www.who.int/csr/resources/publications/deliberate/WHO_HSE_EPR_2008_4/en/index.html> (viewed 7 February 2012); World Health Organisation (WHO) 2012, Informal Consultation on Dual-Use Research of Concern (DURC), World Health Organisation, Geneva, 26–28 February 2013, <http://www.who.int/mediacentre/events/meetings/2013/durc/en/> (viewed 2 September 2013).

5   This section draws upon the following WHO guidance document: WHO, 2010, op. cit.

- Pillar 2: Ethics. This involves the promotion of responsible and good research practices, the provision of tools and practices to scientists and institutions that allow them to discuss, analyse and resolve in an open atmosphere the potential dilemmas they may face in their research, including those related to the possibility of accidents or misuse of the life sciences.

- Pillar 3: Biosafety and laboratory biosecurity. This concerns the implementation and strengthening of measures and procedures to: minimise the risk of worker exposure to pathogens and infections; protect the environment and the community; and protect, control and account for valuable biological materials (VBM) within laboratories, in order to prevent their unauthorised access, loss, theft, misuse, diversion or intentional release. Such measures reinforce good research practices and are aimed at ensuring a safe and secure laboratory environment, thereby reducing any potential risks of accidents or deliberate misuse.

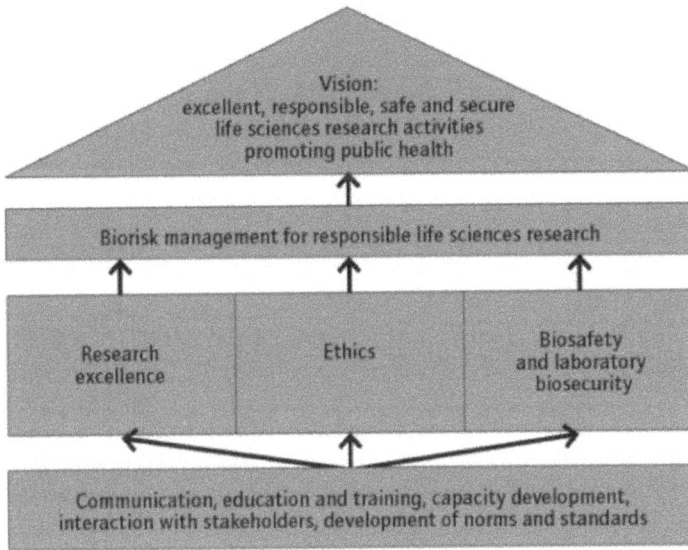

Figure 19.1 Bio-Risk Management Framework for Responsible Life-Sciences Research

Source: World Health Organisation (WHO) 2010, *Responsible Life Sciences Research for Global Health Security. A Guidance Document*, WHO/HSE/GAR/BDP/2010.2, World Health Organisation, Geneva.

This approach recognises that there is no single solution or system that will suit all countries, institutions or laboratories. Each country or institution that assesses the extent to which it has systems and practices in place to deal with the risks posed by accidents or the potential deliberate misuse of life-sciences research will decide which measures are most appropriate and relevant according to their own national circumstances and contexts.

Likewise, this approach also acknowledges the fact that there is a multiplicity of stakeholders developing different measures. The public health sector is one stakeholder among others (for example, national academies of science, ethicists and security communities), and coordination at national and international levels is therefore crucial for addressing those types of risks.

One important advantage of this integrated framework is that it recommends investing capacities in public health areas that either are already in place in many countries or are being developed for strengthening research capacities and for addressing laboratory bio-risks. This approach therefore promotes a sustainable and effective way to address such bio-risks. Indeed it emphasises that one of the most effective ways to prepare for deliberately caused disease is to strengthen public health measures for naturally occurring and accidentally occurring diseases.[6] In a similar manner, the promotion of a culture of scientific integrity and excellence is considered as one of the best protections against the possibility of accidents and deliberate misuse of life-sciences research and offers the best prospect for scientific progress and development. A culture of scientific integrity and excellence can be strengthened through the development and reinforcement of capacities in the three aforementioned public health pillars. In adopting the bio-risk management framework for responsible life-sciences research, countries and institutions are encouraged to consider

- reinforcing public health capacities in terms of research for health, biosafety and laboratory biosecurity management and ethics
- investing in training personnel (laboratory staff and researchers) and students in ethics, the responsible conduct of research, and biosafety and laboratory biosecurity
- ensuring compliance with biosafety and laboratory biosecurity
- identifying multi-stakeholder issues, with different layers of responsibilities and encouraging coordination among stakeholders
- using existing mechanisms, procedures and systems and reinforcing local institutional bodies (if they exist).

While recognising the above, countries and institutions may consider drawing on a bio-risk management framework for responsible life-sciences research, which has been developed to help countries and institutions identify where public health resources can be used for addressing the above risks. To facilitate such analysis, a self-assessment questionnaire on responsible life-science research has been developed to support health researchers, laboratory managers and research institutions to identify strengths and to address weaknesses in

---

6   World Health Assembly Resolution WHA55.16 of 18 May 2002, *Global Public Health Response to Natural Occurrence, Accidental Release or Deliberate Use of Biological and Chemical Agents or Radionuclear Material that Affect Health*, World Health Organisation, Geneva.

each of the three pillars. Among these three pillars, ethics is an important public health area where countries and institutions can invest capacities to address such bio-risks.

An ethics framework

Context and related initiatives

The potential duality of research and science, and related moral concerns, is not a new issue. Following the invention of atomic weapons, many scientists, including Einstein, Oppenheimer and Russell, voiced concern about the potential dangers to humanity arising from scientific discoveries.[7]

Ethical issues arising in research and the corresponding obligations of scientists have for a long time been one of the key areas of work on ethics of the UN Educational, Scientific and Cultural Organisation (UNESCO). Already in 1974, UNESCO's eighteenth general conference adopted the *Recommendation on the Status of Scientific Researchers*,[8] which enumerated ethical responsibilities and rights of scientists.

In 1999, UNESCO organised the World Conference on Science, in Budapest, which gave special attention to ethical principles and responsibilities in the practice of science.

Since 2005, UNESCO has led a new project, with regional consultations, to survey the field of science ethics and to evaluate the need for an update of the 1974 document.[9]

Until recently, however, the bioethics community concerned with research has focused on the protection of human subjects in clinical trials, and other kinds of ethical issues raised by genetics and related disciplines. It is only in the past few years that ethicists have become more engaged with the scientific community and security experts regarding the ethical questions raised by the potential misuse of science.

---

7   See, for example: 'Russell-Einstein Manifesto and the Pugwash Conferences on Science and World Affairs', <http://www.pugwash.org/about/manifesto.htm> (viewed 7 February 2012).
8   United Nations Educational, Scientific and Cultural Organisation (UNESCO) 1974, *Recommendation Adopted on the Report of the Commission for Science at the Thirty-Eighth Plenary Meeting on 20 November 1974*, Paris.
9   Scholze, S. 2006, 'Setting standards for scientists: for almost ten years, COMEST has advised UNESCO on the formulation of ethical guidelines', *EMBO Reports*, vol. 7 (SI) (July), pp. S65–7.

Ethics is now recognised in the field as providing a useful framework for the identification of dilemmas, and the discussion and evaluation of responsibilities of different stakeholders. Ethics can help promote understanding of the nature of decisions that researchers and other actors have to make.[10] The discussion of ethical issues can contribute to consensus-building among the stakeholders, who sometimes have competing interests.

Ethics framework in the WHO project

WHO has, since its establishment in 1948, given great importance to ethical considerations in its programs. In 1967, the World Health Assembly (WHA) resolved that 'scientific achievements, and particularly in the field of biology and medicine—that most humane science—should be used only for mankind's benefit, but never to do it any harm'.[11] A WHO report in 2005 discussed in more detail the ethical questions raised.[12]

Some of the basic dilemmas addressed in the current WHO project, 'Responsible life science research for global health security', are

- how to weigh the potential benefits of research against the risk of misuse
- how to weigh the individual interests of researchers against the common good of public health
- how to best manage the risks associated with research, without hindering its beneficial application to public health
- what are the responsibilities of individual researchers, and of the scientific community as a whole to society?

For the development of an ethics framework, it is helpful to analyse the responsibilities and dilemmas of different stakeholders involved in life sciences, and to develop corresponding tools to support them in making ethical choices.

Researchers, science societies and codes of conduct/ ethics

Individual scientists have been at the centre of the debate on ethical responsibilities in dual-use research. On the one hand, they will aim to carry out and publish beneficial research, while on the other hand they want to refrain from research that could lead to the potential malevolent use of the results. They

---

10 WHO, 2007, op. cit.
11 WHA20.54.
12 World Health Organisation (WHO) 2005, *Life Sciences Research*, WHO/CDS/CSR/LYO/2005/20, World Health Organisation, Geneva, <http://www.who.int/csr/resources/publications/deliberate/WHO_CDS_CSR_LYO_2005_20/en/index.html> (viewed 7 February 2012).

have a widely acknowledged obligation to carry out their research in accordance with good research practices and laws and regulations, including codes (see below). In addition, the debate on dual-use research has made clear that there are also some expectations on the individual scientists to make judgments about the outcomes and potential security implications of their research projects; however, often researchers will not be security experts and may face difficulties in making such assessments. Therefore, while education can help to raise awareness among scientists, in case of doubt, they should be able to turn to mechanisms for reporting and consultation, including for whistleblowing, provided by their institutions.

Codes for researchers have for a long time been proposed as potentially useful tools to address the potential misuse of life-science research at the level of the individual scientists. In fact, codes for professional behaviour date back at least two millennia (the Hippocratic Oath, Maimonides, and so on). Two types of codes are usually distinguished

1. codes of conduct: these provide specific guidelines with respect to what is considered appropriate behaviour

2. codes of ethics: these are more aspirational, setting forth the ideas to which practitioners should aspire, such as standards of objectivity or honesty.

Some prominent examples of codes for scientists include those proposed by scientific and academic societies (such as the American Society for Microbiology, the Chinese Academy of Sciences, The Royal Society in the United Kingdom and the US National Science Advisory Board for Biosecurity), and medical associations (such as the World Medical Association and the American Medical Association).[13] Recently, the International Centre for Genetic Engineering and Biotechnology has been undertaking a project to review codes of conduct.

There has been considerable debate on the effectiveness of codes of conduct/ethics for researchers.[14] Critics have insisted that there is little evidence about their effectiveness in practice. For example, the Hippocratic Oath has not hindered medical doctors from doing horrendous things to their patients. Scientists may simply not comply with such non-binding instruments as codes, and terrorists will not be deterred from using biotechnology with bad intentions. Even with the best intentions, and fully aware of their duties, scientists are not security experts and may thus not foresee the consequences of their experiments, which might have dual-use potential.

---

13 For more examples, see WHO, 2010, op. cit., pp. 58–9.
14 Rappert, B. 2007, 'Codes of conduct and biological weapons: an in-process assessment', *Biosecurity and Bioterrorism: Biodefense Strategy, Practice and Science*, vol. 5, pp. 145–54.

On the other side, the proponents of such codes—for example, Margaret Somerville and Ronald Atlas[15]—have argued that they serve their purpose in raising awareness, fostering norms among the scientific community and establishing public accountability mechanisms. They can help sensitise scientists who might be used by malevolent individuals or organisations and the public health community to the risks of dual-use research. Codes can also foster an environment in research institutions in which whistleblowing could be encouraged. There is an ongoing discussion between the proponents of a universal code, arguing that national codes of conduct will have little effect on the global behaviour of life scientists, and those who prefer developing some basic principles that would then be applicable to and implemented at local settings, institutions or countries.

Education and training in biosecurity and ethics are other tools for raising awareness among scientists. A report published in 2008 by the American Association for the Advancement of Science (AAAS) examined such education programs and emphasised their importance.[16]

Research institutions

As most research takes place in institutions, these also have important responsibilities in the prevention of accidents and the deliberate misuse of research. Research institutions have an obligation to ensure that research within their premises is conducted according to national laws and relevant codes of conduct. They should sensitise their researchers to dual-use issues—for example, through education and specific training, promoting discussion and reflection on research practices, and creating a climate that encourages scientists to come forward in case of difficulties. Procedures and standards for whistleblowing should be in place for cases of scientific misconduct.

Another potentially useful approach to address the dual-use dilemma, which has been proposed at the institutional level, is the ethical review of sensitive research by an ethics committee. In practice, however, it is not clear which committees could take on this role. Traditional research ethics committees review research with human subjects, and not bench science, and their members are not usually trained in biosecurity issues. With specific training, the members of research ethics committees could probably be enabled to play this role. Another

---

15 Somerville, M. and Atlas, R. 2005, 'Ethics: a weapon to counter bioterrorism', *Science*, vol. 307, pp. 1881–2.

16 American Association for the Advancement of Science (AAAS) 2008, *Professional and Graduate-Level Programs on Dual Use Research and Biosecurity for Scientists Working in the Biological Sciences: Workshop Report*, American Association for the Advancement of Science, Washington, DC.

option would be to supplement institutional biosafety committees (IBCs) with ethicists. A further difficulty to resolve is the proper identification of the research proposals that should be subjected to such a review.

Publishers and journal editors

Freedom of intellectual inquiry, sharing of knowledge and publication are at the very heart of science. So one of the most contentious debates about the governance of dual-use research involves the extent to which the publication of potentially dangerous information should be restricted, or censored.[17]

Publications enable progress in beneficial research and can inform about new kinds of threats. In particular, sufficiently detailed 'materials and methods' parts of publications are essential for other scientists to reproduce and verify results. Yet, these sections are precisely the ones that could potentially be misused by ill-minded individuals. Publishers and journal editors have a responsibility to prevent the publication of research results that may be misused. In February 2003, recognising this responsibility, a joint *Statement on Scientific Publication and Security* of the Journal Editors and Authors Group was simultaneously published by *Science, Nature*, the *Proceedings of the National Academy of Sciences* and the American Society for Microbiology Journals.[18] While reminding readers of publishers' 'unique responsibility' for the integrity and the advancement of research, the statement accepts that at times censorship of science may be warranted: 'We recognize that on occasion an editor may conclude that the potential harm of publication outweighs the potential societal benefits. Under such circumstances, the paper should be modified, or not be published.'[19]

Governments, international organisations and funding bodies

Finally, governments, international organisations and funding bodies all face dilemmas in managing the risks related to deliberate misuse of research. On the one hand, they play a key role in promoting scientific development, and in creating an environment that is conducive to stimulating life-sciences research. On the other hand, they have important responsibilities in promoting global health security and in minimising the potential harm to public health, at a national and a global level.

---

17   Miller, S. and Selgelid, M. 2007, 'Ethical and philosophical considerations of the dual-use dilemma in the biological sciences', *Science and Engineering Ethics*, vol. 13, pp. 523–80.
18   Selgelid, M. 2007, 'A tale of two studies: ethics, bioterrorism, and the censorship of science', *Hastings Center Report*, vol. 37, no. 3.
19   Journal Editors and Authors Group 2003, 'Statement on scientific publication and security', *Science*, vol. 299, no. 5610, p. 1149.

There are different risk-management options for governing life-science research with dual-use potential. It is essential to find and maintain the right mix of policies in order to allow the public health benefits of life-science research to be maximised while minimising the risks. Besides the promotion of research excellence, biosafety and laboratory biosecurity, an ethical framework is a key component of a public health approach. While no magic bullet, codes of ethics for scientists, as well as for publishers, can make essential contributions. Further discussion about the usefulness and feasibility of an ethics review of sensitive research is needed. Finally, assessments, awareness-raising and training at local and facility levels are key to the implementation of effective policies. The WHO self-assessment questionnaire can help countries and institutions to identify these needs and capacities and take the necessary actions.

Ethics and the self-assessment questionnaire

The process of self-assessment starts with an identification of strengths, weaknesses and gaps, and concludes with action to address the gaps and weaknesses and to build on or consolidate the strengths. Going through this process will first provide an assessment of the extent to which systems are in place in the national public health system, in research institutions and in laboratories to facilitate and ensure excellence in science, ethics, safety and security.

Second, it will identify priority areas where action is necessary to ensure such practices. In coordination with other stakeholders, the knowledge gained through this process will help countries and institutions to align available resources with local needs and circumstances. This is therefore an important step in the identification of appropriate training and a sustainable way to enhance global health security. This self-assessment tool could also be used as a simple and cost-effective way for interested countries and institutions to raise awareness about responsible life-sciences research. Health researchers, laboratory managers and research institutions are the primary audiences for this self-assessment questionnaire. It was not developed to evaluate the adequacy of measures developed by other sectors (for example, security, academia, publishers and editors). Yet, it recognises the importance of collaboration and coordination between different sectors, especially when priority areas have been identified and when action will unfold.

Methodology

The self-assessment questionnaire on responsible life-sciences research lists several statements associated with each pillar, using the Likert scale. A statement for which a respondent answered 'don't know' or 'disagree' may mean that further information is needed in certain areas and may indicate some weaknesses. On the other hand, responses of 'agree' and 'strongly agree' suggest that appropriate measures are already in place.

The strengths and weaknesses of an institution can then be reported by pillars, giving some overview of advantages and gaps. It is therefore strongly advised to fill in the questionnaire as a whole and not in parts. The self-assessment tool is a voluntary, anonymous and innovative tool in this field.

The ethics pillar lists some of the following statements to which respondents are being asked to give their opinion[20]

- education and/or training is offered on research ethics
- appropriate ethical research guidelines and practices have been published and implemented
- adequate mechanisms exist for investigating and responding to non-adherence to ethical standards
- researchers know how to assess whether the risk outweighs the benefit of continuing with their research or activities
- researchers are aware of and informed about national and international conventions, laws and regulations related to their research
- an ethics committee assesses research proposals involving human subjects
- a review process exists to assess ethical issues raised by research proposals not involving human or animal subjects
- information about the national and international conventions and regulations related to all fields of science is easily accessible.

The self-assessment questionnaire was piloted in a South African facility in 2009.[21] A seminar on responsible life-sciences research and dual-use issues was presented to the audience as a way to introduce the questionnaire to the laboratory scientists and managers. The audience was also informed about the purpose and content of the questionnaire, noting that this was an anonymous and voluntary process. Of the 20 scientists present, 18 completed the questionnaire. The respondents were coming from different laboratories, all part of the same facility. Feedback on the questionnaire was provided to the respondents at the end of the seminar.

---

20   WHO, 2010, op. cit.
21   Gould, C. 2011, 'Biosecurity: a public health approach to reducing risk', *Research Report for the Wellcome Trust Project on 'Building a Sustainable Capacity in Dual-Use Bioethics'*, <http://www.brad.ac.uk/bioethics/Monographs> (viewed 22 September 2011).

The pilot study had some limitations. First, the sample size of the pilot study is too small to draw conclusions that would be representative of the whole facility. The collected data are therefore only illustrative of potential needs and capacities. A second limitation of the self-assessment questionnaire is that these questions may be considered subjective. The validity and reliability of the subjective measures have been enhanced through multiple questions measuring the same subjective state. The self-assessment questionnaire has indeed been designed in a way that statements regarding the different pillars can be analysed in combination, thereby reinforcing the data analysis. It should be noted, however, that one important strength of the questionnaire is its ability to identify differences in opinions about existing systems and measures so that the discrepancies can be dealt with—not necessarily through a change in the system itself but just with, sometimes, further discussion with the staff about existing measures.

Results

The data collected by the pilot study regarding the ethics pillar showed the[22] following.

- Seventeen of the 18 respondents agreed or strongly agreed that education and/or training is offered on research ethics. One respondent was undecided. This is reinforced with the fact that 15 respondents agreed or strongly agreed with the statement that ongoing research training takes place.
- Likewise, a majority of respondents (15) agreed or strongly agreed that research proposals raising ethical issues (whether or not human or animal subjects are involved) are subject to review.
- A majority of respondents, however, disagreed with the statement that education and/or training is offered on dual use. Nine respondents disagree or strongly disagreed, and seven respondents were undecided about the statement. Only two agreed with it.
- Eight respondents also disagreed with the statement that researchers are aware of and informed about national and international laws and regulations related to their research, and three respondents were undecided. This can be read in conjunction with the fact that 14 respondents were undecided or disagreed with the statement that information about the national and international laws and regulations related to all fields of science is easily accessible.
- In addition, 13 respondents were undecided or disagreed with the statement that whistleblowing mechanisms exist and make provision for the protection

---

22 Ibid.

of whistleblowers. Related to this, 10 respondents were undecided or disagreed with the statement that researchers have somewhere to turn for competent advice if they have ethical, safety or security questions relating to their research. Yet five respondents agreed or strongly agreed with the first statement and eight respondents agreed with the second one.

- Eleven respondents were undecided or disagreed with the statement that adequate mechanisms exist for investigating and responding to non-adherence to ethical standards, and half of the respondents were undecided or disagreed with the statement that appropriate ethical guidelines and practices have been published and implemented. Yet seven respondents agreed with the first statement and nine agreed with the second one.

Discussion

The pilot study at the South African facility illustrates the type of useful and informative feedback the self-assessment questionnaire can provide to laboratory managers and their staff.[23] In this case, the pilot study may seem to indicate that ethics training is an asset of the facility and that research proposals are being regularly reviewed. But it also seems to point out some areas for action.

For instance, disagreements occurred in four areas

- about the education and/or training on dual-use issues
- about their awareness of and information regarding national and international laws and regulations related to their research
- about the existence of mechanisms for staff to report unlawful or irregular conduct (that is, whistleblowing mechanisms) and where to turn for competent advice if they have ethical, safety or security questions relating to their research
- about the publication and implementation of ethical guidelines and practices and mechanisms for investigating non-adherence to ethical standards.

When a majority of respondents disagreed with a statement or was undecided, this may point to a certain weakness that needs to be addressed. In this case, a majority of respondents noted the lack of training and/or education on dual-use issues.

Disagreement among respondents may indicate that there may be measures in place but these are not sufficiently well communicated to all employees. This may mean that ethical standards for research may be in place in the facility,

---

23 Although the pilot study implemented the whole self-assessment questionnaire, only the ethics data are being reported in this chapter.

as well as mechanisms to deal with non-adherence to ethical standards, but the employees should be better informed about them. Likewise, employees may need to have more accessible information about national and international laws and regulations related to their research as well as additional information as to where to turn for competent advice if they have ethical, safety or security questions associated with their research.

In summary, the analysis of the data associated with the ethics pillar tends to suggest that the facility has important capacities in terms of ethics training and on ethical review of research proposals while some needs may have been identified in terms of dual-use training. As well, additional information may be provided on the availability of competent advice; publication and implementation of ethical guidelines and practices and mechanisms for investigating non-adherence to them; and on national and international laws and regulations.

This process of self-assessment should therefore provide useful and practical information for research institutions and their managers about research excellence and management, ethics, biosafety and laboratory biosecurity practices within their facilities. When comparing the results of the three pillars, the institution can also identify capacities and needs for each pillar. It can then assess which pillar looks stronger compared with the others and then identify priority areas for action. One way to address potential needs might be to organise a seminar for all staff to share the data analysis, to underline the strengths of the institution and to propose measures to address identified gaps.

In this case, the identified needs could be addressed through a seminar that would be aimed at sharing information about national and international laws and regulations related to their research and the setting up of a web link to encourage staff to make themselves aware of any changes in national and international regulations. A seminar could also be held on the mechanisms or processes in place in the institution if staff have ethical, safety or security questions regarding their research and to make clear where the ethical standards can be found and how whistleblowing mechanisms can be accessed. Information regarding ethical standards can be distributed to the staff or could be put on a web site easily accessed by the employees. Staff should also be able to keep themselves regularly informed about any changes in ethical procedures. Another way of addressing needs might be to invite other facilities and research institutions to share information about their best practices, how they train their staff on specific topics, how they have identified ways for making information on certain issues available to staff and how to keep them regularly updated. Finally the self-assessment questionnaire could be proposed again to the facility employees after the implementation of activities that were aimed at addressing needs to evaluate the progress made.

Self-assessment should therefore not be seen as an end in itself but as a process for achieving better research management, enhanced ethics practices, the implementation of biosafety and laboratory biosecurity measures and procedures, and their continuous improvement. This would contribute to fostering responsible life-sciences research activities in countries and institutions and promoting public health. The questionnaire was also used in Kenya, to assess the capacity of research and diagnostic laboratories in Nairobi and surrounds.[24]

Ethics is essential to address the risks posed by life-sciences research. The pilot study on the self-assessment questionnaire on responsible life-sciences research illustrates how ethics needs can be identified and what kind of activities can be put in place to address those needs.

---

24 Kenya E. et al., *An Assessment of the Capacity of Research and Diagnostic Laboratories in Nairobi and Surrounds*, June 2012.

Conclusion

20. Ethics as …

Brian Rappert

The contributions to *On the Dual Uses of Science and Ethics* have extended an invitation to dwell on a matter of much importance: the unity of knowledge. Each chapter has considered the potential for the skills, know-how, information and techniques associated with modern biology to serve contrasting ends. Each has spoken to steps that might be undertaken to prevent the deliberate spread of disease. A recurring message has been that, to date, the discussion about the 'dual-use' potential of the life sciences has been characterised by silences and absences. To be sure, while some researchers and policymakers have devoted attention to this topic for years, overall the regard is patchy and problematic.

But as the title of this volume advocates, as we attend to the multiple purposes served by the life sciences, parallel regard should be directed towards those served by ethics. Another theme of *On the Dual Uses of Science and Ethics* has been that, to date, the ethics discussion about the multiple potentials of the life sciences has been characterised by silences and absences. This book has sought to redress the state of analysis by bringing together individuals with varied backgrounds. Some have been trained as ethicists but had given little prior attention to the destructive applications of science, while others had long mulled over this topic but not directly informed by the discipline of ethics. As part of building the debate, the contributors have asked what role ethics (particularly bioethics) might serve in averting the deliberate spread of disease.

Part reflection on the preceding chapters, part analysis of the wider literature, and part programmatic agenda setting, this concluding chapter addresses how the history, methods and practices of bioethics could enrich current dual-use deliberations. This is conceived as a two-way exchange. The queries, uncertainties and anomalies raised by examining the unity of knowledge will also be approached for what they tell us about the preoccupations of bioethics. Through this the limits of bioethics, how it must adapt and what is at stake in suggesting 'change' is required will be considered.

In asking about 'uses', regard will also turn to possible 'abuses' of bioethics. Efforts to bring to bear the intellectual resources from this tradition come with their own dangers. The assumptions informing normative approaches, how they are positioned to justify courses of (in)action and the blind-spots of analysis are just some of the reasons for holding together worries about the direction of ethics and the life sciences. For instance, one of the common ways bioethics has positioned itself as a discipline is as an intellectual response to new innovations in science and technology.[1] Ethicists then set out to offer needed analytical responses to the latest challenges cropping up. Yet, this 'techno-origin myth'[2] risks becoming blind to the organisation, funding priorities, agendas and purposes of ethics.

This chapter sets about to accomplish the above aims through a four-part analysis. The next section begins by scrutinising assumptions informing the contributions to *On the Dual Uses of Science and Ethics* as well as other commentaries. Undertaking this critical turning back is vital if we wish to set a productive program for the future. Section three asks how normative and empirical approaches can be reconciled in efforts to prevent the destructive application of the life sciences. The fourth section considers the potential dangers associated with promoting greater bioethical attention to dual use. Informed by the analysis that precedes it, the final section sets out a path for future intellectual and practical engagement. That path entails attending to the whys and hows associated with what is 'not': a) what is *not* recognised as posing a concern in the first place or, if recognised at some level, b) what is *not* treated as a problem, and c) if regarded as a problem then what is *not* acted upon.

Questioning beginnings

This section begins by undertaking a central task for robust ethical analysis: scrutinising presumptions.

1. Categorical condemnation

The contributors to this volume have begun with a normative position that has itself attracted little attention: the categorical unacceptability of biological weapons. This overall evaluation is codified in the 1972 Biological and Toxin Weapons Convention (BTWC) that forbids states from acquiring or developing bio-agents and toxins for hostile purposes.

Within international diplomacy, policy deliberation and academic study, the attribution of an 'abhorrent' or an 'inhumane' standing to these weapons is commonplace (as in van der Bruggen's chapter in this volume). While some have impugned the political expediency of the categorical prohibition of bioweapons vis-à-vis the laxer controls for nuclear weapons (which still enable

---

1   See Borry, P., Schotsmans, P. and Dierickx, K. 2005, 'The birth of the empirical turn in bioethics', *Bioethics*, vol. 19, no. 1; and Kushe, H. and Singer, P. (eds), *A Companion to Bioethics*, Blackwell, London.
2   de Vries, R. 2007, 'Who will guard the guardians of neuroscience? Firing the neuroethical imagination', *EMBO Reports*, vol. 8, S65–9.

retention by major military powers),[3] biological weapons are rarely portrayed as anything other than illegitimate today. That recent diplomatic accusations about the existence of offensive programs all refer to clandestine ones illustrates this exceptional status. With this now longstanding global rejection in place, little in the way of justification is treated as necessary for condemnation. When argument is given, it often amounts to a restatement of the cardinal assertion that inflicting death and injury to life through life is qualitatively different than other means of harming.[4]

Given this normative starting *ought*, it is hardly surprising that discussion about the possible contribution of the life sciences to the spread of disease is characterised by a binary language of 'use' and 'misuse'. This stark language would be out of place for many elsewhere. The systematic engineering of cross-fertilisation between civilian science and war-fighting capabilities has been a longstanding mission of many national innovation policies since World War II.[5] When Colwell commented that weaponising bio-agents 'violates the fundamental values of the life sciences that I and my colleagues hold dear: that science is a vital tool for improving life and health of our planet and enhancing our understanding of the natural world'[6] in a report for the US National Academy of Science, the reference to the life sciences would no doubt be widely shared in North America and much further afield. Generalising this sentiment against weaponisation to 'science' as a whole would, however, be far out of line with scientific practice.

In the past though, the moral standing of what would today be labelled as biological weapons was more contested. In a wide-ranging analysis, for instance, Zanders argued that the condemnation of biowarfare over the ages has never been uniform or absolute.[7] Within the twentieth century, scientists, medics and others have found ample justifications for their willingness to participate in large-scale programs. Patriotism, civilian spin-offs and even the belief that (at least some) bioweapons represented more humane alternatives were among the

---

3   See Falk, R. 2001, 'The challenges of biological weaponry', in S. Wright (ed.), *Biological Warfare and Disarmament*, Rowman & Littlefield, London.

4   For an examination of the ethics of biological weapons, see, for example: Sims, N. 1991, 'Morality and biological warfare', *Arms Control*, vol. 8, pp. 5–23; and Krickus, R. 1965, 'On the morality of chemical/biological war', *Journal of Conflict Resolution*, vol. 9, no. 2, pp. 200–10. Even in attempts to differentiate the moral standing of types of biological weapons, the starting assumption is often that they all share dubious qualities, as in Appel, J. M. 2009, 'Is all fair in biological warfare? The controversy over genetically engineered biological weapons', *Journal of Medical Ethics*, vol. 35, pp. 429–32.

5   James, A. 2007, 'Science & technology policy and international security', in B. Rappert (ed.), *Technology and Security*, Palgrave, London.

6   Colwell, R. 2010, *Understanding Biosecurity: Protecting against the Misuse of Science in Today's World*, National Research Council, Washington, DC.

7   Zanders, J. P. 2003, 'International norms against chemical and biological warfare', *Journal of Conflict & Security Law*, vol. 8, no. 2, pp. 391–410.

reasons.[8] Prior to the ratification of the BTWC, in the 1960s scientific societies (such as the American Society for Microbiology)[9] found themselves embroiled in bitter dispute about the rights and wrongs of offensive programs.

Recognition of this historical contestation is not taken as an opportunity to reject the denunciation of bioweapons and the corresponding need to avert the life sciences from aiding their development. In this sense, this chapter begins where others in the volume have begun (and ended).

Noting this contestation though will be taken as an opportunity to reflect on the basis for today's commonsense moral appraisals and thus how ethical analysis can be made to matter. In relation to the topic of this volume, that means questioning how norms, stigmas and taboos can be bolstered (see below).

Noting this contestation also will be taken as an opportunity to reflect on how the boundary is set between what is and is not treated as a 'biological weapon'. As Whitby details in his chapter about plant inoculants, at times the distinction can be fine to non-existent. The contingency associated with categorisations is evident elsewhere. During ancient Greek and Roman times, debate about the morality of deliberate poisoning and the contamination of water supplies took place alongside debate about the morality of driving enemy troops into mosquito-infested areas.[10] Each of these represented a sort of weaponising of nature, even if the last would not fit today under the term 'biowarfare'. And as yet another aspect of boundary drawing, while the use of pathogens as weapons to inflict disease might be widely deplored, just when it is deemed appropriate for biology to serve war-fighting is of much less accord. Whitman's chapter indicated the range of bio-enabling capabilities that nanotechnology offers militaries. The moral status of such applications is likely to be a matter of disagreement in a way it would not be for other areas of nanotechnology given the link to biology.

What such examples suggest is that the descriptive matter of how things become *understood* needs to accompany the normative matter of how things ought to be *judged*.

Take the example of bio-defence. As a number of contributors in this volume have indicated, the acceptability of activities undertaken in the name of protection is not a matter of unanimity. In no small part this has been due to disagreement about what distinguishes the knowledge and techniques necessary for 'defence' from those of 'offence'. As I have argued elsewhere, at stake in past disputes about

---

8   Balmer, B. 2002, 'Killing "without the distressing preliminaries"', *Minerva*, vol. 40, pp. 57–75.
9   Cassell, G., Miller, L. and Rest, R. 1994, 'Biological warfare: role of scientific societies', in R. Zilinskas (ed.), *The Microbiologist and Biological Defense Research*, New York Academy of Science, New York.
10   Mayor, A. 2009, *Greek Fire, Poison Arrows and Scorpion Bombs*, Overlook Duckworth, London.

what is ethically and legally permissible have been thorny questions about how the purposes of activities can be discerned. Commentators have been divided on whether purpose can be discerned from the characteristics of the activities or whether their meaning has to be found in some wider 'context'. In practice, for many who subscribe to a contextual approach a sense of 'the context' is often mutually defined in relation to a sense of activities that are under scrutiny. For instance, expansion of American military funding during the 1980s was one occasion that featured heated dispute about the appropriateness of bio-defence. As part of this debate, attempts to evaluate aerosolisation-testing centres were subjected to contrasting assessments due to alternative characterisations about the overall purpose of bio-defence programs. But on top of this:

> [S]uch characterizations, in turn, [informed] determinations of intent. Many critics of the US 'biodefense' program treated the secrecy surrounding it and the wider political posturing of US administrations as indicators of the dubious intent of the undertaking. Seemingly acceptable individual projects were re-considered for how they might inform offensive weapons development. In turn, the number and character of such ambiguous activities provided justification for concerns about the ultimate aims served by the program as a whole.[11]

It is through such—what might well be deemed circular—reasoning that attempts to assess what is permitted often unfold. Herein, the trust held in a presidential administration informed the evaluation of particular projects and vice versa. In recognising such co-definition dynamics, it is possible to go beyond simply noting the recurring difficulties with making 'decisions about whether biodefense research can be ethically justified as truly "defensive" in nature'.[12] Instead it is possible to examine how boundaries and evaluations are established.

The previous paragraph indicated the need for empirically informed ethical analysis in relation to matters of dual use—a theme that will be taken up again in the next section.

2. Neglected dis-ease

As mentioned at the beginning of this chapter, a motivation for this volume— and a conclusion supported by its contributors—was a sense that dual-use issues had not hitherto been subjected to significant or sufficient ethical analysis.[13]

---

11 Rappert, B. 2005, 'Prohibitions, weapons and controversy: managing the problem of ordering', *Social Studies of Science*, vol. 35, no. 2, p. 231.
12 King, N. 2005, 'The ethics of biodefense', *Bioethics*, vol. 19, no. 4, pp. 432–46.
13 As also voiced in Chadwick, R. 2011, 'Bio- and security ethics: only connect', *Bioethics*, vol. 25, no. 1, p. ii.

There are, however, past and present lively areas of deliberation related to the overall themes of *On the Dual Uses of Science and Ethics*. The rights of research subjects have long been an enduring topic of mainstream medical ethics. Human experimentation in weapons programs has more than its share of flagrant violations of rights.[14] Just how much research and healthcare priorities ought to pay to concerns about the deliberate spread of disease have animated much debate recently, too, with a feared 'biosecuritisation' of priorities.[15] Many have contended that the increasing funding directed at security post 9/11 has led to skewed research and healthcare agendas. Herein pathogens that might kill in warfare are given disproportionate resources compared with those that do kill on a daily basis. What suffer are human welfare, global justice and international security.[16] How many lives, it might be asked, have been lost because of the choice to spend scare resources on non-existent threats? In contrast, others have advocated for the increasing inclusion of security within public health and thereby reconceiving what is meant by both.[17] Herein, not only do healthcare systems need to do more to prepare for the intentional spread of disease, but also security agencies need to recognise the gravity of 'natural' disease for national protection.[18]

These points about 'biosecuritisation' suggest how the destructive use of the life sciences overlaps with healthcare preparations for the outbreak of disease. To the extent this is correct, turning the previous 'disconnect' between ethics and dual use into 'connect' should be entirely feasible.

Affinity was a recurring theme of the chapter by Bartolucci and Dando, who examined: 1) what neuroethics has to say about dual use; and 2) how neuroethics has positioned itself as similar/different to other areas of ethics. The first of these is relatively straightforward: neuroethics is notable for its lack of direct engagement with the dual-use aspects of the life sciences. At the same time, much of it could be pertinent. The second dimension is more complex. Debates about distinctiveness have long been at the centre of positioning about whether there is a need for 'neuroethics' as a field of study in its own right. With the increase of specialised publications and conferences, over recent years many

---

14 Harris, S. 1999, 'Factories of death', in T. Beauchamp and L. Walters (eds), *Contemporary Issues in Bioethics*, 5th edn, Wadsworth, London, pp. 470–8.

15 Fisher, J. and Monahan, T. 2001, 'The "biosecuritization" of healthcare delivery: examples of post-9/11 technological imperatives', *Social Science & Medicine*, vol. 72, pp. 545–52; and Brown, T. 2001, 'Vulnerability is universal', *Social Science & Medicine*, vol. 72, pp. 319–26.

16 Enemark, C. 2009, 'Is pandemic flu a security threat?', *Survival*, vol. 51, no. 1, pp. 191–214; and Choffnes, E. 2002, 'Bioweapons: new labs, more terror?' *Bulletin of the Atomic Scientists*, vol. 58, no. 5, pp. 28–32.

17 See Fidler, D. and Gostin, L. 2008, *Biosecurity in A Global Age*, Stanford University Press, Stanford, Calif.

18 See, for example, de Waal, A. 2010, 'Reframing governance, security and conflict in the light of HIV/AIDS: a synthesis of findings from the AIDS, security and conflict initiative', *Social Science & Medicine*, vol. 70, no. 1, pp. 114–20.

have adopted a language of intellectual uniqueness, at least at times. One of the dangers identified with this is that previous bioethical work becomes lost or reinvented as neuroethicists divorce themselves from mainstream ethics.

Stepping back from neuroethics as a specific example, there is good reason for contending that the likeness across topic areas will be tricky to pin down. Establishing the extent of resemblance is practical judgment that must be argued for, rather than simply being resolved once similarities (or differences) are identified. Determining the extent of overlap between topics of ethics requires identifying the key issues and defining how they should be understood. For emerging or disputed topics though—such as dual-use life sciences—these can be the very matters where accord breaks down.

I want to take the past two paragraphs as suggesting that the question of how much the ethical issues associated with dual-use concerns are similar to those elsewhere should be posed while also attending to implications of claims of 'similarity' or 'difference'. So in pondering to what extent a distinctive 'dual-use ethics' is required, the question of 'required for what?' should loom large. Attempts in relation to neuroscience and nanotechnology to stake out a distinctive sub-branch of ethics, for instance, are part and parcel of building professional identity and boundaries. While there was little in the way of calls in this volume to clear a distinct space for dual-use ethics, the complexion of the conversation might well change over time.

The implications of claims to similarity or difference figure in ongoing ethical debates elsewhere. Take the area of public health again. Leslie Francis and colleagues have argued that infectious disease has been largely absent from the historical development of bioethics as a field.[19] As a result, insufficient attention has been given to the way contagious individuals are both patients with illness and vectors for transmitting disease.[20] Consequently, traditional conceptualisations of the role for, composition of and need for informed consent are inadequate. These fail to recognise how choices about treatment or non-treatment made by some affect others. Once this is acknowledged, an individual patient's desire need not be the prime consideration.[21]

Turning to the topic of this volume, such observations are also relevant in the case of the deliberate spread of disease. A point to draw out is the way Francis and colleagues insist that current deficits in bioethics cannot be alleviated by

---

19   Francis, L. P., Battin, M. P., Jacobson, J. A., Smith, C. B. and Botkin, J. 2005, 'How infectious diseases got left out—and what this omission might have meant for bioethics', *Bioethics*, vol. 19, no. 4, pp. 307–22.

20   See as well Tausig, M., Selgelid, M. J., Subedi, S. and Subedi, J. 2006, 'Taking sociology seriously: a new approach to the bioethical problems of infectious disease', *Sociology of Health & Illness*, vol. 28, no. 6, pp. 838–49; and Selgelid, M. 2005, 'Ethics and infectious disease', *Bioethics*, vol. 19, no. 3, pp. 272–89.

21   Likewise, Francis et al. argued past thinking about justice in health care in bioethics failed to acknowledge the global origins of disease in anything like a robust manner.

a minor refinement of bioethical principles (such as autonomy). Instead they argue that a fundamental rethink of the traditional liberal foundations of bioethics is necessary. The principle of autonomy, for instance, needs to break from a fixation on the preferences of competent and reasonable individuals. Instead acknowledgment must be given to how the vulnerability of individuals and communities derives from our embodied and relational existence. I take the points above as indicating that alongside claims about what is or is not missing from bioethics about dual use should be attention to how similarity and difference are being conceived.

Meeting in the middle

The previous section examined two of the starting points for *On the Dual Uses of Science and Ethics*: the normative condemnation of biological weapons and the lack of robust bioethical engagement with dual-use life sciences.

Still another starting point for this volume was the belief that working towards a bioethical engagement requires bringing together contributors from a variety of disciplines. The composition of the chapters reflects that thinking. No single field of study—be it ethics, social science, biology, and so on—could adequately label the previous chapters. Moreover, a mix of aims has been pursued: clarifying concepts and normative justifications, detailing how problems are recognised, charting the state of development in science and technology, and so on. To some extent, these varied activities have aligned with contrasting tasks: prescription, description and assessment.

Another way of making the points above is to say that *On the Dual Uses of Science and Ethics* has sought to clear a space for bioethical engagement within the fraught terrain of empirical ethics. In recent years the umbrella term 'empirical ethics' has signalled the need to bring closer together research questions and methodologies from multiple fields—such as ethics, sociology, bioethics, law, applied ethics, epidemiology and anthropology. The aim has been to couple normative analysis and empirical evidence. That has entailed finding some way of combining contrasting preoccupations. While much of bioethics strives to produce useful guides for making decisions in problematic-choice situations, much of the empirical research in sociology and anthropology attends to how notions of right and wrong are socially formed, policed and reproduced.

The need for coupling the normative and the empirical has been evident in this volume in a number of respects.

- *Identification*: In no small part, the very attention to dual use stems from claims about the pace of development of biotechnology today and its potential relevance for intentional destructive purposes tomorrow;[22] however, as with other issues, the path from potential concern to publicly recognised social problem is not one of necessity. Social, institutional and cultural considerations are the reasons certain topics become treated as serious problems. In the case of the topic of this book, the year 2001 remains of paramount importance.

- *Specification*: Moving from a general recognition of a problem to specifying what is pressing and why are often informed by detailed argumentation. The chapter by Kelle, for instance, delineates the multiple strands of synthetic biology in order to map variations in kind and severity of potentials. As he also argued, much of the policy attention to the governance of synthetic biology has been limited to particular strands of it—most notably, DNA synthesis. The reasons for and the implications of such narrow framings are part of what empirical research needs to establish.

- *Appreciation*: Part of ensuring that sensible and appropriate measures are devised is understanding how practitioners—such as laboratory researchers and managers—assess the possibilities of their work. As Connell so vividly and honestly shares with readers, even those with longstanding experience in preventing the deliberate spread of disease can lose sight of the implications of their own work. Understanding the organisational and social reasons for such blindness is vital in knowing how to respond. As Chambliss has warned, everyday routines in organisations often delimit ethical scrutiny.[23]

- *Contextualisation*: Benzuidenhout's chapter explored how recognition and responses to dual-use issues are likely to vary systematically due to research environments. Without regarding information about the day-to-day experiences of those in labs, proposals for what needs to be done can be compelling in the abstract but practically irrelevant.[24]

- *Evaluation*: Assessing the measures undertaken to prevent the hostile use of the life sciences requires empirical data.[25] Without this, action can be misplaced. For instance, the review of civilian research experiments, science journal manuscripts and public grant proposals for their security risks has excited much controversy.[26] As the most prominent instance of this, in 2003,

---

22   See Rappert, B. 2007, *Biotechnology, Security and the Search for Limits: An Inquiry into Research and Methods*, Palgrave, London, ch. 1.

23   Chambliss, D. F. 1996, *Beyond Caring: Hospitals, Nurses, and the Social Organization of Ethics*, University of Chicago Press, London.

24   'Context' should not, however, simply be thought about as a static backdrop directing action. Financial and organisational resource constraints motivate the search for different forms of research collaboration that, in turn, structure possibilities for perception.

25   Douglas, T. and Savulescu, J. 2010, 'Synthetic biology and the ethics of knowledge', *Journal of Medical Ethics*, vol. 36, pp. 687–93.

26   King, op. cit., pp. 432–46; Frisina, M. 2006, 'The application of medical ethics in biomedical research', *Cambridge Quarterly of Healthcare Ethics*, vol. 15, pp. 439–41; and Tyshenko, M. 2007, 'Management of natural and bioterrorism induced pandemics', *Bioethics*, vol. 21, no. 7, pp. 364–9.

32 prominent (Western) journal editors committed themselves to enact peer-review procedures to assess the risks and benefits of individual manuscripts. These were meant to determine whether articles needed to be modified or withdrawn because 'the potential harm of publication outweighs the potential societal benefits'.[27] One concern frequently expressed has been that such measures would jeopardise the free flow of information that is vital to civilian science. Such fears though are arguably misplaced for two reasons

1. experience from at least 2004 has indicated that the risk–benefit logic of the formal review processes means that few manuscripts, grant applications or experiment proposals are identified as posing dual-use concerns (let alone are censored; there are no cases of civilian work being categorically withheld)[28]

2. evidence from social studies of science has illustrated how the exchange of resources and information in research is frequently subject to negotiation in practice.[29]

Such observations raise concerns about the ultimate purposes and prospects of formal oversight procedures. They might either miss important developments or impose needless layers of bureaucracy.

- *Operationalisation*: Policy initiatives can also be misjudged if they misconstrue the unit at which ethical decisions are taken and moral standards enforced. As Miller argues in Chapter 12, reducing ethical concerns to individual decision-making would be fundamentally flawed. Science must be treated as a group and community enterprise.

The need for ethical analysis informed by empirical research is readily acknowledged today in bioethics—indeed, many contend it has always been so.[30] The rub is not with *whether* 'empirical' and 'ethics' can be placed side by side, but rather *how*. The central difficulty is that of bringing together 'what is' and 'what ought to be'. The 'facts' of some issue—for instance, how those in a lab regard their professional responsibilities—do not resolve what should be the case.[31] Clearly, it is not possible to justify standards for morality on empirical data alone, even if in practice 'what is' often shapes notions of 'what ought to be'.

---

27 Journal Editors and Authors Group 2003, *Proceedings of the National Academy of Sciences*, vol. 100, no. 4, p. 1464.
28 See Nightingale, S. 2011, 'Scientific publication and global security', *JAMA*, vol. 306, no. 5, pp. 545–6; and Rappert, B. 2008, 'The benefits, risks, and threats of biotechnology', *Science & Public Policy*, vol. 35, no. 1, pp. 37–44.
29 Rappert, 2007, op. cit., ch. 2.
30 Herrera, C. 2008, 'Is it time for bioethics to go empirical?' *Bioethics*, vol. 22, no. 3, pp. 137–46; and Hurst, S. 2010, 'What "empirical turn in bioethics"?' *Bioethics*, vol. 24, no. 8, pp. 439–44.
31 See Bosk, C. 2000, 'The sociological imagination and bioethics', in C. Bird, P. Conrad and A. Fremont (eds), *Handbook of Medical Sociology*, Prentice Hall, Upper Saddle River, NJ, p. 403.

Recognising this though is just the start of unpacking how empirical study can contribute to the normative. As de Vries and Gordijn argue, 'the conclusion of an empirical-ethical inquiry is not necessarily a moral judgment or principle. Nor are the conclusions of these studies necessarily based solely on empirical results.'[32] In other words, the tasks undertaken as part of empirical ethics are many and varied. Seeking some way to move from description to prescription is just one possibility. Others include detailing conduct, understanding conditions and processes of meaning making, acknowledging alternative ethical reasoning, testing the feasibility of moral precepts, and refining ethical theory.

Ethics for ... and for ...

One succinct gloss of the previous sections would be this: ethics is contested space and a contested label. Whether some concern is treated as 'ethical' and how, for instance, are part and parcel of the way issues become identified as problems that demand action. With this recognition, the vocabulary of 'use' and 'abuse' requires scrutiny. The previous sections included consideration of how framing the potential of the life sciences in terms of destructive and peaceful purposes is indebted to particular ways of thinking.

This section takes the 'use and abuse' in a different direction. Instead of focusing on the *life sciences*, it turns to the contrasting potentials of *bioethics*. So rather than just treating research and innovation as the sources of problems to which bioethics offers some sort of antidote, this section asks how pursuing ethical analysis is associated with dangers. If dosage can be the only thing that distinguishes a medicine from a poison then the same applies to ethical prescriptions. Each of the possible positive utilities for bioethics given below is questioned as to how it contains the seeds for dubious outcomes. In this sense, seeking to 'input' bioethics into current dual-use debates is treated as dilemmatic.[33]

Advising

Bioethicists are often called upon to provide advice in controversies. The 2010 *New Directions: The Ethics of Synthetic Biology and Emerging Technologies*[34] report of a US presidential commission is one such instance. It recommended various

32   de Vries, R. and Gordijn, B. 2009, 'Empirical ethics and its alleged meta-ethical fallacies', *Bioethics*, vol. 23, no. 4, pp. 193–201.
33   Billig, M. 1996, *Arguing and Thinking*, Cambridge University Press, Cambridge; and Billig, M., Condo, S., Edwards, D., Gane, M., Middleton, D. and Radley, A. 1989, *Ideological Dilemmas*, Sage, London.
34   Presidential Commission for the Study of Bioethical Issues 2010, *New Directions: The Ethics of Synthetic Biology and Emerging Technologies*, US Department of Health and Human Services, Washington, DC.

reforms to oversight and reporting procedures for research that were said to be congruent with widely shared ethical principles. Through such advice giving, ethics has a chair at the table of high-level public policy.

The need for considered ethical advice has been made before. Green called for a portion of the multi-billion-dollar per year funding for bio-defence in the United States to be set aside to examine its ethical, legal and social implications (ELSI).[35] This recommendation was based on the precedent of the Human Genome Project. And yet, despite seeking to build on this example, Green also acknowledged various problems with the Human Genome Project ELSI program: its policy consequences, coordination and focus.

At a more theoretical level, attempts to advise can be misguided if they are based on a faulty understanding of how analysis—be it ethical or empirical— is made to matter. In an idealised model of policy, analysis acts as an input into rational decision-making processes through establishing facts and selecting between options. Medical ethics and bioethics are arguably particularly aligned with this 'inputting' model given the centrality placed on decision-making. In contrast, students of policymaking have forwarded a complex path between knowledge and action, one characterised by iterative movements without clearly demarcated beginnings or endings. Indeed, some have gone so far as to question whether it would even be desirable for policymaking to be solely or largely based on analysis.[36] Instead of portraying it as some sort of authoritative input, the role of analysis has been defined in more limited terms—such as helping to identify concerns not widely recognised.

Therefore, a danger with greater bioethical scrutiny about dual use would be that it invests a greater significance in the former than it can bear. In practice, the arguments of ethics might well give a false air of justification to decisions taken for altogether different reasons.[37] Another potential problem is that of misconstruing the nature of the choices faced. Hoffmaster and Hooker have argued ethical dilemmas are often radically open-ended and indeterminate. Herein,

> not only is the solution unknown, but the problem itself is initially not well defined, and the values that ought to drive its investigation and the

35 Green, S. 2005, 'E3LSI research: an essential element of biodefense', *Biosecurity and Bioterrorism*, vol. 3, no. 2, pp. 128–37.
36 Lindblom, C. and Cohen, D. 1979, *Usable Knowledge*, Yale University Press, New Haven, Conn.; Palumbo, D. and Hallet, M. 1993, 'Conflict versus consensus models in policy evaluation and implementation', *Evaluation and Programme Planning*, vol. 16, pp. 11–23.
37 For a consideration of this point, see Winner, L. 1992, 'Citizen virtues in a technological order', *Inquiry*, vol. 35, nos 3–4, pp. 341–61.

valid methods to do so are unknown, unclear, or in dispute, as are the set of applicable theoretical models, the solution set, and the criteria for successful resolution.[38]

The danger of trying to 'apply' bioethics to determine the proper course of action in such situations is that the conventional concepts of bioethics can provide a sterile and static understanding of the issues at stake. Any recommendations that follow then are likely to have a stunted potential.

Guiding

If bioethics might not be thought of as wielding the decisive hand in policymaking, a more modest goal would be to suggest that it provides guides for assessing the morality of acts. Along this line, some have sought to derive general guidance for scientists and research organisations from ethical principles.[39] Kuhlau and colleagues, for instance, outlined various criteria for identifying 'harm' within the moral responsibility of scientists.[40] In line with the ethical principle of non-maleficence, for instance, they argued that '[r]esearchers should be responsible not only for not engaging in harmfully intended activities but also for research with harmful implications that they can reasonably foresee'.[41]

The ability of ethics to derive moral obligations has been a topic of debate since its inception as a field of study. Much of that discussion has turned on the prospects for high-level principles to inform situated action. A frequent refrain against principlism is that it is flawed because what it means to adhere to a principle (such as autonomy or justice) is always indeterminate at some level. As instances of ethical concern are never identical, the future application of normative standards cannot be set out once and for all. Individuals must manage the relevance of principles, what it means to follow or deviate from them, and what consequences are likely to follow from violations. For instance, while the obligation to 'prevent harm' often figures as an ethical bar to bioweapons development, this prescription does not have the same import for conventional (read: commonly accepted) forms of weaponry.[42] Enabling harm is a goal in many areas of research. As such, dual-use discussions are often coloured by a normative starting orientation about which principles matter and why, rather than deriving a sense of the right course of action from the principles themselves.

---

38   Hoffmaster, B. and Hooker, C. 2009, 'How experience confronts ethics', *Bioethics*, vol. 23, no. 4, pp. 214–25.

39   See Ehni, H. J. 2008, 'Dual use and the ethical responsibility of scientists', *Archivum Immunologiae Et Therapiae Experimentalis*, vol. 56, pp. 147–52; and Green, S., Taub, S., Morin, K. and Higginson, D. 2006, 'Guidelines to prevent malevolent use of biomedical research', *Cambridge Quarterly of Healthcare Ethics*, vol. 15, pp. 432–9.

40   Kuhlau, F., Erikson, S., Evers, K. and Höglund, A. 2008, 'Taking due care: moral obligations in dual use research', *Bioethics*, vol. 22, pp. 477–87.

41   Ibid., p. 481.

42   Rappert, B. 2012, *How to Look Good in War*, Pluto Press, London, ch. 6.

A concern with principlism then is that it has failed to acknowledge this indeterminacy in favour of adopting a sense of certainty and inevitability. In response, leading proponents have acknowledged the need for the specification and interpretation of principles in relation to specific situations.[43] Yet, whatever the sophistication and subtlety within academic texts about how to do these tasks, some have voiced concern about the blunt way in which principles figure within day-to-day institutional practice.[44]

Deliberating

A different way of thinking about the utility of ethics is to focus on process. If general ethical analysis has a limited ability to definitely determine the proper course of action then it can help structure democratic deliberation.[45] Herein, as individuals schooled in conceptual analysis and argumentative logic, ethicists can ensure the quality of discussions (even if they do not occupy elevated moral positions). This procedural expertise might be exercised through clarifying reasoning, ensuring consideration of neglected topics, making hidden values explicit, providing comparative examples, and so on.

Each of these notionally 'procedural' contributions though could be questioned regarding how they import in (unrecognised) value commitments. The prescriptive and procedural dimensions of bioethics are exemplified in current disputes about the merits of 'the precautionary principle'.[46] Any discussion of this notion needs to begin with the recognition that it comes in a multitude of versions. While in general terms precaution is aligned with not requiring conclusive demonstration of harm to prompt concern or even action, many formulations of it exist.[47] It is possible, for instance, to distinguish between

43   Beauchamp, T. 1995, 'Principalism and its alleged competitors', *Kennedy Institute of Ethics Journal*, vol. 5, no. 3, pp. 181–98.

44   Marshall, P. 2001, 'A contextual approach to clinical ethics consultation', in B. Hoffmaster (ed.), *Bioethics in Social Context*, Temple University Press, Philadelphia.

45   Boniolo, G. and di Fiore, P. P. 2010, 'Deliberative ethics in a biomedical institution: an example of integration between science and ethics', *Journal of Medical Ethics*, vol. 36, pp. 409–14.

46   In relation to how the prescriptive and the procedural meld, a concern for some is that attention to procedure is a means of decision-making, at least by default. A criticism made of precautionary approaches—but one that could be made of any attempt to promote and conduct deliberation—is that it ends up significantly delaying decisions, and that such a delay thereby favours some over others. It is notable in this regard that precaution has become a byword for inaction and hesitancy in many areas of science policy. See Ledford, H. 2011, 'Hidden toll of embryo ethics war', *Nature*, vol. 471, p. 279.

47   Stirling, A. 2008, 'Science, precaution, and the politics of technological risk converging: implications in evolutionary and social scientific perspectives', *Annals of the New York Academy of Sciences*, vol. 1128, pp. 95–110.

argumentative, process-orientated kinds that establish guidelines for what sorts of arguments are legitimate, and those prescriptive decision-orientated kinds that resolve what action should be taken.[48]

As a *decision rule* for making political deliberations, this principle has its detractors. While claiming the need to safeguard against harms in the face of uncertainty and ignorance is reasonable, specific enactments of precaution are seen as going too far. Clarke's chapter, for instance, offers a sustained critique of the so-called 'strong version' of the principle that exclusively considers potential costs of a particular action (such as the 1994 *Final Declaration* of the First European 'Seas at Risk' Conference). Within this way of thinking, any doubt about severe consequences is treated as providing adequate grounds for stopping an activity from going ahead—the result being a paralysing conservatism that can create larger risks than those forgone.[49]

Whatever the merits of formulations of the precautionary principle for making decisions, as *process* guides they can structure how troublesome questions are approached. For instance, with whom the burden of proof rests to substantiate claims of harm or benefit, to whom and with what level of certainty are important issues for any topic of controversy. Box 20.1 details some of the procedural proposals for how evidence and onus could figure in responses to the destructive potential of the life sciences. The relative merits of these alternatives could be further informed by ethical analysis regarding their procedural dimensions.

Box 20.1 Handling Risks and Uncertainty in the Review of Research

The National Science Advisory Board for Biosecurity (NSABB) was in large part formed following the recommendations of the US National Academies report *Biotechnology Research in An Age of Terrorism*. A central task of NSABB is the development of recommendations on 'guidelines for the oversight of dual-use research, including guidelines for the risk/benefit analysis of dual-use biological research and research results'[a] for the US Federal Government. The guidelines for risk/benefit analysis and oversight represent attempts to define, evaluate and handle concerns about the dual-use potential of research through the creation of formal bureaucratic procedures.

The split NSABB has offered between research that might have some sort of dual-use potential and that which is 'of concern' has been of paramount importance. For the board, the term 'dual-use research' is used 'to refer in general to legitimate life sciences research that has the potential to yield information that could be misused to threaten public health and safety and other aspects of national security such as agriculture, plants, animals, the environment, and materiel'.[b] In contrast, 'dual use research of concern' refers to the 'subset of life sciences research with the highest potential for yielding knowledge, products, or technology that could be misapplied to threaten public health or other aspects of national security'.[c]

---

48 Sandin, P., Peterson, M., Hansson, S., Rudén, C. and Juthe, A. 2002, 'Five charges against the precautionary principle', *Journal of Risk Research*, vol. 5, no. 4, pp. 287–99.
49 See as well Harris, J. 2001, 'Introduction: the scope and importance of bioethics', in J. Harris (ed.), *Bioethics*, Oxford University Press, Oxford; and Douglas and Savulescu, op. cit.

Because it is imagined that few experiments will need to be given security review, the emphasis has been with devising a non-demanding 'tick-box' first stage that should exclude the majority of research from further formal consideration.[d] In this regard, NSABB has proposed that the initial review of whether or not research is 'of concern' be undertaken by the principal investigator (that is, the senior project leader). Herein, this person would ask of their work, 'based on current understanding, can [it] be reasonably anticipated to provide knowledge, products, or technologies that could be directly misapplied by others to pose a threat to public health and safety, agricultural crops and other plants, animals, the environment, or materiel'.[e]

To state that assessors must be able to *reasonably anticipate* a direct threat based on current understanding sets a high threshold for proof. At this initial stage of the review process, the determination of the status of research is not intended to impose significant demands on principal investigators. Should research be found to match the criterion then it would be subjected to institutional risk review.[f]

Such an approach can be contrasted with an alternative oversight model proposed by the Center for International and Security Studies at Maryland (CISSM). This is envisioned as an international legally binding system requiring the licensing of personnel and research facilities. The Maryland system also involves independent peer review. An oversight body needs to approve work going ahead, rather than the investigators making the initial determination. This was justified on the basis that '[i]n addition to having a self-interest in seeing their research proceed, such individuals are also unlikely to have the security and other expertise necessary to recognize the possible dual use risks of their work'.[g]

The criteria proposed as part of the risk–benefit analysis in the Maryland system also go further than the NSABB proposal. As part of assessing research, for instance, individuals are required to consider whether the same experimental outcome could be pursued through alternative means, whether the research is being done in response to a validated (credible) threat, and whether it will yield results definitive enough to inform policy decisions. Such questions place additional demands on those taking part in the assessment process to those as part of NSABB recommendations. They also require forms of knowledge that the average principal investigator is unlikely to posses. As another contrast to the NSABB proposals, the Maryland one provides a metric for evaluating research based on the responses given to the criteria mentioned.

At the heart of such alternative policy options is the matter of expertise and how this should be exercised. While NSABB devolves much of the decision-making down to senior individual scientists who are aided by others, the CISSM proposal places much more emphasis on a diverse range of expertise structured through mandatory requirements.

---

[a] *Charter—National Science Advisory Board for Biosecurity*, 16 March 2006, p. 1.

[b] National Science Advisory Board for Biosecurity (NSABB) 2007, *Proposed Framework for the Oversight of Dual Use Life Sciences Research: Strategies for Minimizing the Potential Misuse of Research Information*, National Science Advisory Board for Biosecurity, Bethesda, Md, p. 4.

[c] Ibid., p. 16.

[d] As in comments made during National Science Advisory Board for Biosecurity, 20 March 2006.

[e] NSABB, 2007, op. cit., p. 17.

[f] Ibid., Appendix 4.

[g] Harris, E. 2007, 'Dual use biotechnology research: the case for protective oversight', in B. Rappert and C. McLeish (eds), *A Web of Prevention: Biological Weapons, Life Sciences and the Governance of Research*, Earthscan, London, p. 120.

Legitimating

If legitimacy is a function of being in accordance with rules and procedures justified by shared societal beliefs[50] then one of the roles often sought of ethics is to legitimate policies and practices. Whether through contributing expert moral reasoning or facilitating deliberation, bioethicists (and others) are often called up to ensure support for core social institutions. An often-voiced reservation with such a purpose is that the basis for accord might be less than warranted or genuine. In other words, rather than legitimacy being positively secured, it is skilfully manufactured through a language supplied by ethics.

Chambliss, for instance, offers a highly critical evaluation of how medical ethics and bioethics training figure within hospital practice. One of the starting points for this is the view that organisational hierarchies are often the source of the dilemmas experienced by professionals.[51] Nurses—with relatively little formal power in relation to doctors or administrators—often struggle with the tension between doing what they think is right and doing what conforms to official policy. As a result, it is not enough for them to receive high-minded instruction about moral principles in order for them to do what is right. For Chambliss, ethical training in the form of abstract, principle-based talk cannot only be irrelevant to lived experience, but it can also serve to give a false impression that dilemmas and challenges are being acknowledged and addressed while, in practice, ethics instruction reinforces relations of hierarchy. Winner too cautions against the way moral categories and ethical arguments are often forwarded without attention to the roles and institutions needed to enable notions of the good to translate into deeds.[52] In the absence of such opportunities, ethical analysis all too easily validates the status quo.

More subtly than this, ethics can also act to reinforce the perception of shared beliefs and values. Take the case of the previously mentioned dual-use review procedures initiated by civilian journals, funders and research organisations since 2004. While they differ in specifics, each proposes a risk–benefit calculation. Expected societal gains from research are to be measured against possible security threats, the balance between the two indicating what should be done. As such, the review procedures are forwarded as embodying core characteristics of rational decision-making. Ethicists have adopted and endorsed this rationalistic framework of balancing risks and benefits.[53]

---

50 To paraphrase Green, P. and Ward, T. 2004, *State Crime*, Pluto, London, p. 3.
51 Chambliss, op. cit.
52 Winner, op. cit.
53 See, for instance, Krohmal, B. and Sobolski, G. 2006, 'Physicians and the risk of malevolent use of research', *Cambridge Quarterly of Healthcare Ethics*, vol. 15, pp. 441–4; and Kuhlau et al., op. cit.

What has not been elaborated is how risk could be calculated (see by way of contrast the aforementioned CISSM system for an alternative framework). Certainly no form of detailed calculus has been set out. Even putting to the side a demand for exactitude, it is not at all clear, for instance, how reviewers could specify the risks from the destructive use of fundamental scientific knowledge. The users, situations, time frame, specific contribution of that knowledge, and so on, are not well defined. The absence of significant past experience of the deliberate spread of disease that might at least provide data points for assessing risks and benefits also frustrates undertaking reviews. In the exceptionally few instances of experiments where some science bodies have judged risks in excess of benefits—as in the initial (but subsequently revised) NSABB recommendations for publication redactions with research on the transmission of a modified H5N1 virus in a ferret model—detailed cost–benefit calculations have not been forwarded. Instead, general appeals to notions such as precaution have been given.[54]

Perhaps most fundamentally, what positive values or negative implications might be relevant for weighing are not clear. From one defensive standpoint, it might be vital to identify and publish research that raises concerns (so as to confirm such fears and to devise countermeasures). This would seem to be the rationale that informed the Defense Advanced Research Projects Agency decision to fund the synthesis of poliovirus.[55] In Nancy Connell's chapter, similarly, demonstrating hyper virulence in a mouse model was seen as providing the grounds for future follow-on funding. Thus, what should count as a 'problem' is not simply poorly defined, but also open for radically opposed interpretations in the first place.

Legitimation dangers do not just include giving undue credence to certain outcomes. Another danger is perpetuating faith in the presumptions, beliefs and competencies that underpin the proposal for what should be done. Rockel, for instance, has offered a trenchant critique of the doctrine of double effect as applied to modern warfare.[56] As contended, its underpinning distinctions related to foreseeable effects and intentions have provided the material for papering over systematic deficiencies in military action. In relation to the themes of this volume, it should also be noted that within the discussion of risks and benefits, the non-destructive applications are often assumed to fall wholly on the plus side. This thinking is in line with many public portrayals of science; however, the extent to which biomedical research is linked to improvements in human

---

54 Cohen, J. and Malakoff, D. 2012, 'NSABB members react to request for second look at H5N1 flu studies', *Science*, 2 March; and Imperiale, M. 2012, Presentation to 'Dual-Use Research and Biosecurity: Implications for Science, Governance and the Law', The Hague, 12 March.

55 Selgelid, M. and Weir, L. 2010, 'The mousepox experience', *EMBO Reports*, vol. 11, pp. 18–24, <http://www.nature.com/embor/journal/v11/n1/full/embor2009270.html> (viewed 23 April 2003).

56 Rockel, S. 2009, 'Collateral damage: a comparative history', in S. Rockell and R. Halpern (eds), *Inventing Collateral Damage: Civilian Casualties, War, and Empire*, Between the Lines Press, Toronto, pp. 1–96.

health can be questioned. It can even be counterproductive when it 'obscures socioeconomic reasons for health problems [and] creates boundar[ies] to other types of action that are more effective, efficient, and equitable'.[57]

In considering the legitimating role of ethical analysis, it should be borne in mind that the designation of a matter belonging to the domain of 'ethical' (or not) is part of what needs to be examined. As a case in point, Houtepen examined how controversy about euthanasia in the Netherlands shifted over time.[58] Prior to the late 1960s physicians debated what should be done largely among themselves. It was only in the late 1960s that others found a recognised voice. Part and parcel of this was the explicit redefinition of euthanasia as a matter of 'ethics', and, as such, a matter where varied perspectives from the public at large had to be brought in. Later, with the introduction of formalised clinical routines and policies informed by ethicists, the overt 'ethical' framing of practice waned. Beyond this specific case, what is deemed 'technical' versus 'ethical' is a product of social negotiation that can readily work to exclude some voices from being recognised. To what extent the dual-use review of research activities is deemed a matter of 'ethics' (which it largely has not been to date) is tied to who needs to conduct such reviews. Whitman makes a related point in this volume in recounting the consequences associated with how presenting issues as 'dilemmas' refracts our understanding of what is at stake.

Stigmatising

As contended previously, the categorical condemnation of bioweapons is historically contingent and collectively produced. It is because of—not in spite of—this thoroughgoing social basis that resistance would be offered to attempts to sanction the employment of life-sciences knowledge and techniques for destructive ends.[59] Looking towards the future, given the numerous possibilities for the malign applications outlined within the pages of this volume as well as the difficulties of trying to enforce the prohibition through national security and policing measures, stigmatisation is likely to be essential.

Taking this to be the case—and agreeing with the need to work against the deliberate spread of disease—implies a certain agenda for bioethics: it should work to find ways of strengthening and renewing stigmatisation. This needs to take place in a manner sensitive to different possible belligerents: states, sub-state groups and lone individuals. How to prevent current efforts to develop

---

57   Sarewitz, D. 1996, *Frontiers of Illusion*, Temple University Press, Philadelphia, p. 150.
58   Houtepen, R. 1998, 'The social construction of euthanasia and medical ethics in the Netherlands', in R. de Vries and J. Subedi (eds), *Bioethics and Society*, Prentice Hall, Upper Saddle River, NJ, pp. 117–44.
59   In other words, the categorical nature of the prohibition in the BTWC and the 1925 Geneva Convention is laudable not because it reflects an objective and 'essential' truth, but rather because of the choices that buttress it.

next-generation 'incapacitants' (as discussed in the chapter by Crowley) from leading to a wider normalisation of drugs as weapons, for instance, should be a high priority.

While norms, stigmas and taboos have been subject to significant consideration within the field of international relations in recent decades, much of that has been of a classical, scholarly variety. Herein the relevance of positive normative positions on the part of that writing is downplayed. By and large, international relations scholars have sought to explain the formation of norms rather than elaborating the practical skills necessary for bringing about reform. Therefore, bioethics in its more applied forms could offer significant contributions to the future of the prohibition against bioweapons.

Seeking such positive engagements for bioethics does not amount to promoting blind faith in existing moral standards. This chapter has not sought such a blind faith, even if it is always in danger of taking for granted certain normative positions; however, doing so does require challenging the conventional way stigma is regarded. In fields such as public health, stigma is routinely associated with acts of prejudice and discrimination.[60] HIV/AIDS would be a classic example of where stigma leads to negative consequences. As a result, attempts to make some scope for it as a tool in public health have been subjected to heated criticism.[61] Of course, one of the things that distinguishes talk of stigmatisation in public health from international relations is the typical object of study: those with illness versus state functionaries. Adopting the latter as the bearers of negative distinction might well not animate fears about the victimisation.

Seeking to employ stigma within the international community in relation to preventing the use of biological weapons comes with its own dangers though. One is that the prohibition of these weapons is not neutral vis-a-vis the power relations between nations. Insisting on the outright objectionable status of one type of weapon while patchy controls exist for many other weapons—and doing so in a world with starkly unequal distributions of power—serves some more than others. Moreover, the shunning and integrating dynamics associated with the stigma raise concerns about the importance attached to renunciation. Undoubtedly, Libya's abandonment of its 'weapons of mass destruction' programs in 2003 helped secure a large measure of re-entry into the international community. In the hindsight of 2013, the justifications for what was bought from this act of disarmament seem questionable.

---

60 Meyer, I. and Stuber, J. 2008, 'Stigma, prejudice, discrimination and health', *Social Science & Medicine*, vol. 67, pp. 351–7.
61 See Bayer, R. 2008, 'Stigma and the ethics of public health: not can we but should we', *Social Science & Medicine*, vol. 67, pp. 463–72; as well as follow-on commentaries such as Burris, S. 2008, 'Stigma, ethics and policy: a commentary on Bayer's "Stigma and the ethics of public health: not can we but should we"', *Social Science & Medicine*, vol. 67, pp. 473–5.

Educating

Who needs to be educated, about what, how and by whom are longstanding matters of commentary in ethics. The charge that those involved in medicine and the life sciences are somehow lacking with regard to an appreciation of the implications of their work is hardly unique to the topic of this volume.[62] Yet moving from such an appraisal to proposals for what needs to be done often proves contentious. Approaches to ethics tuition differ—notably, between prescriptive, procedural and virtue-based varieties. Each of these is aligned with distinct ways of thinking about individuals as moral agents, what count as appropriate learning techniques and how value disagreement ought be handled.

Elsewhere I have considered the dilemmas, tensions and pitfalls of education about the destructive use of life science.[63] In this subsection I want to extend that work by placing the issue of education within a wider political framework. Cribb offers a valuable inroad into this by distinguishing types of health education.[64] For him, a *medical* model treats education as a way of achieving health outcomes through providing information that affects patient behaviour. A danger with this is that healthcare workers assume a highly paternalistic role. Against the medical model another would be to conceive of education as enabling people to make *informed choices* based on their own values and preferences. A danger with this model is that it treats individuals' values and preferences as deriving only from them as autonomous agents. An *empowerment* model, by contrast, starts with asking about the factors that constrain individuals from realising their preferences and then envisions education as part of overcoming those barriers. In other words, the emphasis is with change rather than edification for its own sake. A *social-action* model goes one step further by asking what sort of structural changes in society (for instance, with regard to poverty and social welfare) are necessary to achieve sought-for health gains. As with the empowerment model, this one necessarily involves posing wider questions about what needs reform. Individuals and groups require skills for participation to affect change. In endorsing the social-action model, as with Chambliss, Cribb counsels against divorcing education and training from institutional and organisational conditions.

Such a typology offers a way of classifying the work that has taken place to date with regard to dual-use education. The overwhelming orientation has been in

---

62   As, for instance, in the case of Sales, C. and Schlaff, A. 2010, 'Reforming medical education: a review and synthesis of five critiques of medical practice', *Social Science & Medicine*, vol. 70, pp. 1665–8.

63   See Rappert, B. 2007, 'Education for the life sciences', in Rappert and McLeish, op. cit.; and Rappert, B. 2010, 'Introduction: education as ...', in B. Rappert (ed.), *Education and Ethics in the Life Sciences: Strengthening the Prohibition of Biological Weapons*, ANU E Press, Canberra.

64   Cribb, A. 2005, *Health and the Good Society: Setting Healthcare Ethics in Social Context*, Oxford University Press, Oxford.

line with the medical model. Herein, researchers or the public are expected to take on board messages about the potential of the life sciences. The intention is to help achieve certain thinking or behaviour—such as the competency to identify research of concern. Situated within the international community of states, the danger of paternalism typically associated with this model is compounded by that of neo-colonialism. With much of the recent attention to dual use emanating from North America and Europe, an obvious concern is that the agenda as well as the ethical approaches employed to understand it (for example, individualist and principle based) are indebted to strains of Western thinking.[65] With the effort dedicated to 'education as instruction', much less attention has been given to 'empowerment' or 'social-action' models. Exceptions include the *Kampala Compact: The Global Bargain for Biosecurity and Bioscience* and related *DNA for Peace* report.[66] Both proposed holding together biosecurity measures with social/international development agendas. Brian Martin, too, spoke of the importance of individuals and group empowerment in whistleblowing and dual use.[67] In the absence of such practical skills training, a danger of education is that it does not enable positive reform. Such an outcome can lead to feelings of irrelevance, indifference or frustration on the part of educators and the educated.

A 'non-' research agenda

The previous sections have counselled the need for caution regarding the commitments of our analysis: its starting points, the use of evidence and argument, and the purposes to which it is put. In more or less direct ways, in those sections I have suggested that among the prime challenges for ethics include identifying moral issues and formulating them as problems in need of redress.[68] Both are inextricably tied to processes of categorising, labelling and boundary-setting that help define but are also defined by social routines, institutions and structures.

This section considers how these goals of identification and formulation can be taken forward in a way that enriches bioethical engagement with dual-use life science. The starting move in this is a shift, in a sense, backwards. Rather than suggesting detailed engagement with this or that topic of controversy, the proposal is to attend to the why and the how of what is 'not': what is *not*

---

65   For a more general discussion of this danger, see Widdows, H. 2007, 'Is global ethics moral neo-colonialism? An investigation of the issue in the context of bioethics', *Bioethics*, vol. 21, no. 6, pp. 305–15.
66   Available at: <http://www.utoronto.ca/jcb/home/documents/DNA_Peace.pdf> (viewed 4 April 2010).
67   Martin, B. 2007, 'Whistleblowers: risks and skills', in Rappert and McLeish, op. cit.
68   In line with Clouser, K. D. 1978, 'Medical ethics: some uses, abuses, and limitations', *New England Journal of Medicine*, vol. 293, pp. 384–7; and Hoffmaster, op. cit.

recognised in the first place or, if recognised at some level, what is *not* treated as a serious problem; if regarded as a problem then what is *not* acted upon. In short, the subject for scrutiny is the one of what isn't happening.

A rereading of the Australian IL-4 mousepox experiment illustrates the varied relevancies of the 'non-': the potential for IL-4 to enhance virulence was recognised prior to the mousepox publication by some experts, yet seemingly it was not subject to much professional (let alone public) debate. In this respect, what distinguishes the 2001 mousepox controversy from the majority of other work with a dual-use potential is that the researchers actively voiced their concerns beyond a closely knit expert coterie. Although counterfactual, it seems doubtful that this work would have garnered anything like the same attention in the absence of Ramshaw's communication with *New Scientist*. The fraught path to media agenda item is suggested by what happened subsequently. Despite the profile of the researchers from the mousepox experiment, 9/11, the anthrax attacks and much else besides, the publication of results of a follow-on study indicating IL-4 modified mousepox resisted treatment with an antiviral agent for smallpox garnered little notice. As another instance of the 'non-', only years after the initial controversy did Ramshaw regard his research as entailing a sort of weaponisation of sterilisation for rodents.

As well, on a different level, the way that mousepox has been one of only a handful of so-called 'experiments of concern' that are repeatedly put forward speaks to the narrow, individual case-based approach that has come to dominate framing how the life sciences might aid destructive purposes.[69] Herein what are held as mattering are the choices taken at critical 'ethical moments' (for example, should these results be published? Should experiments be approved? And so on).

This framing has characterised recent debate regarding the justifications for the proposed redaction of research undertaken by Dutch and American-based researchers who mutated the H5N1 virus in a ferret model in such as way as to enable it to transmit between mammals. While this case was a matter of much controversy in early 2012, as with other 'experiments of concern', what is perhaps most notable is its *exceptionality*. With this regard to a very limited number of cutting-edge experiments, much less attention has been cast on what the mundane commercialisation of science means for new biowarfare capacities.

69   As in World Health Organisation (WHO) 2011, *Responsible Life Sciences Research for Global Health Security*, WHO/HSE/GAR/BDP/2010.2, World Health Organisation, Geneva, Part 2.

The 'non-' as a non-issue?

Asking about what is 'not' speaks to many of the themes raised previously in this chapter: moral blindness, taken for granted meaning, the social construction of moral reasoning, the lack of ethical scrutiny to the conditions that produce ethical issues, and the need to move beyond a preoccupation with specific decision points.

Noting such affinities prompts the wider question: to what extent has the 'non-' been addressed within bioethics? On the one hand, the case-based scenario reasoning prevalent in bioethics is typically directed towards manifested dilemmas and choices. Given the widespread technique of posing hypothetical and real-life cases to ask 'what should be done', what is not taking place can be sidelined. Attention rests with possibilities for action and agency in specific scenarios. Certainly some approaches in bioethics, such as the case-based casuistry ethics,[70] are ill suited for directing themselves towards what is not happening.

On the other hand, absences are also prevalent. Bioethical analysis generally seeks to question the basis for what is treated as natural, inevitable, just so, and so on. In some ways, the history of bioethics can be read as a history of seeking to doubt prevalent moral conventions and priorities.

Within bioethics, 'non-action' is often at the centre of dispute about what is justifiable. In a review of ethical analysis of euthanasia, Holland mapped out some of the contrasting orientations given to the distinction between undertaking and refraining from action.[71] As he argued, one common means of differentiating between what is morally permissible and what is not is through the language of 'killing' versus 'letting die'—or so-called active versus passive euthanasia.[72] Often the latter is treated as more justifiable than the former because active euthanasia requires directed intervention. While passive euthanasia can itself entail some sort of action (such as turning off life-support equipment), that this is not the direct cause of death is held by some as justification for a moral distinction between it and (the more problematic) active forms. Tooley has countered such attempts to distinguish between killing and letting die. For him, passive euthanasia should be regarded as morally equivalent. As such, it is the lack of willingness on the part of doctors to intervene to *hasten* death that should be seen as the prime problem. As Holland recounts, this analytical

70   Arras, J. 1991, 'Getting down to cases: the revival of casuistry in bioethics', *Journal of Medicine and Philosophy*, vol. 16, pp. 29–51.
71   Holland, S. 2003, *Bioethics*, Polity, London.
72   See Rachels, J. 1975, 'Active and passive euthanasia', *New England Journal of Medicine*, vol. 292, pp. 78–90; and McLachlan H. V. 2008, 'The ethics of killing and letting die: active and passive euthanasia', *Journal of Medical Ethics*, vol. 34, pp. 636–8.

argument about moral equivalence has itself been disputed through the contention that the widespread belief that a moral difference should be made between killing and letting die itself provides an adequate basis for judging which one is preferable.

In contrast, other ethicists have sought to move away from direct reference to action/non-action as the basis for evaluating morality. One way that has been done is by distinguishing between positive and negative duties. Herein transgressing negative duties (such as refraining from killing) are treated as more serious than positive ones (such as not intervening to prevent death). Yet this is problematic because some actions (such as preventing someone from being saved) cut across the starting distinction between killing and letting die. Still other ethicists have advocated replacing the focus on action in debates about euthanasia with that on agency and responsibility.[73]

The manner in which action/non-action is varyingly configured as relevant speaks to the importance of how issues are identified and how they are formulated. Action and inaction have been the locus for moral argument while also being deemed somewhat beside the point. Thus when attending to dual use vis-a-vis what is missing, in addition to considering the many ways (in)action is said to matter, how the debate is framed must be considered: what is taken as counting as (in)action, what evidence supports such claims, what implications are said to follow.

This complicated picture of the coverage and place of the 'non-' in ethics is mirrored in the empirical social sciences. Again while fields such as sociology and political science are generally preoccupied with what is taking place, what is not happening has also figured as a subject of study. A recurring undercurrent of the commentary by sociologists on bioethics is that it does not attend to the structural and institutional conditions that delimit the possibilities available to individuals. As such, sociological analysis often purports to attend to what bioethics systematically ignores.

Another facet of the study of the 'non-' in social research is the examination of how social concerns about science are nullified. Much of this work starts with Gieryn's observations about how the boundaries between objectivity/ subjectivity, natural/social realms and expert/lay knowledge are routinely managed within the practice and portrayals of scientists.[74] Such 'boundary work' is part and parcel of how control is maintained over the goals and standards of science. Along these lines, Cunningham-Burley and Kerr examined how adept boundary work enabled geneticists to secure the cognitive authority necessary to secure funding, while placing themselves as authority figures for speaking

---

73 Coggon, J. 2008, 'On acts, omissions and responsibility', *Journal of Medical Ethics*, vol. 34, no. 8.
74 Gieryn, T. 1999, *Cultural Boundaries of Science*, University of Chicago Press, Chicago.

about the social consequences of genetics, while also distancing themselves from the responsibility for negative consequences.[75] Specifically in relation to the 'non-', Firth et al. have sought to chart how boundary work was part of the creation of a 'settled morality'[76] in infertility clinics.[77] Boundaries management was central to securing agreement over many issues that were contentious outside the lab. Ethical concerns were not identified with day-to-day practices within clinics as part of what was referred to as the 'no ethics repertoire'. The result of both aspects of settled morality is that those in the clinics rendered their practice immune from outside interference.

Non-groundings

As suggested in the previous subsection then, within both ethics and social science, uneven regard is given to what is not taking place. As a result, attempts in relation to the dual-use life sciences to combine normative justification with empirical analysis face two types of problems: how the 'non-' is treated within fields of study and how these fields can be brought together. In relation to what is absent, what really needs attention is the status quo and therefore what is likely to foil efforts to move on from it. As part of this, the normative and the empirical must seek to identify each other's underlying assumptions. The 'non-' as a topic of study in this respect proves advantageous. This is so because attending to what is absent requires not just clarifying thinking, but instead also inquiring about the conditions under which quandaries arise and are structured.

If the 'non-' has advantages as a topic for study in fostering this dialogue between the empirical and the normative, it also has drawbacks. A prominent one is its open-endedness. In de-anchoring analysis from something definite to something that could be happening, the range of relevant considerations multiplies manifold. Appeals to research-community interactions, time and organisational constraints, awareness, professional socialisation,[78] widespread cultural myths and narratives,[79] and so on, are among the reasons that could be cited to explain the lack of professional attention to dual-use issues. And since those considerations relate to what is not happening, proving their counterfactual relevance is not straightforward. This in turn makes choosing

75   Cunningham-Burely, S. and Kerr, A. 1999, 'Defining the "social"', *Sociology of Health & Illness*, vol. 21, no. 5, pp. 647–68.
76   From Hoffmaster, B. 1990, 'Morality and the social sciences', in G. Weisz (ed.), *Social Science Perspectives on Medical Ethics*, Kluwer Academic, Boston.
77   Frith, L., Jacoby, A. and Gabbay, M. 2011, 'Ethical boundary-work in the infertility clinic', *Sociology of Health & Illness*, pp. 1–16.
78   As suggested in Sture, J. 2009, 'Educating scientists about biosecurity: lessons from medicine and business', in Rappert, 2009, op. cit.
79   Gordon, D. and Paci, E. 1997, 'Disclosure practices and cultural narratives', *Social Science and Medicine*, vol. 44, no. 10, pp. 1433–52.

between explanations a demanding task. While empirical and normative argument can be marshaled to make a case that is persuasive for many, this is a case that will need to be made. The 'non-' in relation to dual use faces a similar demand. Barring recourse to an objective sense of social problems that could be read back to determine exactly how much of a concern is really posed by the destructive use of science,[80] normative and empirical questions can be raised about claims regarding what should be recognised as a problem but is not.

Methodology

From the previous argument of this chapter it is possible to draw some important conclusions for the study of what is not going on in relation to concerns about the destructive use of the life sciences. First, while some of the existing literature in bioethics and social sciences speaks to how issues are not recognised or what actions are not taken, more systematic thought is needed. Second, with the somewhat inevitable reliance on the counterfactual and the speculative, arguments about the 'non-' require careful justification. Although it might be possible to make a case for why something is not happening that is persuasive to many, it is likely to be disputed too. Third, a dialogue must be established between the normative and the empirical. Locating a discussion of the 'non-' of dual use within the emerging literature about 'empirical ethics' could provide additional (normative) analytical resources for probing compared with those typically called upon by social scientists.

In thinking in more specific methodological terms about how to study non-issues and non-actions, in this subsection I want to advance two considerations: 1) comparative examination, and 2) interventionist inquiry.

Comparative examination

One of the strategies used in researching the exercise of power has been to match up situations that shared pertinent similarities in order to account for their variations. For instance, the responses of communities affected by air pollution have been juxtaposed to inquire about the reasons for those differences. In a related vein, using time as the variable, periods of major social disturbance have been examined for the opportunities created (and closed) for unconventional ideas and practices.

---

80   For a consideration of objectivist and constructivist orientations to dual use as a social problem, see Rappert, 2007, op. cit., ch. 1.

With respect to dual-use issues, it is possible to take inspiration from this comparative strategy. For instance, a given line of experimentation—such as the insertion of IL-4 on pox viruses—could be examined regarding why some *failed* to raise dual-use concerns while others did.

To take comparison in a different direction, another tack would be to put side-by-side efforts to *dissociate* science from bioweapons concerns actively. For instance, in recent years an international do-it-yourself (DIY) community has emerged, encouraging the formation of small-scale open-access biological labs. While much of the DIY bio community emphasises the democratisation of science in order to address problems ignored by corporations and universities, the proliferation of capabilities beyond accredited labs has been repeatedly associated with fears about bioterrorism.[81] Leaders in the DIY bio community are seeking actively to distance their work from such fears, in part by establishing codes of conduct.[82] Likewise, through innovations in art, the Critical Art Ensemble has sought to debunk the link between biology and bio-threats.[83]

Such attempts to move from 'is' → 'is not' could be compared with attempts to move from 'is not' → 'is'. Along these lines, civilian researchers who have raised dual-use threats could be studied with a view to the reactions they experienced from colleagues, funders and others. The pushback and resistance faced by organisational whistleblowers would likely prove a salient comparison.[84] In the military area, the aim of US Defense Advanced Research Projects Agency funding to link basic research to the protection against bioterrorism (as through the synthesis of poliovirus that brought its own dual-use fears) would be one such effort to move from 'is not' to 'is'.

Another comparative tact would be to juxtapose the evaluations made by different communities. For instance, security experts could be enlisted to identify lines of research they believe pose security risks but which have *not* been the subject of much scrutiny to date. Then those working in the identified areas could be approached to determine whether practitioners agreed with the assessments of security experts, the extent to which researchers have identified dual-use concerns with their work, and the reasons those concerns have (and have not) been communicated. Much the same could be done for ethicists and how their ranking of what ought to be a matter of concern compares with those in the life sciences.

---

81   Ledford, H. 2010, 'Garage biotech: life hackers. Amateur hobbyists are creating home-brew molecular-biology labs, but can they ferment a revolution?' *Nature*, vol. 467, pp. 650–2; and Nature 2010, 'Garage biology: amateur scientists who experiment at home should be welcomed by the professionals', *Nature*, vol. 467, p. 634.

82   As in the DIYbio Continental Congress held on 8 May 2011 attended by the author.

83   Critical Art Ensemble n.d., *Bodies of Fear in A World of Threat*, <http://www.critical-art.net/mp.html> (viewed 4 April 2010).

84   Martin, B. 2007, 'Whistleblowers: risks and skills', in Rappert and McLeish, op. cit.

Interventionist inquiry

That non-issues are *non*-issues and non-actions are *non*-actions speak to the way in which a spirit of intervention needs to infuse their study.

To expand, the process of questioning practitioners about matters of dual use is likely to be an act of questioning assumptions, priorities and world views. For instance, since 2004 Malcolm Dando and I have conducted seminars for university faculties and other public research centres in order to inform participants about current life-science security developments as well as to generate debate about how research findings should be communicated, whether experiments should be subject to institutional oversight and what research should be funded. More than 130 seminars have been undertaken in 15 countries (ranging from the United Kingdom to Uganda and Japan to Argentina), with more than 3000 participants. While it has been possible to generate lively (but bounded)[85] discussion about dual-use issues at these events, as interactions they required careful management because they asked participants to think about their work anew. As a result, what was said, how, to whom and when were all subject to lengthy methodological consideration. The decision to run seminar discussions akin to focus groups in which attendees were encouraged to deliberate with each other was itself the result of the limitations experienced in a one-to-one interview format.[86]

The manner in which probing about non-issues de facto amounts to a form of intervention suggests the need to consciously attend to how this intervention is conducted. Overall, what is required is a systematic process of planning and execution that allows for learning and experimentation. As part of this, inquiry should be thought of as a practical, intellectual, action-orientated and consequentialist form of action.[87] Kurt Lewin's often quoted suggestion that 'if you want to truly understand something, try to change it' indicates the potential for deliberate intervention to yield insights not readily obtainable through unobtrusive means.

In fields of social science such as 'action research', such practical inquiry is linked to the aim of transforming social relations.[88] As with the 'empowerment' and 'social-action' models noted in the previous section, the factors that frustrate change require attention. This practical step entails incorporating

---

85   See, for instance, Rappert, 2007, op. cit., ch. 5.

86   Ibid., ch. 2.

87   Dewey, J. 1929, *The Quest for Certainty*, George Allen & Unwin, London.

88   See, for example, Ospina, S., Dodge, J., Godsoe, B., Minieri, J., Reza, S. and Schall, E. 2004, 'From consent to mutual inquiry', *Action Research*, vol. 2, no. 1, pp. 47–69; and Winter, R. 1996, 'Some principles and procedures for the conduct of action research', in O. Zuber-Skerritt (ed.), *New Directions in Action Research*, Taylor & Francis, London; and Winter, R. 1998, 'Managers, spectators and citizens', *Educational Action Research*, vol. 6, no. 3, pp. 361–76.

positive normative goals into the design of inquiry. Central to a robust process of transformative intervention is to ensure that a conversation takes place between the methods of inquiry and its normative aims. The latter should promote scrutiny regarding the fallibility and commitments of the methods employed. Also, the methods should enable the refinement and revision of what is held as necessary and desirable. Doing so not only requires a certain kind of intellectual understanding, but also practical skills.

What seems essential in studying what is absent is to find means of questioning taken-for-granted assumptions about what counts as an ethical or a social 'problem' in the first place. Rather than going out and probing straightforwardly overt, recognised issues widely labelled as 'ethical' or 'contentious', research techniques and strategies must help to cultivate thinking afresh in order to avoid confirmation bias, to encourage alternative hypotheses and to embrace negative evidence.

Inquiry about non-issues and non-actions then cannot be conceived simply as an attempt to reveal holes in understanding. Instead, it must be a project of questioning the historical, political and situational bases for how understandings are formed and thereby what is counted as 'missing' in the first place. As such, ensuring inquiry interrogates its own starting points is vital. One interesting direction to take the 'non-' is to consider the hows and whys regarding ethicists' lack of engagement with dual-use life science. Some preliminary reflections have already been given on this matter;[89] however, undertaking systematic empirical research might inform an understanding of not only the priorities and presumptions of ethicists, but also therefore the likely limits of existing academic disciplinary resources. Moreover, such a line of empirical investigation provides an opportunity for bringing to the fore questions about how the normative and the empirical can be combined.

At stake is the question of how facts, figures, concepts and arguments should be made sense of in order to assess whether ethicists themselves have been remiss for their past level of regard. The contention that some issues are being neglected compared with others is commonplace in bioethics, with its attentiveness to distributive justice. At times this extends to commentary on bioethics' own agenda.[90] For both, the justifications for claims are open to disagreement in relation to their underlying ethical assumptions. Appeals to consequences versus duties versus rights, for instance, can result in far different assessments about what is lacking from the agenda of bioethics. Each appeal is also reliant on

---

89 Selgelid, M. 2010, 'Ethics engagement of the dual-use dilemma: progress and potential', in Rappert, 2010, op. cit.

90 For examples of this, see Selgelid, M. 2008, 'Ethics, tuberculosis and globalization', *Public Health Ethics*, vol. 1, no. 1, pp. 10–20; and Selgelid, 2005, op. cit.

different types of 'evidencing' to substantiate neglect. While many or even most bioethicists might agree that a particular topic is being relatively 'neglected', the basis for this might well vary.

Complicating the situation further, varying appraisals of what counts as a neglected 'non-' will likely be part and parcel of meaningfully alternative ways a given topic is framed. That then raises important questions about whether commentators are orientating to neglect in 'realist' or 'non-realist' terms—that is, whether they are assuming some definitive sense of what 'topic X' is and how it should be understood. There may well be meaningfully different topics X_1, X_2, X_3 and so on at play.[91]

In short, an empirical examination of how ethicists contend about whether and how the dual-use aspects of the life scientists are being 'neglected' could be a way into examining the in-practice reasoning and bounds of bioethics.

If inquiry is linked to the aim of transforming social relations then another reflexive dimension of the relationship between the normative and the empirical opens up, one that again calls into question conceiving of the study of non-issues as the cataloguing of knowledge gaps. To make this point I want to return to where this chapter began, with the categorical condemnation of biological weapons. As contended earlier, assents to this denunciation are typically accompanied by little explicit justification. As with many stigmas, the one against biological weapons is often portrayed as self-evident.

To engage in explicit empirical-normative inquiry about what should count as a problem might well cast doubt on this orientation. That could happen through questioning arguments that biological weapons are really more 'inhumane' or 'indiscriminate' than other weapons. If empirical-normative analysis were to undermine the often visceral, intuitive reactions against biological weapons, some would no doubt regard this as a deplorable outcome. As a result, *not* undertaking analysis might well be judged as a wiser course of action. Whether or not this appraisal holds depends on the goals sought from analysis. As ever, then, the uses of science and ethics are topics for considered inquiry.

---

91 Further, then, whether some definitive sense of the topic needs to be established (and what that should be) is a choice that needs to be addressed.

Appendix A

Who is working in neuroethics? Where are they?

In a preliminary overview of the field, it was noticed that 'neuroethicists are not saying enough about the problem of dual-use',[1] a conclusion promptly validated by Peter B. Reiner, Professor in the National Core for Neuroethics, who wrote:

> The truth is that other than Jonathan Moreno, few neuroethicists have applied serious scholarship to the issue of dual use. Of course, it is a simple matter to just say no: neuroscience should only be used for improving the quality of human life. But frankly, that is too simplistic.[2]

This appendix provides the reader with an overview of where neuroethics is carried out, who is working in the field and what neuroethicists say about dual use. Only some of the main institutions and networks as well as leading neuroethicists are mentioned as representatives of the wider field.

Where

Center for Neuroscience & Society, University of Pennsylvania

The stated mission of the centre is to increase understanding of the impact of neuroscience on society through research and teaching, and to encourage the responsible use of neuroscience for the benefit of humanity. <http://neuroethics.upenn.edu/>

Center for Cognitive Neuroscience, University of Pennsylvania

Penn's Center for Cognitive Neuroscience is a multidisciplinary community dedicated to understanding the neural bases of human thought. Their current research addresses the central problems of cognitive neuroscience, including perception, attention, learning, memory, language, decision-making, emotion and development. Methods include functional neuroimaging, behavioural testing of neurological and psychiatric patients, transcranial and direct-current

---

1   Dando, M. 2010, 'Neuroethicists are not saying enough about the problem of dual-use', *Bulletin of the Atomic Scientists*.
2   Reiner, B. 2010, *Neuroethics at the Core*, 19 July, <http://neuroethicscanada.wordpress.com/2010/07/19/neuroethics-of-dual-use/>.

magnetic stimulation, scalp-recorded event-related potentials, intracranial recording, computational modelling, candidate gene studies and pharmacologic manipulations of cognitive processes. <http://ccn.upenn.edu/>

Sage Centre for the Study of the Mind, University of Santa Barbara, California

The Sage Center for the Study of the Mind at the University of California, Santa Barbara, is designed to be a catalyst for interdisciplinary study of the relationship of brain and mind. The centre integrates a wide range of scholarly endeavours and technologies in the humanities, social sciences and the sciences. <www.sagecenter.ucsb.edu/intro.htm>

Dana Foundation

The Dana Foundation is a private philanthropic organisation that supports brain research through grants and educates the public about the successes and potential of brain research. Dana produces free publications; coordinates the International Brain Awareness Week campaign; supports the Dana Alliances, a network of neuroscientists; and maintains a web site: <www.dana.org>. The Dana Foundation's science and health grants support clinical research in neuroscience and neuro-immunology and their interrelationship in human health and disease.

European Neuroscience and Society Network

The European Neuroscience and Society Network (ENSN) is the leading European network for interdisciplinary discussions of the social implications of the neurosciences. Funded by the European Science Foundation and convened by researchers at the BIOS Centre, London School of Economics, the ENSN has been established to serve as a multidisciplinary forum for timely engagement with the social, political and economic implications of developments in the neurosciences, a field that has experienced unprecedented advances in the past 20 years.

Neuroethics Society

The Neuroethics Society is an interdisciplinary group of scholars, scientists, clinicians and other professionals who share an interest in the social, legal, ethical and policy implications of advances in neuroscience. Their stated mission is to promote the development and responsible application of neuroscience through interdisciplinary and international research, education, outreach and public engagement for the benefit of people of all nations, ethnicities and cultures. <www.neuroethicssociety.org>

National Core for Neuroethics, University of British Columbia, Vancouver, Canada

The National Core for Neuroethics hosted by the University of British Columbia is an interdisciplinary research group dedicated to tackling the ethical, legal, policy and social implications of frontier technological developments in the neurosciences. The objective is to align innovations in the brain sciences with societal, cultural and individual human values through high-impact research, education and outreach. <www.neuroethics.ubc.ca>

Brain Research Centre, University of British Columbia, Vancouver, Canada

The Brain Research Centre, located in Vancouver, Canada, is a unique partnership between the Vancouver Coastal Health Research Institute and the Faculty of Medicine at the University of British Columbia. The hospital has combined forces with broad, multidisciplinary research expertise at the University of British Columbia to advance knowledge of the brain and to explore new discoveries and technologies that have the potential to reduce the suffering and cost associated with disease and injuries of the brain.

The Oxford Centre for Neuroethics, University of Oxford

Established in January 2009, the Oxford Centre for Neuroethics aims to address concerns about the effects neuroscience and neurotechnologies will have on various aspects of human life. Its research focuses on five key areas: cognitive enhancement; borderline consciousness and severe neurological impairment; free will, responsibility and addiction; the neuroscience of morality and decision-making; and applied neuroethics. It is the first international centre in the United Kingdom dedicated to neuroethical research. It is founded by the Wellcome Trust's Biomedical Ethics Strategic Awards Program. <www.neuroethics.ox.ac.uk>

National Institute of Neuroscience, Italy

The Institute of Neuroscience of the National Research Council considers itself to be one of the top institutions in the field of neuroscience in Europe. It concentrates many of the most important Italian scientists involved in the study of the nervous system, who are organised in research groups, located in Milan, Padua, Pisa, Rome and Cagliari. The Institute of Neuroscience of the National Research Council addresses all the principal topics in the study of the nervous system, investigating the development and plasticity of the nervous circuitry; vision and cognitive sciences; the mechanisms of memory and learning; as well as those involving cellular transmission and neuronal communication;

neuromuscular and neuronal-glia interactions, and the neurobiological bases of alcoholism and drug dependence. <www.cnr.it/istituti/Descrizione_eng. html?cds=061>

Neuroscience Centre, Dartmouth (NCD)

The establishment of the Neuroscience Center at Dartmouth (NCD) in 2002 produced a new and unique interdisciplinary group whose mission is to foster collaborative and interactive research and education in the neurosciences. The NCD draws from its strengths in three key areas: clinical; cognitive and behavioural; and molecular/cellular/systems neuroscience. It is the vision of its researchers to produce and disseminate new knowledge, and in doing so train and educate the next generation of neuroscientists. Interactions among members of the neuroscience community are enhanced and foster a highly interactive atmosphere through the development of this integrated centre. By promoting multidisciplinary efforts in both basic and applied research, the centre's scientists will contribute to human health and wellbeing by increasing our understanding of the mechanisms underlying nervous system function, both in health and in disease. This will lead to valuable discoveries that translate into novel pharmaceutical agents and therapeutic approaches for the treatment of a variety of central nervous system diseases and disorders. <http://dms. dartmouth.edu/>

International Neuroethics Network

The International Neuroethics Network (INN) was launched in 2005 at the Society for Neuroscience Annual Meeting in Washington, DC. The network's vision is to foster international collaboration in neuroethics through the identification of common priorities and joint funding opportunities. The INN's objective is to serve as a means of communication and support among neuroethicists all around the world. <www.neuroethics.ubc.ca>

Stanford Program for Neuroethics, Stanford University

The Stanford Program for Neuroethics is a research team devoted to the new field of neuroethics, with an initial focus on issues at the intersection of medical imaging and biomedical ethics. These include ethical, social and legal challenges presented by advanced neurofunctional imaging capabilities, the emergence of cognitive enhancement neurotechnologies and pharmacology, self-referral to healthcare and imaging services, incidental findings, and fetal MRI. New initiatives are under way in regenerative medicine, neurogenetics and pediatric neuroethics. Several program members are also involved in the John D. and Catherine T. MacArthur Foundation's Law and Neuroscience Project, including Hank Greely, Emily Murphy and Teneille Brown. The project seeks to address issues that neuroscience raises for our legal system through its three Research Networks—Diminished Brains, Addiction and Decision Making—and its Education and Outreach Program. <http://neuroethics.stanford.edu>

MacArthur Law & Neuroscience Project

The MacArthur Law and Neuroscience Project investigates the impact of modern neuroscience on criminal law and in particular the diverse and complex issues that neuroscience raises for the criminal justice system in the United States. <www.lawneuro.org>

The Neuroethics Research Unit

The Neuroethics Research Unit is committed to training a new generation of students in neuroethics through the conduct of collaborative interdisciplinary research. Research interests include the ethical application of neuroscience in research and patient care, empirical bioethics research, and pragmatism in bioethics. It is based at the Montreal Institute of Clinical Research. <http://www.ircm.qc.ca/microsites/neuroethics/en/index.html>

Novel Tech Ethics

The team at Novel Tech Ethics is committed to public discussion of the ethics issues that affect all human beings. Key focal points of the Novel Tech Ethics Research Team include

- how are psychopharmacologies creating a 'new normal' for human behaviour—and what is normal for humans when 'regenerative medicine' mixes and matches cells across species
- how will the results emerging from neuroimaging studies and behavioural genetics affect our understanding—and social and legal enactment—of free will and responsibility
- what are the particular kinds of 'harm' and 'benefit' offered by neurological treatment, neurological enhancement and neurological control, and how do these challenge traditional notions and practices of risk assessment?

<www.noveltechethics.ca/site_events.php>

University of Wisconsin: Neuroscience and Public Policy Program

The program emerged from the recognition that the rapid advancement in neuroscience demanded research neuroscientists to be trained to think critically about both neuroscience and the making of public policy, and to have appropriate skills, experience and networks to facilitate an effective integration of the two. The program is hosted by the University of Wisconsin–Madison. A central element of the program is the weekly Neuroscience and Public Policy Seminar, which challenges students to synthesise information across neuroscience and policy research. <http://npp.neuroscience.wisc.edu/index.html>

The Neuroethics New Emerging Team (NET)

The Neuroethics New Emerging Team (NET) is based at Dalhousie University, Canada. Launched in 2003, the NET aims to undertake an interdisciplinary study of, and disseminate their findings on, the ethical issues posed by advances in neuroscience technology. It is funded by the Canadian Institutes of Health Research. <http://www.neuroethics.ca/Neuroethics.ca>

Who

Colin Blakemore

Colin Blakemore is a neurobiologist at Oxford University. He is specialised in vision and development of the brain. He also holds professorships at the University of Warwick and the Duke University—National University of Singapore Graduate Medical School, where he is chairman of Singapore's Neuroscience Research Partnership. His research has been concerned with many aspects of vision, the early development of the brain and plasticity of the cerebral cortex.

Turhan Canli

Turhan Canli is Associate Professor at Stony Brook University. The work in Dr Canli's laboratory focuses on the hormonal and neurogenetic bases of individual differences in emotion and cognition. The research addresses these questions: what are the biological mechanisms that can explain human personality? What is the mechanism by which life experience, in interaction with genetic variation, influences brain function to generate behavioural patterns that we associate with certain personality traits? Do men and women differ in how their brains respond to these genetic and experiential influences? Can this information be used to identify healthy individuals at risk for psychopathology? To address these questions, Dr Canli's team uses a number of different technologies: functional magnetic resonance imaging (fMRI); transcranial magnetic stimulation (TMS); molecular genetics; and hormone assays.

Arthur Caplan

Arthur Caplan is Emanuel and Robert Hart Professor of Bioethics and Philosophy at the University of Pennsylvania. His research interests focus on transplantation research ethics, genetics, reproductive technologies, health policy and general bioethics.

Jocelyn Downie

Jocelyn Downie is Professor at Faculties of Law and Medicine, Dalhousie University. She works at the intersection of law, ethics and health care. Her research interests include women's health, assisted death, research involving humans, and organ transplantation. Her work is interdisciplinary, collaborative and geared both to contributing to the academic literature and to affecting change in health law and policy at federal and provincial levels.

Martha J. Farah

Martha J. Farah is a cognitive neuroscientist at the University of Pennsylvania, who works on problems at the interface of neuroscience and society, including

- the effects of childhood poverty on brain development
- the expanding use of neuropsychiatric medications by healthy people for brain enhancement
- novel uses of brain imaging in, for example, legal, diagnostic and educational contexts
- the many ways in which neuroscience is changing the way we think of ourselves as physical, mental, moral and spiritual beings.

Kenneth R. Foster

Kenneth R. Foster is a professor of bioengineering. His research interests relate to biomedical applications of non-ionising radiation from audio through to microwave frequency ranges, and health and safety aspects of electromagnetic fields as they interact with the body. For example, he examines the prospects of workers in electrical occupations and the possibility (or lack of) cancer risk. Another and somewhat broader topic of interest is technological risk and the impact of technology (principally, electro-technologies) on humans. His goal in this area is to examine technology, putting into perspective its relative risks and benefits to society. What he hopes to impart is a better perception of the social use of science.

Michael Gazzaniga

Michael Gazzaniga is a professor of psychology and the Director of the Sage Center for the Study of the Mind at the University of California, Santa Barbara. He oversees an extensive and broad research program investigating how the brain enables the mind. Over the course of several decades, a major focus of his research has been an extensive study of patients who have undergone split-brain surgery that has revealed lateralisation of functions across the cerebral hemispheres.

Henry Greely

Henry Greely is Deane F. and Kate Edelman Johnson Professor of Law. A leading expert on the legal, ethical and social issues surrounding health law and the biosciences, Hank specialises in the implications of new biomedical technologies, especially those related to neuroscience, genetics and stem-cell research. He frequently serves as an advisor on Californian, national and international policy issues. He is chair of California's Human Stem Cell Research Advisory Committee and served from 2007 to 2010 as co-director of the Law and Neuroscience Project, funded by the MacArthur Foundation. Active in university leadership, Professor Greely chairs the steering committee for the Stanford Center for Biomedical Ethics and directs both the law school's Center for Law and the Biosciences and the Stanford Interdisciplinary Group on Neuroscience and Society. Professor Greely serves on the Scientific Leadership Council for the university's interdisciplinary Bio-X Program.

Ronald M. Green

Ronald M. Green is Eunice and Julian Cohen Professor for the Study of Ethics and Human Values. Ronald M. Green has been a member of Dartmouth University's Religion Department since 1969; he also directs Dartmouth's Ethics Institute, a consortium of faculty concerned with teaching and research in applied and professional ethics. Professor Green's research interests are in genetic ethics, biomedical ethics and issues of justice in healthcare allocation. He is the author of six books and more than 130 articles in theoretical and applied ethics.

Judy Illes

Judy Illes is Associate Professor of Pediatrics (Medical Genetics) and Director of the Program in Neuroethics at the Stanford Center for Biomedical Ethics. She also co-founded the Stanford Brain Research Center (now the Neuroscience Institute at Stanford), and served as its first executive director between 1998 and 2001. Today, Dr Illes directs a strong research team devoted to neuroethics, and issues specifically at the intersection of medical imaging and biomedical ethics. These include ethical, social and legal challenges presented by advanced functional imaging capabilities, the emergence of cognitive enhancement technologies and pharmacology, the commercialisation of cognitive neuroscience, and clinical findings detected incidentally in research.

Jonathan Moreno

Jonathan Moreno is the David and Lyn Silfen University Professor of Ethics and Professor of Medical Ethics and of History and Sociology of Science at Pennsylvania University. He holds a courtesy appointment as Professor of Philosophy. He is also a Senior Fellow at the Center for American Progress in Washington, DC, where he edits the magazine *Science Progress*. He was a member of President Barack Obama's transition team for the Department of Health and Human Services. In the course of his career, Professor Moreno has applied serious scholarship to the issue of dual use in neuroscience, revealing that several of the new technologies are 'potentially applicable to medical therapy or other peaceful purposes as well as combat, riot control, hostage situations, or other security problems'.

Stephen J. Morse

Stephen J. Morse is Professor of Psychology and Law in Psychiatry. Stephen J. Morse is an expert in criminal and mental health law whose work emphasises individual responsibility in criminal and civil law. Morse was co-director of the MacArthur Foundation's Law and Neuroscience Project and he co-directed the project's Research Network on Criminal Responsibility and Prediction. Morse is a diplomate in Forensic Psychology of the American Board of Professional Psychology; a past president of Division 41 of the American Psychological Association (the American Psychology-Law Society); a recipient of the American Academy of Forensic Psychology's Distinguished Contribution Award; a member of the MacArthur Foundation Research Network on Mental Health and Law (1988–96); and a trustee of the Bazelon Center for Mental Health Law in Washington, DC (1995–present).

Peter B. Reiner

Peter B. Reiner is Professor in the National Core for Neuroethics and a member of the Kinsmen Laboratory of Neurological Research, Department of Psychiatry and the Brain Research Centre at the University of British Columbia. Dr Reiner has a distinguished track record as a research scientist studying the neurobiology of behavioural states and the molecular underpinnings of neurodegenerative disease. Dr Reiner also has experience in the private sector, having been president and CEO of Active Pass Pharmaceuticals, a drug discovery company that he founded to tackle the scourge of Alzheimer's disease. Upon returning to academic life, Dr Reiner refocused his scholarly work in the area of neuroethics, with interests in neuro-essentialism, the neuroethics of cognitive enhancement and the commercialisation of neuroscience.

Paul R. Wolpe

Paul R. Wolpe is Associate Professor of Psychiatry in the Department of Psychiatry at the University of Pennsylvania, where he also holds appointments in the Department of Medical Ethics and the Department of Sociology. He is President of the American Society for Bioethics and Humanities and is Co-Editor of the *American Journal of Bioethics*. Dr Wolpe serves as the first Chief of Bioethics for the National Aeronautics and Space Administration (NASA).

www.ingramcontent.com/pod-product-compliance
Lightning Source LLC
Chambersburg PA
CBHW050100220326
41599CB00049B/7201